Cover Illustrations:

Frank Anders	Carbon fibers in tissue	Justice Harry Blackmun	Albert Szent-Gyorgyi	Cell phones
Philip Handler	Milton Zaret	Kirlian photograph of a leaf	Patty Ryan	Marguerite Hays
Chauncey Starr	Stunted growth in mice from EMFs	Fluorescent bulbs lit by EMFs	Mike Wallace	Morton Miller
Stone quarry	Richard D. Phillips	Secondary glaucoma from EMFs	Capt. Paul Tyler	765,000-volt powerline
Launch of Sputnik	Dr. Robert Becker	"Baby" (NASA computer)	Fred Brown	Oscilloclast-like device

◆

Going Somewhere
Truth About a Life in Science

◆───────

Books by Andrew A. Marino

Electromagnetism & Life [Facsimile Edition]
Electromagnetism & Life Revisited
Electric Wilderness (with Joel Ray)
Modern Bioelectricity

Further publications available at *andrewamarino.com*

GOING SOMEWHERE
Truth About a Life in Science

Andrew A. Marino

Cassandra Publishing, Belcher, LA

Published by Cassandra Publishing
P.O. Box 26
Belcher, LA 71004

© 2010 by Andrew A. Marino

All rights reserved

No part of this book may be used or reproduced in any manner whatsoever without written permission except in the case of brief quotations embodied in critical articles and reviews.

For information: cassandra@cassandrapublishing.net

Library of Congress Control Number: 2010937390

Publisher's Cataloging-in-Publication
Marino, Andrew A.
Going somewhere: truth about a life in science /
Andrew A. Marino.
p. cm.
Includes index.
LCCN 2010937390
ISBN-13: 978-0-9818549-1-5
ISBN-10: 0-9818549-1-5
1. Marino, Andrew A. 2. Scientists--United States--Biography. 3. Electromagnetic fields. I. Title.
Q143.M2265A3 2010 509'.2
QBI10-600200

To
Joel Ray,
Friend, and muse
For this book

To
Linda,
Wife, and partner
For life

———————◆———————

Contents

Preface		11

I Growing Up: 1956–1974

1	Starting Out	21
2	Incubation	33
3	Warrior	43
4	A Legal Turn	59
5	Beginner	69
6	Knowing	77
7	EMFs	85
8	A Long Day	91
9	Wagging Tails	99

II Wandering: 1974–1980

10	The Commission	111
11	Monsters	127
12	Risk	139
13	Dangers	149
14	Zaret	159
15	Bone EMFs	169
16	Turning Point	181
17	Southbound	191

III Detour: 1980–1994

18	Floating	207
19	Carbon Fibers	219
20	Power	239
21	Experts	247
22	Masters of Energy	273
23	Razzle-Dazzle	291
24	*Ipse Dixit*	309

IV Understanding: 1994–1998

25	Szent-Gyorgyi	337
26	Looking at EMFs I to I	341

V Return: 1998–2010

27	Christopher	369
28	Young Richman	393
29	Measure for Measure	411
30	Mind War	425

Epilogue — 443

Index — 451

Preface

This is a story about what happened in the world I experienced and in me during my life-long journey through a part of science-land, the part that relates to the ability of electromagnetic fields to make people sick. What initially appeared to be a simple problem turned into something immensely complicated, and then turned again to reveal something I never expected, that the truth of the matter depended on values and assumptions. Not completely, but enough to guarantee that there will never be final answers, only differences in perspective. I don't mean something real that can be looked at in different ways, I mean different perspectives regarding what *is* real. Does my discovery apply to other areas? I think so, because power, fear, self-interest, and solipsism, which were at least as important as scientific laws and observations in determining what was generally accepted as the truth, aren't unique to questions involving health hazards of electromagnetic fields.

I wrote about a search for an explanation of why the issue of health risks has been contentious, not about a search for truth within or about myself. Nevertheless, I could not omit me. What went on inside me is a central issue because I was the protagonist in a conflict involving the culture of science and the interests it serves. So I included the tensions that I experienced and my attitude about things, but I stayed on the periphery as much as possible, consistent with my purpose.

Herodotus wrote his history of the Persian wars "to show why two peoples fought with each other." John Milton wrote Paradise Lost to "justify God's ways to man." I tried to express my motivation succinctly, but all I could come up with was "to give information" or "to tell a story about how science really works," which I decided were inadequate explanations of my motivation. So I reached into my heart and asked myself why I actually did it, and I found that I had many motivations. The most frivolous was that I was sick and tired of books in which the scientist's accomplishments were portrayed as heroically unemotional, and science itself was depicted as the highest form of knowledge, like Plato's forms. It had been my experience that scientists were no more noble or unbiased than anybody else. Sometimes they didn't really know what they thought they knew, and when they did know something and could prove it, the knowledge was always a mixture of blessing and curse because it helped some people but hurt others. Occasionally there was a book that captured something that looked familiar, for example, James Watson's *The Double Helix*. But his story involved a relatively simple problem, the mechanics of a molecule, so his book could provide only a small window into how science actually worked. I wanted to write a story that told a larger version of the truth. By the time I recognized this motivation I had already decided that there *are* versions of the truth – in fact I had decided that, in science, there were *only* versions of the truth, some better than others.

The second motivation is more complicated. As a professional scientist, I write for a living. Sometimes I describe experiments that I performed, but I also write grant requests and modify manuscripts to accommodate reviewers. I love to write about what I understand to be the truth. I dislike writing what I think grant reviewers think is the truth, so they will fund my work, or what the reviewer says is the truth, so that the editor will publish the manuscript. I have written several hundred grant applications and publications about electromagnetic fields, also called EMFs. Each dealt with only a tiny piece of the overall picture, and most were compromised to a certain extent by the forced inclusion of the truth of others. So I said to myself, "I'm going to write a book about EMFs that includes the whole truth as I understand it, and only the truth that I understand," and I set out to see if I could do it.

Other motivations came more directly from my life. My concern with EMFs had always been from multiple perspectives. As a physicist, my gaze was objective because the conventions of physics required it, and I saw EMFs as something to be harnessed. When I learned biology it became possible for me to see the mystical side of EMFs, that they make life possible. After I became a lawyer, I saw how EMFs could be a means of injustice, like pliers in the hands of a torturer. I didn't set out to write a book that extolled or condemned EMFs, or maintain that one frame of reference was more fundamental than another. Rather, I wanted to write a story that integrated my different perspectives to see if they added up to something coherent. My hypothesis was that they did, and I did the experiment to test the hypothesis.

Another motivation was to tell a story about having a career in science while still remaining free. A recurring theme in my conversations with students involves how they would go about making a living doing research. In graduate school, the direction of the work is chosen by the student, based on his choice of mentor, and it proceeds under their joint control. After graduation, however, the student becomes an employee and must do the bidding of management. When the research is controlled by others, the process is far less exhilarating. Students realize this, and they worry whether their academic pursuits are nothing more than preparation for life as a slave. Traditionally, stories about scientists who escaped being controlled by others, and made a life pursuing goals that they chose, involved larger-than-life figures like Paul Dirac or Albert Szent-Gyorgyi. I have not achieved their fame, but I have led a successful life, based on what I counted as important. I wanted to fill a need that I perceived by telling a story about a man in which it is perfectly clear that he takes charge of his destiny by means of the choices he makes, none of which are heroic, makes a life in science, and remains free.

Another important motivation was the need for a more inclusive approach regarding the intended consumer of knowledge about EMFs. I didn't make headway by writing only scientific articles that were intended for experts. Whenever I wrote about an experiment in which EMFs produced some kind of a cellular change, an industry expert soon appeared and wrote about how he had performed an improved version of my experiment but had not observed the cellular change. This state of affairs seemed cacophonous to those outside the EMF orbit, as if the science had not yet ripened to the point where the remedial steps needed to protect human health were clear. So people would say, "More research is needed to resolve the issue," and then move on to something else. I had been writing for a small polarized audience consisting of truth seekers and truth deniers, and I realized that a better tack would be to explain the cacophony to a larger audience by telling the story that was behind the controversy rather than to try to resolve it on the basis of facts presented to scientists.

"You should write a book that's factual," I was told, one that "proves EMFs really cause cancer." But I wanted to write about something much bigger than that. I wanted to write about what "causes cancer" means, and how we know whether there is any truth to the phrase when we apply it to a particular situation. I learned a lot about these issues in the context of EMFs. I came to see that when we try to understand what EMFs do to people the same way that we understand physical forces such as gravity, we are like a man looking near a lightpost for something that he lost somewhere else. No sane person thinks about breaking the law of gravity, but many apparently sane people smoke. That alone tells you that there's something fundamentally different between physics and the laws of disease causation. I wanted to explore that difference in the most honest way possible. I decided that way would be to describe particular events I experienced so that you could see and hear what took place, as if I had a video camera and tape recorder, to explain what the events meant to me at or around the time they actually occurred, and then to place all this in a narrative that would allow the deep themes to emerge.

When people read about EMFs in a newspaper or magazine and start to worry about the antenna or powerline near their home, what they often do is call the company and ask about safety. Invariably they are told, "Our equipment exceeds all applicable state and federal rules and regulations." Often that ends the matter, but some people remain skeptical and so they contact a federal or state agency. They are told that the company is in compliance, just as it said. What happens in all of these cases is that the question "Is it safe?" is metamorphosed into the question, "Does it meet the applicable standards?" which is an entirely different question. Power companies love state regulatory commissions, and cell telephone companies love the Federal Communications Commission. Why? Because companies can use their political influence to set standards so low that they can be easily and economically met. Then the companies can honestly say, "Our equipment meets all applicable standards." I wanted to make it possible for people to see the situation in a new, more realistic way. EMF standards represent a political choice, not an objective scientific determination. That's a big part of the truth that I wanted to write about. This was not because I wanted to serve as the vanguard for a movement to raise the standards. I don't know if they should be raised; maybe they should be lowered. It was because the new way of looking at the situation is more truthful.

The law doesn't know what to do with science. It just doesn't. It allows any Tom, Dick, or Harry with an M.D. or a Ph.D. to testify in court as an "expert," and whatever he says is "true" is regarded as such. The only gatekeeper for the process is the trial judge. But how does he know what is reliable? Like the lawyers, the last time he studied science was in high school. This situation couldn't be more perfect for the power companies and cell telephone companies, for the law firms that represent them in court, and for the law firms that sue the companies. What these law firms call "science" is a distortion specifically designed to promote victory, not truth or justice; the firms are to science what Rush Limbaugh is to dialogue. The scientific testimony offered on all sides in the EMF cases for the last thirty years has been nothing more than eristic, and not the principled version as in Protagoras or Gorgias, but the kind we get from Euthydemus or Dionysodorus. I finally reached the point where I decided – another motivation – that I had a responsibility to show how our legal system often prevents reliable science from entering the mainstream of society. To tell this story you have to be a working scientist, a lawyer, and a person who is free to do what he thinks best. I met those criteria, and I don't know of anybody else who has, so I felt qualified and responsible. I'm not trying to solve the underlying problem – nobody commissioned me to do that – just to illuminate it.

EMFs were an obscure, uncharted area when I started to study them. Perhaps I found the area attractive because I had little competition, or perhaps because I sensed that EMFs were important, I don't know which. I did my first experiments in 1972 and found EMF-induced effects in rats and mice. At that time, there was no ready explanation in terms of prevailing theories for what I observed, nor any obvious application. The experts told me that there was no "persuasive" evidence that such effects could occur and therefore that there was something wrong with my experiments. For a long time I thought "persuasive" was the most hideous word in science because it was a presumably objective way of expressing a completely subjective sentiment. Then I came across "junk science," and I had to admit that was even more hideous.

I began to keep notes of who said what regarding EMFs, and why. Around the time that I was interviewed for *60 Minutes* I began to suspect that my experiences were important. I taped Mike Wallace the day he came to our lab to tape me. No one told me, "You ought to write about that," I just started to do it, and I began collecting a vast amount of material, letters, reports, minutes of meetings of state and federal agencies, and documents that had never been intended to see the light of day but which I obtained as a result of Freedom of Information requests. Sometimes the information came directly from the agencies, other times it came as a result of appeals from decisions denying information; my biggest single haul was four thousand pages of correspondence and reports from the Department of Energy, which arrived after I appealed to Vice-President Mondale. So, I didn't have to trust too much to memory. All I had to do was decide what to put in and what to leave out.

First, what to put in? Well, the book is about a man on a journey through the world of science. In the beginning he doesn't even realize that he's on a journey. Instead, his metaphor is one of science as a beautiful object, to be possessed. He sees that science can be selfish and tawdry, but that doesn't diminish his love for it. One day he chooses a destination and begins heading toward it, not knowing if he will ever arrive, but completely content with simply the process of traveling. In essence, then, the book is about a man's growing understanding of the world of science. So I put in the moment in my youth when I realized that the train of life was barreling by and I had to jump on somewhere, and I picked physics. I put in Sam Ensor, Father Wallner, and Nietzsche because they taught me important things about how to travel. I put in Dr. Becker, of course. I never really knew whether he liked me, but he was the greatest man I ever knew; it was during the seventeen years I worked in his laboratory that I learned to be a scientist, and discovered my destination. I put in stories about people who helped me along the way, and about those who tried to stop me from going where I was going. Trying to stop me from doing what I wanted to do, it turned out, was roughly like trying to stop a bull by waving a red flag.

What to leave out? Well, I'm not writing a textbook so I left out all of the deep science and math of EMFs. I'm not writing a legal treatise so I left out almost all of the legal mumbo-jumbo of the court cases in which I was involved. I didn't include anything unless I had a good reason to put it in, and I refrained from indulging in sentimentality.

At one level, this is a book about who was right about the health hazards of EMFs, and it turned out that I was right. I don't see the point of being right if you don't write about it, and now I'm mature and confident enough to do that. But it's dangerous to describe yourself as a victor because you run the risk of sounding arrogant or impudent. I did the best I could to leave out hubris, or at least to deflect attention from it.

My greatest achievement is that I've been married to the same woman for forty-five years, all four of our children earned advanced degrees in law, science, or business, have happy families, and we all still love one another very much. I put in almost nothing about this because it did not directly serve my story even though my family was as essential to my life as the air I breathed.

I intended to write a work of nonfiction literature about what scientists actually do in the laboratory, why, and how lawyers use them in courts. I felt that my most serious responsibility to the reader was to tell the truth. Second-most, was my responsibility to avoid trivial matters; otherwise, the story would drown in a sea of countless insignificant details. The distinction that really mattered to me was between what was true and important, and what wasn't.

One narrative consequence of my sense of the truth is the use of compression (time, characters, or both) to express in a single story the meaning of a related series of experiences. For instance, I saw many testifying experts exhibit facets of a particular kind of scientific disorder. Combining their behaviors (in Anthony Erdgas) made possible a coherent presentation of how such people think and talk. Similarly, Patty Ryan was the unitary embodiment of a species of defense lawyer against whom I fought, and the chapter on risk assessment was based on contacts with many power-industry contractors. In these cases, compression required me to alter names in order not to misrepresent the words of real people. The reader can see in the Index those characters who have been given invented names.

When I look back on my life I am amazed at how fortunate I've been. I'm the son of a working-class Italian immigrant, yet there are few people whose education has been as broad and deep as mine. I can't remember a day when it wasn't a joy to go to work in the laboratory. While working there I discovered something new and important about nature. I also learned something about myself. I'm not weak. I'm not a coward. I'm not afraid to speak to power about my understanding of the truth. I did it. With each controversy I became even happier and more energized, and I survived to write this story. I hope you find it edifying.

Part I

Growing Up: 1956-1974

*Forget the pastimes of a child: You are
a boy no longer. Or have you not heard
what fame Orestes gained when he avenged
the murder of his father?*

(*Odyssey* 1: 295-298)

I
Starting Out

◆

Nobody knew for certain but everybody had an opinion. Some guys in my neighborhood thought that pitching overhand was best for stickball, but others favored underhand. One day I proposed that we play two games in which both teams pitch underhanded in the first and overhand in the second. Somebody suggested we play only one game in which each team used a different method, but I pointed out that we wouldn't be able to tell whether the pitching technique or the ability of the players accounted for the results. Somebody else reasoned that the teams would be tired after the first game so the second game should be played the next day. But that wouldn't work because Jimmy Lynam had basketball practice and Dickie Gilpin had to visit his grandmother. After further heated discussion we decided to try my idea. The teams split the games, and fewer runs were scored in the second. I took the results to mean that the overhand method was better, but many disagreed with my interpretation. That's the way it was on the block of Greenway Avenue in West Philadelphia where I grew up. There were always arguments but nothing ever seemed to get settled.

The big guys were at the top of the hierarchy. A few had graduated from high school but most had quit or been thrown out. They chose the sides for our street games, played cards and dice in the alleys, smoked, drank beer, and passed around pornographic pictures. One night they gave Gilpin a cherry bomb which he lit far from the light on the power pole so the neighbors couldn't see. But the fuse stopped hissing, and as he crawled around looking for the cherry bomb it exploded and blew out his left eye. I thought a lot about who caused the accident, whether it was the big guys, Dickie himself, me for not stopping him, or God, but I didn't reach any conclusion.

The power poles were on my side of the block, at more or less equal intervals. Three wires spaced about a foot apart ran from pole to pole. I never knew whether the electricity came in from 60th Street and went down Greenway Avenue to 59th Street, or whether it was the other way around. There were no power poles on the two streets that dead-ended on Greenway Avenue from the north; instead, wires were strung across Greenway Avenue and ran along the backs of the homes on both streets. I had thought that electricity being so close to the houses was dangerous, but one day I noticed that there were wires attached to the back of our house. After that I stopped worrying about the electricity.

The nearest basketball court was in a colored neighborhood, so we just nailed a rim and backboard on a power pole and played in the street. If a shot hit the back of the rim just right the ball would shoot downwards, catch the edge of the curb, and ricochet into traffic. I once lost a basketball under the wheels of a truck on my first shot after I took the ball out of its box.

We played boxball in the street because there was no place large enough in our neighborhood to play baseball. The pitcher would deliver a hollow rubber ball on a bounce and the hitter would swing with his hand. A miss was an out, as were two fouls. Every time I missed I claimed "tip," which was a foul and therefore afforded a second chance.

We rarely played tackle football because the only available patch of grass was no bigger than my grandfather's victory garden. During one game I went low to make a tackle and caught the runner's knee squarely on my nose. I heard nothing and saw only flashes of red and blue light for I don't know how long before I came to my senses. I wondered where those lights had come from.

We played touch football in the street. Sometimes the ball would hit the wires that went across the street. If the pass was caught the play stood, otherwise the offensive team claimed "interference" and we replayed the down. One day when I was on defense, a pass hit the wires square and fell to the street like a dead bird. As usual the other team cried "interference," but I argued the point. "'Interference' is really something that someone shouldn't have done," I said, "like when a catcher sticks his mitt in the way and stops a batter from hitting the ball. The wires didn't interfere with us, we interfered with the wires." I thought I was making sense. "You're claiming interference only because you didn't catch the ball *after* it hit the wires. Whether it's 'interference' shouldn't depend on what happened *after* the contact occurred." The guys on the other team mostly ignored me, but I went on anyway. "Suppose the ball had hit a parked car. Nobody would have called that interference. What's the difference between wires and a car?"

In the end we replayed the down, but I thought I was right and I felt good about putting up a fight.

Starting Out

• • •

Early in the fall of 1956 Denny DeMarco and his family moved to Greenway Avenue. He told me he wanted to be a singer like Elvis Presley, and asked me what I wanted to do. I had no idea. The only things my father had ever said about that were "Work with your mind, not your back," which I took to mean that I shouldn't work in a factory, like him, and "In America you can be anything you want." I knew a lot of guys who had quit school and gotten a job, but my sense was that they weren't intending to be anything in particular, and what they wanted was to make money to buy a car and get girls. Bob Charles was typical. He was skinny and had a hook nose and long black hair that sparkled because dust stuck to his pomade. His typical outfit was dungarees and a white t-shirt with the sleeve on the right side folded back to form a pocket for cigarettes and matches. He had a fast '52 Mercury with skirts and chrome trim, and a different girl each week. One evening he told me how much he made pumping gas and changing tires and what the payments were on his car, and it turned out that he was working almost entirely just to keep the car. I too wanted to get girls but what Bob was doing seemed like a dead end.

Denny could spell any word. One night I tested him with a dictionary and the only words he misspelled were the ones I mispronounced. Almost as amazing, he knew what the words meant, and why. His explanation of the meaning and origin of "antediluvian" took my breath away. He knew the names of the Greek gods, the twelve tasks of Heracles, the adventures of Odysseus after he left Troy, and many other stories from Greek mythology. He knew the difference between Socrates, Plato, and Aristotle, and he knew about Alexander the Great. He could describe details of Civil War battles and he knew the names of the four Japanese aircraft carriers sunk at the battle of Midway. My father had bought an Encyclopedia Britannica from a door-to-door salesman that for six years just sat at the bottom of the stairs in our living room. I began reading the encyclopedia, and I bought my own copy of Bullfinch.

Denny invented a game called whiz-bang; he said he got the idea from a television show that featured a professor from Columbia University who answered questions about anything. None of the guys in my neighborhood were interested in playing except Dave McCarron and me. Denny would ask a question on any topic; if I wanted to answer I would say "whiz" and if Dave wanted to answer he would say "bang." You got one point if you answered correctly and lost a point if you answered incorrectly, and whoever got 10 first won. After a while I would routinely be ahead 9-0 and Denny would be reduced to asking questions like, "Name three of Dave's aunts," or, "What did Dave have for breakfast this morning?" But I soon learned as much about Dave's family and personal habits as there was to know, so not even that artifice created competition. As time went by, Denny and I got into a lot of arguments regarding whose version of the facts was correct, and we drifted apart.

• • •

In high school, there were smart guys, dumb guys, and a middle group. I thought of myself as being at the top of the middle group, ready to bust into the smart-guy group. I had taken college preparatory courses to preserve my options, but by the time I finished my third year of high school I still had no idea what I wanted to be. In a physics class early in my senior year my teacher dropped a ball from different heights and showed he could use an equation to predict exactly when the ball would hit the ground. He also did experiments with meters and light bulbs in white porcelain sockets connected by a network of wires. Whenever he unscrewed a bulb he could predict which of the others would also go out. I wanted to do experiments like that, so after class I took a meter he had used to measure the current.

That same month the Russians launched Sputnik. It kept circling around the earth day after day without coming down, which I had thought was impossible. A newspaper article about Sputnik quoted the director of the Yerkes Observatory, an Army general, and a professor from a university in California. I wrote and asked them what someone who wanted to learn more about Sputnik should study in college. They answered that I should study physics, and that's what I decided to do.

A few days before I graduated from high school, a classmate named Joe Spaeder died from cancer. I hadn't been aware that students in high school died from cancer. I didn't want to get cancer, so I asked people who I thought might know, "How do you get cancer?" I soon learned that the answer my mother had for so many questions was as good as any other answer. "It's a mystery, nobody knows."

• • •

Soon after I started at St. Joseph's College I was in big trouble. Everything was a struggle. In Theology I couldn't understand how the teacher knew what Christ said or what He meant. In German I penciled the English translation in my copy of *Drei Komeraden* because I couldn't remember the vocabulary words and I always got caught. I had no chance of learning German grammar because I didn't understand English grammar. My chemistry teacher didn't like me. He gave me a D on every test even though, I thought, my answers were the same as my classmates who got C's.

My biggest problem was calculus, which was taught by Sam Ensor, a short, balding man who wore a frozen smile that showed no teeth. He carried a textbook in one hand and a pointer in the other, and he entered the classroom with a flourish, like Loretta Young. He gave four tests during the semester, plus a final examination. In addition to grades of A, B, C, D, and F, he had a grade called F_0. You got an F_0 if, while he was marking the examination, you had done so poorly that even if you got everything right from that point on you would still get an F. I got an F_0 on my first test, which fed my growing fear of failure and the uncertainty of its implications. In grammar school I had always been terrified by the nuns, who somehow knew that beating me would make me better in the eyes of God. Once while I was in high school I was picked up by the police for loitering in front of a hoagie shop. I had sat for hours on a wooden bench at the police station, worrying that my father would have a heart attack when they called him. Then there was the night I went drinking with friends before a dance. While driving from the bar to the dance we sideswiped about a half dozen parked cars. Those emotions weren't even a patch on the fear that my calculus class generated in me, which was that I would flunk calculus, get kicked out of college, and wind up working in a factory.

On my second calculus test I got another F_0 and at the top Ensor wrote, "I don't believe physics is for you." He couldn't tell that I had flunked the third test until he marked the last problem, so I got an F, but on the fourth test I got another F_0. I went to see him and asked, "What should I do?"

"What do you want to do, Andrew?"

"I want to study physics. I want to study physics so bad I can taste it; I love physics; I can't imagine living life without being able to study physics." I laid it on pretty thick, I guess, but I was sincere. Physics *was* the most exciting thing I'd ever seen.

"Well, you need to pass calculus," he said. "If you flunk it, take it again. If you flunk it nine times, take it a tenth time. There's no alternative if you want to study physics."

Things must have looked bad from the outside. The other physics majors had more or less stopped talking to me after my second F_0 thinking I'd soon be gone. But inside, there was really no problem. The conversation with Ensor had crystallized things for me. I expected to flunk and then take the course again, and again if I had to, until I passed, which, one day, I would do. I felt at peace, and that I had control of my life.

I took the semester final on a cold December morning. From the moment I touched the examination paper I was in a zone all by myself. The first problem was to solve an equation, which I immediately saw how to do. I also had no hesitation or uncertainty with the second problem, or the third. There wasn't a problem on the test I couldn't solve. That morning I felt as if there wasn't a calculus problem in the world I couldn't solve.

Three days later I went to Ensor's office to get my test grade and learned it was an unprecedented 99%. But what about my course grade? He sat behind his desk with his arms bent at the elbow and his palms horizontal, alternately moving them up and down slowly as if to simulate the scales of justice. In his right hand sat D, and in his left sat F. D was heavier.

I didn't breathe until I saw that I got four C's and two D's that semester, but I survived, and the 99 confirmed to me that I really was a smart guy.

• • •

Near the end of my first year in college my uncle Larry offered me a summer job working on a maintenance crew at a paper mill about seventy miles north of Philadelphia. I needed a car to take the job, so my father gave me our family car and he took the bus to work. The car had a deep significance for him, and in ordinary times I was lucky to get it one night a month. He had once said to me, "I would never have had a car if I had stayed in old country." But he gave me the car for the summer without the slightest bit of resentment. That was the way my father told me he loved me. He never said it in words.

The paper mill was a dirty, noisy, dangerous place. The men who worked there were glad they had jobs but they always looked forward to quitting time. The boss of my crew was an old man named Walter who had no formal education, but who knew how to do things. He carried a dog-eared copy of the Bible in his back pocket which he read on his breaks. One day we had to replace a rotted railroad tie. Walter studied the situation for a few minutes and then gave orders for the placement of pry bars at particular places on the track, and when we leveraged them the track elevated at exactly the right place. He didn't know anything about physics, but he could solve any problem he faced at the mill. Another time, he positioned the bucket of a huge payloader against the front face of a storage bin. When he raised the bucket, the bin tilted up and men went into the gap to reposition wooden blocks. If the bucket had slipped the men would have been crushed. But Walter throttled the payloader engine perfectly so the bucket remained motionless. He went to work every day wearing overalls and boots and a dirty cap, unlike Larry who wore a white shirt and tie and polished shoes. Larry had an education and could move from job to job in different parts of the country. Walter had no education, so he had worked at the mill for forty-two years.

One night as I drove back to the mill after visiting my family in Philadelphia, I pulled over because I could feel myself falling asleep at the wheel. Suddenly an intense beam of light flashed over my head. About a minute later what sounded like an explosion started coming toward me. I thought I was having a vision. Then I saw a train that, apparently, had stopped before crossing the road in front of where I had parked and had started up as I sat there. The next morning I told Walter about the incident and he said it meant I was going to be "the St. Paul for an idea." I didn't know what that meant, if it meant anything.

During one of my many conversations with Larry I told him how rough my first year in college had been, and that I didn't think my teachers liked me.

"Do you ask questions in class?"

"Yes."

"Maybe you don't ask them the right way."

"How many ways are there to ask questions?" I asked.

He said I should phrase my question as if it were based on my inability to understand rather than the teacher's shortcoming in explaining the point.

"You should say, 'Did I understand you to say such-and-such?' If you're low-key and non-threatening, the conversation will last longer and finish better."

I resolved to do that.

• • •

Physics had equations for mechanics, heat, sound, electricity, light, and the atom, and they all yielded rock-solid, certain knowledge about the real world, not just about x's and y's as in Ensor's class. If I memorized the equations I could control what happened, like a master controlling a slave. Where did the equations come from? Maybe it had happened this way. God told archangel Michael that He intended to create a world where everything would be subject to laws. Michael asked God how He would implement this plan. "Equations," God answered, and at that instant He created equations in their countless numbers and infinite diversity and said, "All the events in my world will unfold in accordance with equations. An endless number of events needs an infinity of equations."

I learned to see patterns in the problems at the ends of the chapters in the physics texts, so when I solved one problem I knew how to solve others like it; many of my classmates didn't see the patterns. I got an A in my final math class with Ensor. At the top of my last test he wrote "I remember when you were a struggling fish, and I almost threw you back."

A gap progressively developed between me and ordinary people who didn't know how to use the equations and, more and more, I felt special. Even so, some things about physics were disquieting. Fudge factors, for example. I had first seen them in high school when we were supposed to predict when a falling ball would hit the ground. I never got exactly the number predicted by the equation, although I always came close. If a ball was supposed to obey an equation and it didn't, obviously there needed to be some other equation that the ball obeyed. I had a conceptual problem with why the ball obeyed any equation in the first place, and a bigger problem with how the ball chose which one to obey. There was always some excuse for why our numbers were never exactly what was predicted, and why each of us got a slightly different number. But we were systematically taught to focus on what could be predicted, and the great utility of that was clear to me. So I accommodated myself to the use of fudge factors, which I came to understand simply as devices that turned observations into answers that I knew were correct even before I did the experiment.

I was uncomfortable with "force." It was supposed to explain why the motion of things changed. Force was said to be the "cause," and the change in motion was the "effect." But what caused the force? It couldn't be another force because, by definition, forces caused changes in motion, not other forces. I could accept that gravity was a force, but what caused gravity? And what caused motion that wasn't changing, like Halley's comet? The book said "nothing," which I was expected to believe because that's what Newton had said.

I had another problem—at that time, really just a question. Physics books never mentioned the most important thing in the world, life. If all you read were physics books, you would never know that there was such a thing as life. Why?

The other side of my education consisted of standard Jesuit courses in philosophy and the humanities. Initially I resented those courses because they took time away from physics, but then I began to see that they were actually teaching me to think. Father Wallner's class in logic did that.

"Joe can swim, fish can swim, therefore Joe is a fish; anything wrong with that, Marino?"

I didn't think it was right, but I couldn't tell him why. After a while, however, I learned why. It was really just common sense, and I could figure it out if I made little drawings that represented the various possibilities and then asked whether something would or would not be true, given that something else was or wasn't true.

Starting from a particular fact and figuring out its implications was exciting, and seemed to be a good way to deal with everyday life. The first time I tried was with my uncle Angelo. We had begun arguing about politics. I was sure I was right, and I had a copy of *Newsweek* that proved it. He was passionate about everything, and when he argued about something you got the feeling that who won or lost really mattered. In the middle of our argument he had to go somewhere, but he wouldn't concede that I had won. I followed him out the door and down the sidewalk toward his car, all the while arguing that I was right, and that he would realize that logic and truth were on my side if he would just look at the article. But he said he didn't care what was in *Newsweek*, and that I shouldn't believe everything I read.

Father Wallner was also passionate, but in a more controlled way. He was middle-aged, of medium build, with a full head of black hair mildly streaked with gray. He would pace back and forth in front of the class, with one hand tucked into the sashlike girdle that was part of the Jesuits' standard dress. His usually serious demeanor would soften a little and a hint of a smile would cross his face just at the moment he made his point. The Church had banned many books as heretical but he wanted to read them. He had asked for permission but been turned down, so he kept appealing the decision through the Church's labyrinthine hierarchy. Magnificent independence and tenacity, as I saw it. I wanted to be like him.

The Jesuits would typically raise a problem and then solve it in a way that gave you the impression that there was really no reasonable alternative. For a while I was just swept along. I bought their argument that there had to be a Natural Law, and that the complexity of life proved that there was a God. Then I started questioning what they were saying. I wasn't disagreeing because I didn't know enough to disagree, I was only trying to understand what was meant or how they knew. "The little Green Man on the wall" in my epistemology class was an example. Various lectures were peppered with that reference whenever the teacher wanted to distinguish between what something "really" was, and what an ordinary observer saw. The little Green Man was supposed to be able to see "the thing in itself," which was the underlying reality you couldn't see but whose existence could be inferred with certainty, using reason. I wanted to know how anyone could know about a world they couldn't see or experience. I pressed the point one day until my teacher called me "an agnostic" and told me to be quiet.

I felt I probably wasn't the first person to have doubts about the unseen world, but the books in the philosophy library dealt only with the Jesuits' scholasticism, which had no such doubts. When I went to the library at the University of Pennsylvania, however, I was shocked to find many other flavors of philosophy, most of which, unfortunately, were impenetrably obscure. Nietzsche was one of the worst. There was a lot written by and about him, but it made little sense to me. One day, in metaphysics, when the teacher was talking about prime matter and substantial form which, according to scholasticism, were what everything was made of, I asked, "What would Nietzsche say about that?" No answer. In ethics I asked, "Why did Nietzsche say that the Judeo-Christian tradition is a form of servitude?" Again, no answer. In epistemology I wanted to know what the teacher thought Nietzsche meant when he said truth was "a mobile army of metaphors." No reply.

I came across existentialism; it dealt with "the problem of existence," which seemed important. But the end of my college career was approaching and there were no courses left where "existence" was to be a topic. So I tried to identify what the "problem" was by reading Kierkegaard, who supposedly started existentialism, but his books were dull and they put me to sleep. I thought I might understand what existentialism was if I found someone who disagreed with Kierkegaard, and explained why. That led me to Hegel, whom I had first heard of in Dr. Burton's history class. Dr.

Burton was an impressive man who made me see events and institutions in a way I hadn't previously imagined. A story he told about the Korean War had that kind of effect on me. For weeks he had been fighting in the rain and mud on a hill that he identified with a three-digit number, and finally he was ordered to the rear. He trudged down the hill, walked for several miles, and came across a Red Cross station in a tent where he got coffee and doughnuts. At this point in his story Dr. Burton paused, rolled his lower lip, and drawled, "...and they charged me 20¢."

The fact that Dr. Burton had mentioned Hegel suggested to me that he was important. But a book about Hegel said he thought that "physics was seriously defective because it really doesn't explain anything," which I thought was absolutely ridiculous. In the end, I just forgot about existentialism.

On the last day of my senior year Father Wallner gave us some parting advice: "Always do the right thing. No one can tell you what that is. You must figure it out for yourself. You now have the tools to do that."

2
Incubation

When I finished college I accepted a job with the RCA Corporation and enrolled part-time in graduate school at the University of Pennsylvania. The company's training program involved a series of assignments at different locations. I was sent initially to a lab that was developing the use of lasers to measure distance. Whenever I pushed the "on" button, the laser emitted an intense pulse of red light that instantaneously traveled across the room and struck a target on the far wall. The energy that bounced back to where I stood was used to calculate the distance to the target. Some of the engineers had developed cloudy vision, but their major concern was whether the Army would renew the contract.

At my next assignment one of my tasks was to measure microwaves radiated by an antenna. The Ph.D.'s decided what measurements I would make and how I would make them. The technicians mostly just soldered circuits. One day I soldered something. Thirty minutes later a man handed me a notice that said, "Warning – you are engaged in activities that are to be carried out solely by union personnel as provided in the contract between the International Brotherhood of Electrical Workers and RCA. Cease and desist immediately or you will be subject to disciplinary action." I crushed the paper into a ball and threw it away, but my boss told me not to do any more soldering.

The trainee whose desk was next to mine objected every time I lit a cigarette.

"It's bad for your health," he said.

"How do you know?" I said.

"Common sense," he said. "The lungs weren't designed to cope on a day-to-day basis with the smoke of burning leaves."

"Well," I said, "every day we stand in a microwave beam and measure the radiation pattern. Our skin wasn't designed to stand in a microwave beam. Does that mean microwave exposure is a health hazard?" He had no reply to that.

At Penn I took a course on electricity and magnetism. A special lecture one night was "Nonthermal Biological Effects of Electromagnetic Fields." Well-groomed Air Force officers in the audience contrasted sharply with the students, who mostly wore jeans and hadn't shaved. The visiting professor was Herman Schwan, a handsome man with wavy gray hair and a youthful face. He spoke with a heavy German accent as he filled the blackboard with equations about the effects of microwaves on red blood cells. I had no way of knowing that a day would come when Schwan and I would clash in court over whose notion of science was correct, or that concerns regarding the biological effects of electromagnetic fields would turn out to be my life's work.

During my third assignment I was asked to write a report about possible methods for miniaturizing electronic circuits, and to recommend the best method. One possibility was to make tiny versions of the individual circuit components and then solder them together under a microscope. Another approach consisted of controlling the chemical composition of a solid block of material so that different areas performed different functions, leading to what were called "integrated circuits." I didn't believe integrated circuits would work, so I concluded that miniature soldered circuits were the wave of the future and that integrated circuits would not be of any practical importance.

During my last assignment I had an opportunity to do my own experiment. The goal was to determine which of three brands of light bulbs was best. They had been designed to operate at three volts, but the test would have taken too long at that voltage so my boss, Mr. Schmidt, told me to use twelve volts because the bulbs would fail much faster. I set up a panel containing 150 light bulbs of each brand, and for the first few days all 450 bulbs burned continuously. On the third day they started to burn out, and when I ended the test seven days later, as planned, 87% of the Westinghouse bulbs had burned out, compared with 42% of the General Electric bulbs, and 21% of the Sylvania bulbs. I concluded in my report that the Sylvania bulb was the best.

After Schmidt read the report he called me into his office and asked, "Could you have made a mistake? Could it really have been that the Westinghouse bulb was the best?"

"That's impossible," I told him, "because the bulbs had different shapes and each bore its brand name."

He told me that equipment containing the Westinghouse bulb had already been delivered to the Air Force, and that I should change my report to say that the Westinghouse bulb was the best. When I refused, he got angry and said I would never work in his department. His threat didn't bother me because I had no desire to work permanently in Camden, NJ, which was where I thought God would have stuck the nozzle if He ever decided to give the United States an enema. But I wondered what I would have done if I had needed that job. Then I wondered how many things that I had read and accepted had been written by people who needed their jobs.

When my training period ended I was assigned to work on a project called magnetohydrodynamics. My boss, a Ph.D. physicist named George Frasier, said it would revolutionize how electrical power was generated. Each day he did calculations, searching for the equation that would change the world, and I tested them by measuring the amount of energy required to heat gases in a large tank, and the amount of energy we could extract from the gases. Unfortunately, on average, we always lost energy.

Even though Frasier struggled, I could see he was in charge of the project and had freedom to make choices. I decided I had to have a Ph.D. in physics so I too would be free. But RCA lost its funding for Frasier's project and in February I was assigned to a lab in Boston. I quit RCA and took a job at Westinghouse with the intention of leaving in September to study physics full-time somewhere other than Philadelphia.

At Westinghouse I worked in the Nuclear Power Department, which had a Navy contract to build nuclear steam generators for submarines. There were no Ph.D.'s at Westinghouse, just engineers and draftsmen who sat side-by-side in a large room adjacent to the factory. All the men wore white shirts but only the engineers wore ties.

I calculated the stresses that would occur where the steam generator was connected to the submarine. Using equations I got from a book and data about different kinds of steel, I figured out that the generator connections wouldn't break during a dive even if the force were twenty times that of gravity. It seemed as if everything in the world followed a mathematical law, and that if you knew the law you could predict what would happen. People were made of atoms, just like steam generators, so I wondered why there wasn't a law that predicted how people would behave. I saw no sense in believing that an atom removed from a steam generator and placed in a person would no longer follow a law. On the other hand, I didn't believe there was a law that made the atoms in me do the things I freely chose to do. I did not understand even how to think about the relation between equations and human behavior.

A draftsman named Vince Carbala had the desk next to mine. He spent many hours patiently answering my questions about how to read blueprints. When a machinist in the factory had a problem he would call me, and I would talk to Vince and then call back with the answer. I earned almost twice what he earned but he never showed any resentment toward me. He had been taking engineering courses in night school for almost twelve years, and he expected to finish in September. I confided in him my plan to leave, and we both hoped he got my job.

I was accepted by Notre Dame and Syracuse University, and was ready to accept the Notre Dame offer when I learned it was an all-boys school. Half the students at Syracuse University were women, so I chose Syracuse. At my going-away party Vince learned that Westinghouse had a policy not to promote draftsmen to engineers, and someone else was hired to replace me.

• • •

In Syracuse I rented a one-bedroom apartment. The day I met my landlord he was lifting a puppy over his head. I asked why, and he said if he could lift the dog over his head today, then he could do it tomorrow, and so forth. He didn't think the dog could grow enough in any given day to affect whether or not he could lift it. I knew that day would come, even though I couldn't prove I knew or predict exactly when.

He had a TV repair business. His strategy was to replace each tube, one by one, and see if that fixed the problem. If it didn't, he put the original tube back in and repeated the process until he found the bad tube. He was unable to fix my TV because, I learned later, it had two bad tubes.

The physics department was housed in an old stone building that looked as if it belonged in a Frankenstein movie. The department's recruiting poster claimed there was "a close faculty-student relationship," but the faculty were cold and aloof, at least toward the first-year graduate students. Dr. Trischka, whom we called "Trashcan," was the worst. We invented imaginary tortures to balance out the psychological hegemony he exercised over us in Experimental Physics. The winner by general acclaim envisioned Trischka sliding down a board lined with razor blades into a vat of iodine. My most frequent contact in the department was with Paul Gelling, who was the man in charge of the equipment we used in the physics labs. He hated the president and was the only person I saw smiling on November 22.

Each of my professors had an "interest." For some it was "general relativity," a theory of gravity and time pioneered by Einstein. Some of my classmates thought general relativity was exciting, but not me. Other professors specialized in "high-energy physics," which involved the use of the atom-smashing machine under the stands of the football stadium. I couldn't tell what the point was, except I knew atomic weapons were involved. I just didn't care about that kind of work. Another faculty interest was "solid-state physics." Some professors specialized in its theory, others in measurement techniques. I decided I wasn't interested, and that I wanted to work on something more meaningful than the next generation of transistors, lasers, or radar.

My classmate William was my best friend. We talked about politics, religion, sports, almost anything. Our longest-running debate was whether mankind had made any progress. He argued that people had once lived in caves and hunted in order to survive. Now, they lived in houses and bought their food at the supermarket. We have medicines, books, and myriad creature comforts. He took all this to mean that there had been progress, but what I saw was that as the physical stresses had decreased, the psychological stresses increased. People lived in tiny boxes in crowded cities, drove cars on congested highways, were fearful of criminals, and worried about what would happen to them when they got old. "Where is the progress?" I asked, more or less rhetorically.

Science was William's shining example of human progress. In the beginning, he said, humanity understood essentially nothing about nature. Fire was considered a gift from the gods and people carried an ember from place to place so that the miracle would not be lost. Eventually the secret of fire was discovered, the wheel was invented, and weapons were developed to give humans the edge over animals. The laws of planetary motion were discovered, so the rising and setting of the sun were no longer seen as miracles. Science led to the invention of the telephone, the computer, the automobile, the airplane, spacecraft, the discovery of blood circulation, and to knowledge of how the heart and brain work.

I thought William's argument begged the question. Certainly science had developed and was producing increasing amounts of information. The question, however, was whether that development evidenced progress. I was dubious because I didn't think science had elevated humanity as a whole. Science had helped some but hurt others, not like a tide that lifts all ships.

• • •

Father Charles ran the Newman Center, which was the focal point for Catholic students at the university. He was a round, animated, friendly man, half bald with a big nose. Our arguments were frequent but friendly. I always picked something he had to defend because he was a priest, but really couldn't because he had no evidence. Using the method of logic I had learned from the Jesuits, I bested him on indulgences, limbo, transubstantiation, purgatory, and St. Thomas Aquinas' proofs for the existence of God.

One Friday evening someone at the Center began showing a French film. The scene opened with a man and a woman sitting on a couch. First he unbuttoned the top button on her blouse, and then the second. Father Charles walked to the back of the room and sat near the projector. The actor unbuttoned the third button, and moved the dark blouse to the sides, revealing a white bra. Father Charles put his hand on the projector's on/off switch. Then, in a lightning-fast series of events, the man popped out her left breast, the camera zoomed in so that the nipple covered the entire screen, and Father Charles switched off the projector. I was outraged at what I thought was rank censorship, and I walked out. A graduate student who had just arrived said to me, "What the hell was that all about?" "Didn't you see what he did?" I asked. "All I saw was a big pimple on the screen," he said.

There was a dance that evening at a local church. When I arrived I saw a bunny-hop line was snaking its way around the dance floor, so I clamped my hands around the waist of a slender, attractive woman with red hair. When the dance ended I learned her name was Linda Lavin. Her father was mostly German but she identified with the Irish on her mother's side. She had graduated from LeMoyne College, which surprised me because I didn't think Jesuit colleges accepted women. LeMoyne did, she told me, but the entrance requirements were five points higher for women. She was teaching first grade in the western suburbs of Syracuse, where she lived with her parents.

Linda was impressed by my status as a physicist-in-training, and I tried to exploit the opening. "We discovered the omega-minus particle today," I said, using "we" to include myself in a group of physicists on Long Island who found the particle in an atom-smasher. The discovery had been in the news that day, and I thought my line was one of the all-time greats.

"What's it for?" she asked.

I didn't know quite how to answer. "What do you mean what is it for?"

"I mean, what does the omega-minus particle do?" she said.

"It doesn't really do anything. It can't because it lives for only a millionth of a millionth of a millionth of a second."

"You mean it's alive?" she asked.

"Well...no, not exactly. When two atoms smash into one another the particle appears and can be seen using special instruments, and then an instant after it's seen it's gone."

"Where did it come from and where did it go?" she asked, but I changed the topic.

Despite my inartful approach, in the subsequent months we fell in love and made plans to marry.

• • •

Near the end of my first year at Syracuse the question of my "interest" became acute because I was expected to pick a lab and begin my doctoral research. Dr. Charles Bachman's interest was "biophysics," which intrigued me even though it was unpopular with most of my classmates. I went to his lab to learn what biophysics was. Soon after I arrived I saw a student measuring electrical current in a rat, using a wire connected to the rat's tail. Another student was analyzing data from bones that Dr. Bachman had gotten from a doctor who worked across the street from the University, at the Veterans Administration hospital. I asked Dr. Bachman whether I could do some biophysics research. He pointed to a tank containing goldfish and told me to measure the voltage between the fish's nose and tail. I wondered what I had gotten myself into, and where it would lead.

• • •

Soon after I began working in the lab, the doctor at the Veterans Administration hospital got a government research grant and began looking for a physicist to work on the project. Dr. Bachman recommended several of his graduate students for the job, including me. On the day of my interview I went to the hospital, knocked on the door of a research laboratory on the ninth floor, and heard an authoritative "Come in." I did, and met the man who was to be my boss for the next seventeen years, and a dominant influence on my life.

"Hello, Andy, I'm Dr. Becker," he said. My family had always called me "Andrew" and I preferred it, but it didn't seem like a propitious time to say so. He was a wiry man in his forties. Between puffs on his pipe, he told me in a businesslike way about his experiments involving animal magnetism, and electric currents in the brain and in bone. He said his goal was the regeneration of missing limbs.

"When people lose an arm or a leg," he said, "I want to be able to treat them so they grow a new one."

"Is that possible?" I asked.

"Why not? If you grew an arm the first time, why not a second time?"

The following Sunday there was an article about him in the newspaper. The headline said, "Human Transistors — A Medical Adventure," and photos showed him measuring the voltage from a salamander and doing other scientific things. The article said that he and Dr. Bachman worked together on "a journey into hitherto unknown areas involving the electric fields within living organisms." According to the article, they had discovered that "billions of tiny transistors operate in our bodies' bones and nerves to flash out signals for growth," which was a "discovery that could have unprecedented implications for curing human ailments."

I thought that the kind of work Dr. Becker was doing was wonderful, and I decided to take the job if it were offered. In the meantime I finished my first year in graduate school and got a summer job with the National Aeronautics and Space Administration at Langley Field in Hampton, Virginia. As I drove into Hampton I entered a traffic circle and was almost rear-ended by a pick-up truck. The screeching of the truck's brakes caught the attention of a cop who gave me a ticket.

I was assigned to work for an engineer named Max. He had a huge analog computer which we used to simulate rendezvous that would be attempted between the Agena and Gemini-7 spaceships. He called the computer "Baby," and we programmed her by using long cords, like an old-time switchboard operator. I felt awed standing in front of Baby, and privileged to be able to plug cords into her. How amazing that Baby not only obeyed equations, like everything else, she could actually solve them.

Against Max's advice I went to court to contest my traffic ticket. I thought I wouldn't get anywhere if I contradicted the cop, so I took a different approach. I told the judge, "Your Honor, I got the ticket because I didn't yield, but I believed I had the right-of-way. That's the law in New

York, Your Honor. I got the ticket the first day I was in Virginia. I had no chance to learn that Virginia law was different from New York law." I didn't know what the law in New York was, but neither did the judge, and he dismissed the ticket.

During my last week in Virginia, as I drove through Langley Field I passed the test facility for the lunar excursion module and saw a prototype of the module. It seemed amazingly frail, and my first thought was of the courage of the men who would fly in that machine. Then I realized something else – that they must have been preparing their whole lives for the trip even though they couldn't have known that was what they were preparing for.

Just before I returned home Dr. Becker notified me that I had the job. It was a sweet deal, a full-time position as a research biophysicist in his laboratory at the Veterans Administration hospital. He had arranged with Dr. Bachman to allow my work to count in the physics department as my doctoral research, and he had adjusted my duty hours so that I could complete my last year of coursework.

3
Warrior

♦

The day I started work in Dr. Becker's laboratory I felt as if I were setting sail for someplace wonderful. Many things in the lab were new to me: pieces of bone, x-rays, syringes, elaborate microscopes, surgical instruments, and specialized electronic equipment. Something else was new, although I didn't recognize it then. My life had always been completely controlled by others. Whatever distinctions my teachers had made between correct and incorrect, true and false, right and wrong, or important and unimportant, were what I had accepted. That uncritical acceptance had superimposed structure and order on what otherwise would have been for me a chaotic world, but that acceptance had also relieved me of responsibility to think for myself. It had been much the same in industry, where I drew graphs or did calculations as instructed. I had gone to graduate school because I wanted to be free to make decisions, but I had no insight into the responsibility that freedom entailed, or how difficult it was to think rather than let others do it for me.

A biologist named Fred Brown worked in the lab. The first time I saw him he was inside an electrically shielded cage making voltage measurements on a small cube of human bone. Wires connected it to a rack of measuring equipment and a chart-recorder. From time to time the needles on the meters slammed into the stops at the end of their allotted ranges, plainly indicating that he was doing something wrong. I thought to myself that he needed help. Later in the day, as he talked about working at the hospital, he happened to say something about a urine test on a patient. I said, "You mean they analyzed his urine to see if he was sick?" He nodded, so I said, "That's the last place I would think of looking," to which he replied, "That's because you don't know anything about biology."

Dr. Becker didn't tell me to do anything specific so I began reading his published papers in the hope of getting a hint about what to do. In his first publication he had described the similarity between the neuroanatomy of salamanders and the pattern of voltages that he measured on their skin. His next paper had reported on experiments in which he amputated salamanders' limbs and measured the voltage at the amputation sites; there was a big change immediately after he cut off the limb, but the voltage returned to normal as the limbs regrew. For comparison purposes, he had also amputated the limbs of frogs, which don't regenerate their limbs, and found that the pattern of voltage change was quite different from that in the salamander. He had concluded that the nervous system of the salamander somehow regulated limb regeneration, and that the frog didn't regenerate because its nervous system couldn't make the proper electrical signal.

In a paper published in *Science*, he had presented evidence that a flow of electrons in nerves gave rise to the voltage that controlled the regeneration. In a paper in *Nature* he had concluded that the brain was like a battery and the nerves were like wires; some types of nerve cells carried the current away from the brain and other types formed a return path to the brain. In a second paper in *Science*, written with an orthopaedic surgeon at Columbia named Andrew Bassett, he reported that bone emitted electrical signals when it was bent or squeezed. How the signal arose and what its purpose was remained a mystery.

In a book chapter he had described studies by scientists who experimented with the effects of magnetic fields on people and animals, and he concluded that living things could somehow detect magnetic fields. The same year the chapter had appeared he had also published the results of a study of the pattern of admissions to psychiatric hospitals; he found that its monthly changes matched the pattern of changes in the earth's magnetic field, suggesting that the changing field might be associated with psychiatric disorders.

In an article that described what made bone grow, he wrote that bone was made up of millions of microscopic transistors formed by the arrangement of tiny crystals of bone mineral on a protein matrix. When mechanical forces were applied to bone, as in walking, the transistors generated an electrical signal that regulated the activity of bone cells and made them build new bone. Among other things, that explained why prolonged bed rest resulted in osteoporotic deterioration of the skeleton. In another paper, he and Bassett had apparently proved the theory that electricity could make bone grow. They built small battery-powered electric circuits and implanted them in the legs of dogs; after twenty-one days, a large amount of new bone had grown at exactly the point where the electricity exited from the circuit into the tissue.

Dr. Becker's research accomplishments were awesome, and each new publication had added to his stature. He had been awarded research grants by the Veterans Administration and the National Institutes of Health, his work had been featured in national magazines, and he had received the Veterans Administration's highest honor for medical research. He lived in a beautiful house that had a studio where he did oil paintings, and another room where he made sculptures out of pecan wood that he had shipped in from Louisiana. He and his wife had a greenhouse where they grew orchids.

The janitors, electricians, and elevator operators who worked at the hospital idolized Dr. Becker. They often came to his office to talk about their aches and pains. The people in the hospital administration, however, had a quite different attitude. They thought he was too independent, and derisively called him "the cowboy" because of the hat and boots he wore and the pick-up truck he drove.

Soon after I had begun work in the lab I asked him how he got started in research. He told me that when he was in medical school he had wondered why some animals could grow a new limb and others couldn't. "After all," he said, "the limb of a salamander is as anatomically complex as yours or mine. It has the same bones, muscles, and nerves as those in a human being; they're just smaller. It occurred to me that the only thing missing in people might be a special signal that turned on the cells and instructed them to build a new limb. I knew from experiments done in the 1940s and 1950s that nerves were somehow involved in the ability of the salamander to regenerate a limb, but I didn't know how. I decided I would do experiments after I finished my medical training, and that's why I took this job at the Veterans Administration."

"Why here?" I asked.

"The Veterans Administration has a program that funds its doctors to do some research. If it didn't offer this perk nobody would work here because it's a crappy job for a doctor."

I began thinking of Dr. Becker's project as my own. No other kind of work that I knew about even came close to providing the feeling of importance that I got when I pictured myself doing what he was doing. But this identification of my "interest" heightened my angst because I didn't know if I had what it took to come up with ideas, like him. I thought about how awful failure would be, and about how Ensor's advice wouldn't work if I did fail.

I wanted to know what Dr. Becker's formula was, so I asked him, "How did you get the idea that magnetic fields and electricity have anything to do with nerves and bone?"

"The thing that impressed me most," he said, "was that the body knew how to heal itself." He stopped, puffed on his pipe as if he were drawing the story out of its bowl, and continued. "But nobody knew how that happened, and nobody was even studying it. What the biochemists did was cut out tissue, dissolve it, and then study the chemicals it contained. They didn't seem to realize that they had destroyed the organization of the tissue, which was something really important. Human limb regeneration seemed so impossible that everyone considered it unscientific to even discuss the subject. Then, the year I finished my residency the Russians launched Sputnik. Do you remember Sputnik, about 10 years ago?"

"I sure do."

"Well, after Sputnik suddenly there was a lot of money available for science. One of the things the government did was to start translating Russian science journals. One day a truck drove up here and deposited a load of them at our library. I came across an article that described the use of electricity to make tomato plants grow faster. That started me thinking about electricity and life."

• • •

I began smoking a pipe, using Balkan Sobranie No. 3 tobacco, which was Dr. Becker's brand. The laboratory was full of modern equipment, but I really didn't know what kind of research I should pursue, like having a Corvette and no place to go. Dr. Becker taught residents how to do orthopaedic surgery, but he wasn't teaching me how to do bioelectrical research, and I began to recognize that I would have to get the job done by myself.

I didn't know anything about experimenting with animals, but making electrical measurements on bone looked like something I could do. The first bone samples I prepared by myself came from the left tibia of a young woman whose leg Dr. Becker had amputated because of cancer; the bone arrived in the lab in a metal pan covered with a saline-soaked rag. I tried not to think about her as I put the bone in a vise and cut away a few inches that were bulbous and irregular because of the tumor. I passed a rod down the central canal to push out the marrow, and I treated the bone with acetone for a week to dissolve the remaining marrow. I used a chisel to remove the fibrous layer on the outside of the bone and smoothed the surface with sandpaper. I cut off ring-shaped sections about 1/2 inch thick, from which I cut small cubes. They were as white as sawn ivory.

I tried to detect the bone transistors Dr. Becker had written about but I couldn't find any. Then I tried to repeat Bassett's measurements that had shown bone made an electrical signal when it was squeezed, but I had many problems. The results were erratic because it was hard to apply a force to the bone in the exact same way over and over again, and any small differences affected the results. In addition, if the humidity in the room changed, then the signal I measured would change. I just couldn't get the same results that he had published, at least not consistently.

Dr. Becker knew that bone picked up water from the humidity in the air, and he had purchased an apparatus that could measure the dielectric constant of bone which he believed showed how much water was in bone. He asked me to make the measurements; that was one of the few times he ever asked me to do anything.

The apparatus consisted of a variable Wheatstone bridge, a null detector, and an adjustable capacitor that held the bone samples. I built equipment to control the humidity at which the bone was stored and measured. At a particular humidity, each bone sample adsorbed a certain amount of water from the air, and then no more. For a particular sample, the dielectric constant was a maximum after equilibration with 100% humidity, and a minimum after the sample was heated which caused all of its water to evaporate. I could measure how much water was adsorbed at a given humidity by weighing the sample after it stopped picking up water at that humidity, heating the sample, and then weighing it again.

I worked out all the details by myself, and at the end of the project I could draw a graph of dielectric constant versus amount of adsorbed water. The graph started off as a slowly increasing straight line and then it registered an amount of adsorbed water at which the slope of the graph increased steeply. I interpreted the change in slope to mean that there were *two* compartments for water in bone, one consisting of water adsorbed directly onto bone and the second that consisted of water that was adsorbed onto the water that was already in the bone. I got that idea from an article about how salt crystals pick up water from the air.

The values I found for where this transition occurred were different for each bone sample, whether or not it came from the same person. I measured ten bone samples and found that the average transition point was 37.2 milligrams of water per gram of bone, which I took to be the correct value because, like everybody else, I assumed that the average of a set of measurements was always closer to the truth than any of the individual measurements. I truly became a scientist when I realized that assumption was not always valid, but fifteen years would pass before that happened.

When I made the measurements of the water compartments in bone my point of view was entirely conventional in another sense. Once in Trashcan Trischka's class I was measuring the wavelength of an electromagnetic field by the method of locating the nodes of a standing wave produced by current in a wire, and I couldn't get what my text said was the correct value. Then I noticed that the ends of the meter sticks I was using had been worn away, so when I placed them end to end they gave an overall length that was shorter than it should have been. That's the way I thought of a measurement — that there was always just one correct value. It didn't depend on who made the measurement; if there were no people on earth, the wavelength would still have the same value. There was really nothing of me in the result. I didn't write, "I measured the wavelength and I observed that it was X" but rather, "The wavelength was X," as if the measurement had been done by a machine. I thought of the bone measurements the same way, as if there was a "right" answer, and I just assumed that everything in science was like that. The possibility that there could be many different right answers to some scientific questions hadn't dawned on me.

•••

Dr. Becker bought an electron spin resonance spectrometer because he understood that it could detect electrons, which were key to his bioelectrical theories. Except for the physics department, no one else in Syracuse had such an instrument. It worked by means of microwaves and magnetic fields, and had a built-in oscilloscope and an array of gauges and knobs that reminded me of the cockpit of the lunar excursion module. Dr. Becker was never comfortable operating the spectrometer and one day he turned it over to me. He had been in the middle of some measurements which he asked me to complete. I could hardly believe my luck.

The purpose was to measure the signal from bone and compare it with those from bone mineral and bone collagen. I never knew exactly why, I just did as I was asked. The bone samples had to be in the form of a powder, which I made by scraping bone with a glass slide. I prepared samples of bone mineral by treating the powder with a strong organic solvent that dissolved collagen, which was the protein that formed bone's matrix. The solvent left a yellow residue which I decided to remove by washing the mineral in running tap water for 24 hours. I made samples of bone collagen by treating bone powder with formic acid, which dissolved the mineral. When I examined the samples in the spectrometer, bone and bone collagen gave similar results, but those from the bone mineral were quite different. Dr. Becker published the results in *Nature* and listed me as an author. It was my first paper.

Soon afterwards, I told Dr. Becker that I had an idea about why the signal from bone mineral was so different. If bone really were made up of millions of tiny transistors formed by bonds between mineral and collagen, as he thought, then the organic solvent should have created new surfaces on the mineral at those places where it had previously been bonded to the collagen. Suppose that the electrons responsible for the signal I had measured in the mineral were in atoms that came from the tap water and stuck to the new surface of the bone mineral. That idea could support the transistor theory and *also* explain why we got the signal from bone mineral. Evidence in favor of my idea would be measurements showing that I didn't get the bone-mineral signal when I washed bone powder in tap water and that I did find it when I washed bone collagen. That's what happened. Dr. Becker was happy, and I felt as if I were really helping him. This time I wrote the paper.

The next thing I tried to do was identify what it was in the tap water that contained the electrons, and ultimately I was able to show it was copper. When we published that paper we emphasized how the results had helped us to understand the structure of bone. We downplayed the fact that we had never intended that something from the tap water would stick to the bone mineral, and that we were only trying to get rid of the residue of the solvent.

I published additional papers based on data I got using the spectrometer. But as I continued the work I began to sense that I wasn't going anywhere because the results weren't really summing up to anything. Dr. Becker was satisfied, however, and I was publishing papers in *Nature* which, according to him, was the world's most prestigious journal in experimental biology. I thought that I was doing well, especially considering that I still didn't know anything about biology.

• • •

Andrew Bassett was a handsome man, particularly when he was wearing his white lab coat. His long wavy hair sat perfectly on his head, like a hat. He had sixty scientific publications, each of which was a gossamer web of arguments, data, and citations to the work of others that was so well organized it made his conclusion seem more than plausible, almost necessary. I most admired the fluidity of his transitions; he used many different words and phrases as vehicles to move smoothly from paragraph to paragraph or idea to idea. I began incorporating his transitions into my writing.

He and Dr. Becker were cordial to one another but they disagreed about many things, one of which was the origin of the electrical signal from bone. Bassett said it was piezoelectricity and Dr. Becker said it came from the transistors that he thought were present in bone. I found an article that described a clever way to measure piezoelectricity and I told Dr. Becker I wanted to use it to study bone. He did not receive my suggestion warmly, and when I thought about it I realized why. I was really proposing to look for evidence supporting Bassett's theory, because no possible result from my proposed experiment could support Dr. Becker. The only direct test of his theory would be to make bone that was exactly the same in all respects as normal bone but had no junctions of the type he said formed transistors, which was impossible.

I did the experiment and found that bone was piezoelectric. More than that, it was robustly piezoelectric. One of the hospital residents had been a Peace Corps physician in Chile, and had brought back bones from a burial ground of a civilization that had disappeared more than 800 years ago. Even that bone was piezoelectric. I worried that Dr. Becker might think me disloyal, but he agreed to be a coauthor on the paper.

Bassett and Dr. Becker had worked together on an experiment in which Bassett passed a weak electric current through a solution of collagen in a dish and observed the formation of fibers that precipitated at right angles to the flow of the current. They interpreted the results as evidence that the electrical signal from bone made collagen fibers line up parallel to the axis of the bone so that they could become mineralized to form new bone. The story seemed like a great scientific advance because it connected an electrical property of bone and the effect that applied electricity had on its growth, and when I duplicated the experiment I got the same results. Nevertheless I was troubled because no one knew how the bands formed in the dish. If the responsible process occurred only in dishes and not in the body, then the story would fall apart.

I read about the chemistry of collagen and learned that it would precipitate from solution if the pH was about 8; otherwise it stayed in solution. I measured the pH of the solution in the dish after electricity had been passed through it and I got drastically different results depending on exactly where in the dish I made the measurement. I soon realized that the pH changed from its initial uniform value of about 6.5 because the voltage I had used caused water molecules to break down. Oxygen formed at one electrode and hydrogen at the other. On the hydrogen side the pH became 12 and on the oxygen side dropped to about 2; in between it had intermediary values and where it was 8 was exactly where I saw the bands. I did the experiment many times and the result was always the same. Unfortunately, pH changes just couldn't happen in the body because the fluid environment of bone was so well buffered that any tendency to change pH would immediately be opposed by substances dissolved in the fluid. It was clear to me, therefore, that even though the phenomenon of band formation in dishes was real, it was not evidence in support of how currents promoted bone growth.

I told Dr. Becker what I had found and he suggested that I write and tell Bassett, and invite him to be a coauthor on a paper describing my results and their implications. He said, "We could say 'we previously showed that collagen precipitated from solution in response to the passage of weak currents, and we interpreted the result as support for the theory that stress-generated electrical signals in bone are part of the growth-control system for bone. Further studies, however, have shown etc., etc.' We need a graceful way out."

I did as he asked and Bassett wrote back saying that he wanted to do further experiments before he made a decision regarding my offer. Months went by, and I didn't hear from him. I sent him a second and then a third letter, but received no reply. Finally, I gave up and we published the paper without Bassett.

• • •

Dr. Becker respected Bassett and found it expedient to cooperate with him, but he didn't trust Bassett. I first understood why after he told me a story. "Just after Bassett joined the faculty at Columbia, Bernard Baruch went to the dean and told him about rumors of a doctor somewhere on a mountaintop in Switzerland who had developed a treatment for restoring youth. Konrad Adenauer and Pope Pius XII had received the treatment, and Baruch wanted someone to go to Switzerland and investigate whether it actually worked. The dean picked Andy Bassett and sent him to Switzerland. After he came back I asked him whether the treatment worked. He said he couldn't tell, but that he had told the dean it didn't work. The dean passed on the information to Baruch who wrote a check to Columbia and then promptly departed. I asked Andy why he told the dean the treatment didn't work if he couldn't tell one way or the other. He told me that his career at Columbia would have been over if he had said that he didn't know, or if he had said it worked and Baruch went to Switzerland and got the treatment and then still looked and felt like an old man. On the other hand, Andy said if he told the dean it didn't work, he would be extremely unlikely to send somebody else to make another investigation."

• • •

Dr. Becker thought his ideas would lead to important improvements in science and medicine, and he was pure and selfless regarding these fruits. He never thought in terms of owning the knowledge he produced, or of using it to start a business. But he wanted to be recognized for his contributions, which was something he felt didn't often happen to pioneers. The story that best illustrated his fears was one that he frequently told about a Hungarian physician named Ignaz Semmelweiss. "Semmelweiss worked in a maternity ward in a hospital in Vienna in the 1840s. The death rate from infection in the public ward was almost zero, but there was a high death rate among both the mothers and the babies in the private ward, which was where all the rich people went and where the treatment was supposed to be much better. One day Semmelweiss noticed that the medical students routinely went to the private wards after they had dissected cadavers in the gross anatomy laboratory, and that their visits to the public ward always occurred prior to dissecting the cadavers. It occurred to Semmelweiss that whatever caused the infection was being carried by the students from the anatomy laboratory into the private ward, and he urged that the students be required to wash their hands before they examined the patients. He even did an experiment in which he got some of the medical students to wash their hands, and the infection rate in those beds on the private wards dropped to zero. The experts of the day, however, thought that his suggestion was absurd, and he died in disgrace in an insane asylum." Dr. Becker typically paused at this point and fiddled with his pipe before delivering the point of the story. "Fifteen years after he died, Pasteur discovered that bacteria cause infection."

Perhaps because of his sensitivity regarding the issue of credit, Dr. Becker was always careful to explicitly recognize the importance of specific ideas of others in the development of his theories. In his articles he said that he got the idea about applying electrical currents from experiments with tomato plants by Sinyukhin, the idea about currents in nerves from Zhirmunskii, the idea about the role of nerves in regeneration from Polezhev and Rose and Singer and, most important of all, the idea about the flow of electrons from a Nobel-Prize-winning Hungarian biochemist named Albert Szent-Gyorgyi. Still, Dr. Becker feared that others would not treat him the same way. "They will steal my ideas," he would say, "I know it."

One morning Dr. Becker picked up his copy of *Scientific American* and saw an article entitled "Electrical Effects in Bone," written by Bassett. It described the work they had done together, but Bassett had taken all the credit and never even mentioned Dr. Becker's contribution. "I knew it," Dr. Becker said as he went into his office and slammed the door.

• • •

Paul Dirac came to the physics department to give a lecture. The graduate students jockeyed for seats farthest from the blackboards to minimize the possibility that he might call on one of us. When a speaker introduced Dirac as the Lucasian Professor of Mathematics at Cambridge and said that the position had once been held by Newton, a jolt went through the room. Dirac had divined an equation that predicted the existence of a subatomic particle that no one had ever seen, and then someone saw it. To a physicist, Dirac's achievement was something sublime.

He filled the blackboard with equations about the deepest properties of electrons and atoms. After his lecture someone asked him how he had decided which equations were correct, and he said that "mathematical beauty" was the ultimate criterion. "Suppose the mathematics were ugly," someone asked, "but gave good agreement with experiments?" Without any hesitation Dirac replied, "Just because the results happen to be in agreement with observation does not prove that one's theory is correct." I had a hot flash because I thought the great man had contradicted himself. If truth was beauty, how could physics be objective?

On my way from the lecture hall to the reception where the graduate students would have an opportunity to meet Dirac personally, one of them said to me, "That's physics, Marino, not the crap you do with bone. You are going to wind up like Rachel Carson." I just gave him the finger, and waited for my turn to talk to Dirac. I considered asking him to explain his comment about "truth" and "beauty." I thought it might have something to do with why Hegel didn't like physicists; maybe he thought we were hypocrites.

Finally I was face to face with Dirac.

"Hello sir, I am Andrew Marino."

"What kind of work do you do, Andrew?"

"Well ... I ... I work with bone."

"And what do you do with bone?"

"I try to understand what makes it grow."

"That's very interesting, Andrew," he said. "Erwin wrote a book about biological matters. Have you read it?"

"No sir, but I will," I said.

When I left the reception I walked down the broad creaky stairway that led to the large wooden front doors of the physics building, exited, turned left and headed back to my lab. A stiff intermittent breeze blew in my face as I thought, Erwin? ... Erwin? ... Jesus Christ! He must mean Schrodinger!

All the books on quantum mechanics started with something like "The wave equation, which was first formulated by Erwin Schrodinger in 1926, is given in equation 1...", and would then go on to use the equation to calculate energy levels in atoms. I had sweated over those kinds of calculations for three semesters in my courses in quantum mechanics, but I had never thought about the human being who had created "Schrodinger's equation."

There was not much about him in the library. He had gotten the Nobel Prize for his equation but hadn't done much after that, except for a book entitled "What Is Life?" in 1945. The copy in the physics library had been checked out only once, for two days in 1953.

Now, *there* was a question. Equations governed radar and lasers and moon rockets, but what about cells and people? Where were the equations that governed cells? Either the behavior of living things was governed by the laws of physics, or it wasn't. Considering that people behaved in more or less reliable ways, I figured something had to account for that. Was there an equation that could tell you who you should marry or who would make the best President? Was there an equation that governed who would get the flu this year or who would get cancer next year? Was there an equation that could tell you when you will die? Even in principle, was it possible to discover such equations?

I had met a graduate student who believed that God was directly responsible for everything, as if life were a puppet on a string. I saw some merit in her perspective because it dealt with the big picture, not with just a few of the parts like gravity, electricity, and atoms. But science had certain rules, foremost of which was that whatever happened in the world was assumed to happen for earth-based, non-transcendent reasons that could be learned by analyzing observations. So if you claimed what you were doing was science, there were only three possible stories for explaining anything. You could tell everyone exactly how the laws of physics governed the observation, you could argue that the explanation for the observation was latent in the laws and would be found someday, or you could claim that the observation was governed by a law that hadn't yet been discovered. You couldn't say "God did it" because that was simply breaking the rules. There was no *scientific* alternative to the conclusion that laws of some kind somehow govern everything.

Schrodinger couldn't explain life on the basis of the laws of physics and he didn't think that there were undiscovered laws, so the only story open to him was that some super-smart physicist would someday show how life was governed by known laws. I didn't think there was anything wrong with that belief, but it obviously wasn't scientific because there were no observations or analyses that warranted it. It was simply philosophy, which was still another perspective on the world, and from all I knew, the least credible. I saw that the sense in which physics "explained" life was simply not helpful because the "explanation" had no predictive power.

I found much about Rachel Carson in the library. To her, life was a beautiful kind of relationship among plants and animals and people and the air and the water and the land; it was the process by which everything was perpetually renewed. In sad and delicate prose, she said that we were inadvertently destroying nature as a consequence of our use of pesticides, insecticides, and chemical fertilizers. I found a long series of attacks against her by the chemical companies, and by the man who won the Nobel Prize for his work in developing new strains of wheat. He called her work "half-science, half-fiction," and blamed it for "instigating a vicious, hysterical propaganda campaign against the use of agricultural chemicals."

What was the difference between Schrodinger and Carson?

Obviously, one dealt in precise facts, and one didn't.

Wait. Homer, Aeschylus, the Bible, Shakespeare, and Goethe are still read today, and that wouldn't be the case if they didn't tell us something about ourselves, some timeless facts.

Perhaps, but they would be esthetic rather than observational facts.

That makes no sense. Can you imagine someone writing about or understanding anything in the absence of observations?

Science enables us to explain and predict phenomena.

Yes, as do the understandings that we can distill from our great books.

Science is certain.

Not all parts of it. Not biology!

Wouldn't it be wonderful if people like Rachel Carson and Erwin Schrodinger talked to one another about the science of living systems. What is the truth? How will we know it when we see it? Unfortunately, such people didn't talk to one another. I saw that most clearly in a dream. I was sitting across the room from Albert Einstein, who was slumped in his desk chair, holding a pipe. His long white hair was rumpled and pressed backwards so that it framed his face and emphasized his high, lined forehead. The two top lines went from one side to the other, and the lowest was interrupted just above his nose.

"What do you want to talk about, Andrew?"

"About science. I want to know what it is."

"What do you think it is?"

"I don't know. There is the world of equations, where everything happens the way it must happen, and there is the world of bone and blood and nerves, where there are no equations to predict what will happen. It's as if there were different worlds."

A sense of sadness seemed to come over him and he said, "The community of the intellect that once united the world of knowledge seekers is dead. Men of learning have become representatives of narrow traditions, each with their own methods and goals. The sense of an intellectual commonwealth has been lost. Nowadays we are faced with the dismaying fact that there are no exponents of general ideas."

• • •

I earned my Ph.D. in 1968. At my commencement Walter Cronkite said, "There are a lot of things wrong with this world we've made: poverty, ugliness, corruption, intolerance, waste of our resources, pollution of our air and water, urban sprawl, inefficient transportation, outmoded concepts of national sovereignty, the secret society of the establishment elite, the power of the military-industrial complex, the atomic arms race, the population explosion, war." As I listened to him it struck me that science had made all those things possible.

When I had begun studying physics I thought it was mankind's most noble undertaking. Over the years that attitude had faded, like a bright color exposed too long to the sun. The whole enterprise had come to seem elitist, something that benefited only some human beings. When Galileo had studied the earth and the sun he hadn't been trying to make money. When Roentgen had discovered x-rays, he refused to apply for a patent because he believed that the knowledge belonged to all humanity. Einstein was a humble man; self-aggrandizement seemed to be the furthest thing from his mind. There didn't seem to be any men like them, at least that I knew about, except Dr. Becker.

The president of the National Academy of Sciences had said that physicists were an elite corps of disinterested experts who were better than everybody else because of the knowledge they had produced. But that knowledge had also caused great pain. It seemed to me that if you weighed the pluses and minuses, perhaps physicists weren't even as good as everybody else. They argued over and over that knowledge generated by atom-smashers would benefit all humanity – absurd. Meanwhile, detergents fouled the water, insecticides poisoned the land, automobile exhaust created smog, and physicists wouldn't acknowledge the impossibility of continuing to exploit nature with impunity, even though anyone with common sense could see it.

• • •

It was hard to get a job the year I graduated, and those that were available had starting salaries below what I was making in Dr. Becker's lab. He never really asked me to stay, but he never told me to go. I had all the equipment and supplies I needed, and I could do any kind of experiment I wanted. Moreover, he had told me about new experiments that would "revolutionize" regeneration research, and I wanted to be there when the results came in. So I decided to stay in his lab and continue to try to find some kind of research pathway that would be meaningful. I hadn't been able to generate the kind of ideas that he did, but I thought that perhaps if I hung around longer, it would happen. I kept looking for a way to tie into his purpose and help him carry the load.

4
A Legal Turn

Linda and I got married, though her parents thought we should wait and mine had hoped I would marry an Italian girl. Our wedding took place near the end of the federal fiscal year, and I took scientific catalogs on our honeymoon because if I didn't soon submit purchase orders our lab would have lost the unspent money.

We rented an apartment next to a park. One day I rescued a bluejay that had fallen from its nest and was about to be attacked by some cats. The bird had a deformed foot and a bad eye, so we kept it as a pet and named it Joe. I decided to house-train Joe because we frequently let him fly around the apartment. Whenever he did his business outside his cage I would squirt him with a water pistol, and continue squirting until he returned to his cage. I did this for months, but all I accomplished was to teach Joe to do his business any place he wanted and then immediately fly back to his cage.

Linda and I had frequently taken drives into the country when we were dating. It was during those times that I saw a cow for the first time. I watched as it walked and ate and swatted at flies with its tail. I also saw different kinds of trees growing in the valleys and on the hills. There were maple, elm, walnut, poplar, pine, spruce, hemlock, and white flowering dogwood which grew profusely on steep banks. On a visit back to Philadelphia I made it a point to see what kind of tree it was that grew across the street from our house. I had gone from not realizing there was a countryside to embracing it as a wonderful place to live, and we bought a house and some land in the country, not far from Skaneateles Lake.

The house needed much work, but for $19,000 I thought it was a bargain, and we had $3,500 saved, so we could pay for improvements. But soon after we moved in I was shocked to learn that the estimate for modernizing just the kitchen was $3,800. I decided I would do the job myself.

I had no clear idea how to build a kitchen because my father had never taught me any manual skills. He saw his job as providing a roof over our heads and food for the table, not as teaching. He made all the repairs around the house. I was neither required nor permitted to do anything. I always knew the reason wasn't that he didn't love me, it was that he did. He had come from Italy as a small boy and had grown up without a mother in a poor neighborhood in south Philadelphia where, mostly alone, he had faced many prejudices and dangers. He sought to protect his children against such adversities, and he did so in the only way he knew, which was to prevent the conditions that gave rise to them.

The first step in modernizing our kitchen was to be the construction of a wall. Linda suggested that I ask her cousin Jack for help, and on the appointed day he walked in wearing a big smile and carrying a toolbox, and said, "Where do you want the wall?" I was struck by how simple a thing is if you know about it and how daunting if you have no clue. He showed me many things; others I learned by trial and error. During the time I was building the kitchen I acquired the confidence that I could do things in the world, not just in a classroom or a laboratory. I realized I could do anything I wanted to.

Sure, there were flaws. I should have put in two electrical outlets instead of only one. The countertop sloped in the direction of its length. I covered it in formica, but I didn't have a router so I had to use metal trim to hide the jagged edge produced by the saw cut, which made the counter look like one I had seen in a White Tower restaurant. Several of the cabinets were attached with fewer screws than desirable because there were fewer studs in the wall than there should have been. There were gaps in several places between the rug and the wall because it bowed and I had cut the rug on a straight line. Still, I finished the project at a fraction of the cost of the initial quote, and I knew that I would do much better next time.

The General Crushed Stone Company bought some land about half a mile from our house. I first learned about it when I saw some men cut down the woods and begin scraping the topsoil off the bedrock with bulldozers. They piled the soil around the edge of the property to form a berm, on top of which they built a ten-foot fence. Our house was in Elbridge; nobody on our town board knew anything about what the company was doing on its land, which was in the town of Skaneateles.

One morning a terrific explosion rattled the picture windows in our living room. Over the next several weeks there were more explosions; some were muffled thuds and others were ear-shattering. Soon the intermittent pattern of explosions was augmented by a more or less continuous rumbling and groaning that went on around the clock. I climbed the berm and the fence and saw huge payloaders scooping up pieces of fractured rock and dropping them onto a conveyer belt that fed an enormous machine, at the other end of which walnut-sized pieces of stone emerged. The crushing process created a dust cloud that was carried northwesterly by the prevailing winds, toward my house.

The Skaneateles zoning enforcement officer told me that the company didn't have a permit to operate a quarry, but he said he wouldn't take any action until he was told to do so by the town attorney. I called the town attorney but it turned out that he was also the attorney for the stone company, and he wouldn't talk to me.

I complained to the town board, the zoning officer, and the lawyer's secretary about the noise and dust and explosions, about the overloaded trucks that hauled the crushed stone away, and about the fact that the stone company didn't have a permit. I also complained about what the company had done to the land. It had been a hardwood forest with a small trout stream, but it had taken the stone company only a few weeks to make it look like a desert. The town board finally decided to hold a public hearing.

The hearing took place in the fire hall on a hot August evening. The board sat at a table, facing the audience. The first ten rows of seats were roped off, so the local residents had to sit in the back of the hall. There were no company representatives in the hall and we began to think that none would show up. Suddenly, the back doors of the hall opened and an entourage of men in suits and ties entered, walked to the front, and sat in the roped-off seats. Their grand entrance and privileged seating reminded me of when I had made my first holy communion.

My plan had been to stand up and say what I had to say right after the meeting started. But when I did the town supervisor told me to sit down, and he called on the company to present its position. One of the men in suits stood up and identified himself as a lawyer from one of the big law firms in downtown Syracuse, and he introduced three other lawyers who he said were part of the team that represented the company. He said it paid its taxes, employed many people, and provided a product that was essential in constructing roads and highways. He said that the company represented progress and economic development, and that it would be a big mistake for the town board to be swayed by those who had romantic notions about preserving forests. He went on and on, and I got angrier and angrier.

About the time I thought he was finally finished, he announced that he would present witnesses to prove his case that the stone company was a good corporate citizen. He placed a chair between the audience and the board and one of the men in suits walked over to the chair and sat down.

"Would you please tell the Board your name and place of employment?" the lawyer said.

"Dr. Franklin D. Hart. I am Professor of Acoustical Science at North Carolina State University," he replied.

"Dr. Hart, in your capacity as professor of acoustical science, are you familiar with the laws of acoustics?"

"Yes, I am."

"At my request, have you done calculations of the sound levels that would be predicted to occur on Dr. Marino's property during various steps in the process of preparing the product for market?"

"Yes, I have done extensive calculations."

"Did you use computers in making those calculations?"

"Yes."

"What results did you find?"

"The calculations clearly showed that it was impossible for any auditory consequences of the activities on the General Crushed Stone property to be heard by someone standing on Dr. Marino's property."

"Did you do measurements to confirm your calculations?"

"Yes. I went onto Dr. Marino's property and measured the sound levels using a B-weighted sound meter, and found that the sound levels were only 65 decibels, which is roughly equal to the sound of rustling leaves."

"Dr. Marino has complained to the board that sounds produced during product extraction processes were unusually loud. Did you find this to be the case?"

"No, what he said is simply untrue," Dr. Hart replied.

I got so angry that I started to see flashes of red and blue light. My senses started shutting down. I could see the lawyer talking to Hart but I couldn't hear anything. Then I couldn't see anything – just the colored light. I couldn't feel anything so I didn't know whether I was sitting or standing. I thought I'd died. I didn't know how long I was in that condition, but after a while I could feel my back against the chair and my feet pressed against the floor. At that point I saw the lawyer talking to another suit, a professor from Cornell. In response to a question I couldn't hear he answered,

"Yes. I positioned the seismograph at several places around the perimeter of Dr. Marino's house and recorded the results when a typical percussion operation was performed as part of the limestone preparative process."

"What did you observe?" the lawyer asked.

"Absolutely no vibrations were detectable on Dr. Marino's property as a consequence of the extractive operations on the General Crushed Stone property."

"In your considered scientific opinion, could the operations that take place on the General Crushed Stone property be responsible for the cracks in the foundation of Dr. Marino's house, as he alleges?"

"It is impossible for that to be the case because the vibrations produced there by the explosions had zero amplitude," the witness replied.

I have no clear memory of what happened next. I know I stood up and started yelling. I saw the town supervisor point to me and motion for two policemen to come forward. They grabbed me by each arm and lifted me off the floor and carried me to the exit, a task made easier by my anger, which had produced a state of near tetany in my arms. My wife started screaming, and I began yelling and kicking, knocking over chairs in the process. The audience was shouting. The last thing I remember saying before they threw me into the street was "Tyranny is not dead – it is alive and well in Skaneateles!"

The next day the Syracuse newspaper reported that the board had issued a permit, "despite the opposition of Andrew Marino who spoke excitedly and criticized the gravel company's operations as a whole."

I couldn't understand how the scientists could have said what they did. From the day I saw the experiment with the light bulbs in high school, I had thought of science as a steely rational activity. It was done by human beings, of course, but in itself it wasn't human, it was divine, it allowed man to look into the mind of God and see how the universe worked, to actually see His plan. Because it was divine it couldn't come in shades of opinion or degrees of meaning. It had to be just one thing. If you were a scientist you saw what that was. I knew that even the greatest physicists hadn't been able to read God's mind about life, but for things like vibrations or sounds, I didn't understand how I could perceive them but the company's experts couldn't. Then, I guess I just had a eureka experience, because I suddenly realized they weren't really looking into God's mind – they were just doing as they were told, like Mr. Schmidt had wanted me to do.

When the company had asked lawyers to fight complaints about its operations, the lawyers didn't say, "Sorry, we can't help you because it's clear what you're doing is wrong." A lawyer doesn't seek absolute truth, but only the truth favorable to his client. The same goal determines which expert is hired. The lawyers for the stone company may have gone through a dozen experts who agreed with me before they found Hart down in North Carolina. Alternatively Hart may have been the first expert the lawyers contacted, and he told them the story he could easily see was what they wanted to hear. I realized that when an expert testifies, the whole process is biased in the sense that the expert is an advocate for the side that is paying him, just like with a lawyer.

It was really the lawyers who were in charge. The scientists were only pawns. The lawyers were the ones who had pushed me around. At first, I thought that I had to make a lot of money so that, in future disputes, I could afford to hire lawyers. Then I realized I could be my own lawyer. I could learn the rules. It sounded like overkill to some of my friends, but they didn't know how intensely I hated being pushed around. The idea also sounded impractical to them, but I investigated the situation and found that wasn't the case. My lab was just across the street from the law school.

Linda worried about where we would get the money and where I would find the time, particularly considering that our second baby was on the way. But I told her I wanted to do it. I didn't want to be pushed around, and I saw that was what happens when your adversary knows the rules and you don't. I had been fortunate in Virginia, but in Skaneateles I lost because only my adversary knew the rules. I expected there would be conflicts throughout my life, and I didn't want to ever again be so woefully unable to protect myself. It wasn't just that I didn't want to live that way, I *couldn't*. It was simply impossible. I didn't expect to come back and beat the stone company. I knew I had lost. I felt more like someone who, in preparing to travel to a strange land, would obtain provisions for the journey. I never had to explain this to Linda. Somehow, she simply knew. She subordinated her life to my ambition, my needs, and my vision of what was necessary for our protection.

All the courses were taught during the day, so I needed Dr. Becker's help to go to law school. He was seated at his desk when I went into his office to ask for it. The way I put it was that I wanted to take some law courses. He raised his eyes, but not his head, and said "Why?" He didn't like lawyers, so I gave him a reason I thought he could accept. "It will be a challenge, and I think it will help me to think more clearly and write better." He approved my request, and signed a memo officially changing my tour of duty so nobody would wonder where I was from 9 until 12 on Mondays, Wednesdays, and Fridays. There were two things I could always count on with Dr. Becker: he would almost never ask me for anything, and he would always give me what I wanted.

Less than three months after I had decided to go to law school I was sitting in my first class, torts. Professor Alexander walked briskly into the room, holding the casebook against his chest; by the time he got to the lectern, the room was silent. "What was the fact pattern in *Palsgraf v. Long Island Railroad Company?*" he asked, and then pointed at me.

"The plaintiff was standing on a railroad platform after buying a ticket. A train stopped at the station, and a man carrying a small package ran forward to catch it. The train started pulling away, and the man jumped aboard the railroad car but lost his balance. The guard on the car, who had held the door open, reached forward to help him in, and another guard on the platform pushed him from behind. In doing this, the package was pushed out of the passenger's hands and fell onto the railroad track. The package contained fireworks, and when it hit the tracks it exploded and the concussion knocked over a metal scale at the end of the platform which fell on the plaintiff, injuring her. She claims that her injuries were the result of the negligence of the guards."

"Did the guards cause the injury?" he asked.

The whole story seemed absurd. I couldn't see how it got to be a court case, so I said, "I think it was just an act of God."

"Let's leave God out of this, Mr. ...?"

"Marino."

"Okay, Mr. Marino. This is law, not religion, so let's leave God out."

He walked from behind his lectern and up the steps to the row where I was sitting and said, "Could the scale have been knocked over by a stampede of passengers who were frightened by the explosion?"

"The explosion caused the concussion that either directly knocked over the scale, or caused the stampede that knocked it over."

"Well," he said, "which one was it?"

"Judge Cardozo said it was the concussion."

"He wasn't there, was he?"

"No."

"So how does he know?"

"There were people there, and they must have testified it was the concussion."

"Is a concussion something you can see?"

"No."

"Then nobody could have testified that they saw the concussion knock over the scale, right?"

"Right."

"Then why do you think the judge said it was the concussion?"

"Maybe he figured that since something had to have knocked over the scale, the concussion was a good bet."

"Bet, Mr. Marino, is that what you think the judge did?"

"I meant there had to be some explanation, so he picked something that could suffice."

"Why didn't he say that a frightened passenger knocked it over?"

"Maybe because nobody saw that."

"How can that be the reason, considering that nobody saw the concussion but he still picked that?"

"I don't know why he chose one explanation over the other."

"Do you mean there were only two explanations?"

"I don't know what else it could be."

"Isn't it possible that one of the legs on the scale was defective, and it happened to break at exactly the same moment the explosion occurred?"

"I think that's highly unlikely."

"Perhaps, but is it more unlikely than the scale being knocked over by a concussion?"

"I think so."

"You haven't made any measurements of those probabilities, have you?"

"No."

"But still you would bet against the defective-leg explanation?"

"Yes ... no ... I don't know."

"Could the concussion have arrived at the scale at the same instant a passenger bumped into it?"

"Yes."

"In that case, what would you say caused the scale to topple?"

"Only God knows," I said as the class broke into laughter. I smiled and nodded a little because I thought that made it look like I had intended to be sarcastic.

"Let's go a little further back," he said. "What caused the explosion?"

"The fireworks that fell on the track."

"Was the explosion of the fireworks the only cause of the injuries that the plaintiff suffered when the scale fell on her?"

"Yes, because she wouldn't have been injured but for that."

"Could Mrs. Palsgraf have sued Babe Ruth because she could prove he was in New York the day the accident happened?"

"No, because he didn't have anything to do with the accident. She sued the railroad company because there was a link between what their employees did and the injury that she suffered."

"What about the company that made the scale? After all, it couldn't have fallen if it wasn't there. Isn't that a link between the company and plaintiff's injury?"

"You can't blame the company because it didn't do anything wrong."

"So the plaintiff couldn't sue Babe Ruth because he wasn't connected with the incident, and she couldn't sue the scale company because it wasn't negligent; but the defendant, at least on the plaintiff's theory, was directly responsible?"

"Yes. If the guard who worked for the railroad company hadn't pushed the man carrying the fireworks, then there wouldn't have been any connection between the plaintiff and the defendant, and the defendant wouldn't be liable."

"Can there be legal liability if there is no causal link?"

"No."

"Can there be a causal link and no legal liability?"

"In a sense, yes."

"In what sense?"

"Well, the company that made the scale caused the accident in the sense that if they hadn't made the scale, it couldn't have been on the platform and fallen on the plaintiff."

"Suppose the plaintiff had taken a cab to the railroad station and the cab driver broke the speed laws. Would the speeding be a cause of the accident?"

"Yes, in the same sense."

"Suppose there was a hole in the street leading to the railroad station that had just been repaired, and that if it hadn't been repaired the taxi driver would have taken a more circuitous route with the result that the plaintiff arrived at the railroad station later than she actually did arrive. Would fixing the hole in the street be a cause of the accident?"

"Yes."

"Well, Mr. Marino, applying these principles to the present case, was the explosion the sole cause of the plaintiff's injuries?"

"No."

"Mr. Marino, can you conceive of an event that doesn't have millions of causes?"

"No."

"And Mrs. Palsgraf's injury wouldn't have happened but for the occurrence of each of those causes, right?"

"Right."

"If that's the case, why does the law allow the plaintiff to pick out the cause for which the railroad company was responsible and hold it liable?"

"The law has a policy against negligence. The company's contribution to the causal chain occurred because of its negligence, so the law regards it as the legal cause."

That discussion opened my eyes to the idea that law and science followed different rules. If you believed in equations, which I did, then whatever happened had a single cause, an "x" in an equation. But in the everyday world the number of causes of anything was enormous, and what people regarded as the cause was usually something they chose on the basis of policies or values. Equations were irrelevant. It didn't occur to me, then, to wonder about whose rules would apply when the two worlds came together. I just continued to try to learn how to function within each separate world.

5
Beginner

♦

My main problem in the lab was figuring out what to do. I desperately wanted to help Dr. Becker but I didn't know what he wanted. I made suggestions about experiments, but often they were ludicrous because I knew little about biology. "You can't do that to living tissue," he would say with a sense of exasperation. Once I asked him "Why can't horses heal broken bones?" He looked at me like I was crazy and said, "Horses can heal broken bones." "Then why do they shoot horses when they break a bone?" I asked. "Because it's too difficult to immobilize them long enough for the bone to heal," he replied in an exasperated tone.

Some scientists were open to Dr. Becker's idea that electrically activated cells made fractures heal and limbs grow back, but I didn't know anybody who approached research the way he did. Others would treat tissue with formaldehyde, cut it into thin slices, and then colorize them with dye so they could be seen under a microscope. Eventually there would be a report that described how the cells looked, and speculated about how they might have functioned when they were alive. Dr. Becker called them "painted tombstones," and said that staring at them through a microscope was no proper way to study cell function.

His way was to do something to cells and then see how they changed. That's how he did the experiment he had told me earlier would "revolutionize" regeneration research. He had passed electricity through dishes containing frog blood cells to prove that they could revert to their primitive form, called stem cells. In his theory, that reversion was the first and most crucial step in the healing process, because stem cells could develop into the muscle, nerve, and bone cells needed to form a new limb. The novel idea that cells could move backwards developmentally was another example of how smart he was.

The stem-cell work was done mostly by Fred Brown. I watched, but only when Dr. Becker wasn't there so that he wouldn't think I was interfering. At the end of each experiment the best Dr. Becker could say was that he saw a cell change into what looked like a stem cell. The only way to really know it was a stem cell was to see it develop into a cell that made bone, muscle, or nerve. But cells in a dish just couldn't make those tissues, so Dr. Becker could never really know that a blood cell had dedifferentiated into a stem cell. I raised this problem with him, but I felt he regarded my comments as a sign of disloyalty. I knew others would raise the issue, and if I hadn't cared about him I would have kept quiet.

He eventually concluded that the changes he saw in the dish supported his theory that electrical signals caused blood cells to become stem cells that could subsequently differentiate into specific cell types needed to grow a new limb. How much of this was really true and how much was only a story I couldn't say. Perhaps he hadn't dedifferentiated the blood cells but only electrocuted them, as some people said. The problem was that you couldn't say for sure what a cell was capable of doing by just looking at it. You had to do more. Otherwise all you were doing was studying unpainted tombstones.

While Dr. Becker was continuing to break new ground, other investigators worked intensively in small regions of his wake to develop commercial electrical devices. An orthopaedic surgeon at the University of Pennsylvania named Carl Brighton was one such man. He used transistorized circuits and stainless-steel electrodes, technical improvements over what Bassett and Dr. Becker had done, to deliver electricity to the bones of animals, and thereby make them grow. One day Brighton put a stainless-steel electrode into the fracture gap of a fifty-one-year-old woman whose broken ankle had failed to heal for almost a year. When he passed negative electricity into the gap from a circuit attached to her cast, the ankle healed in nine weeks. An article describing his achievement appeared in the *New York Times*.

To make his treatment appear to be scientific, Brighton needed a story about why electricity made bone grow. He proposed that electricity increased the oxygen levels in the tissue, which in turn stimulated the bone cells. In essence he was simply saying that the transistorized circuits had resulted in delivery of a drug. I didn't like that rationale because it didn't tap into the primordial bioelectrical essence of living things, as did Dr. Becker's rationale. I could see that either Brighton was wrong or Dr. Becker's work had no meaning.

I wondered what might have motivated Brighton to reduce bioelectricity to the level of drug action, rather than to treat the subject as a new approach of potentially great value. When I had first started working for Dr.

Becker, I got the feeling that there was something dark about the origins of bioelectricity, something that forced potential advocates into a defensive posture, as one might expect of a man in a social situation with people of a higher class. Dr. Becker seemed to act this way, and so did the prominent Dr. Bassett and the rising Dr. Brighton. When I visited a museum in Minneapolis devoted to the historical role of electricity and magnetism in the life sciences, my eyes were opened to a sociological dimension of bioelectricity that I had never suspected.

I saw the earliest machines for making electricity, finely crafted of lacquered mahogany with a crank for turning a leather belt to produce sparks, and shocks. The machines had been used in medicine, with controversial results. A French physician who had treated paralyzed patients reported that "priests had to be called in to convince the neighboring people that the cures had not been wrought by magic." But Benjamin Franklin wrote: "A number of paralytics were brought to be electrified, which I did at their request. ... I never knew any advantage from electricity in palsies that was permanent." Still, electricity was used to treat consumption, dysentery, plague, smallpox, venereal diseases, and cancer.

Some believed that the machines could produce disease, but a practitioner at the University of Rouen wrote: "I spent several years making experiments in my office, living fifteen minutes to twenty-four hours of the day in an environment well-charged with electrical fluid and ... I did not experience any marked effect from living that way."

Another conflict had involved a French physician named Franz Mesmer, who used magnets to treat patients. The Marquis de Lafayette had become an avid proponent of Mesmerism and tried to establish it in the United States. However, Thomas Jefferson, then a United States envoy in France, had effectively blocked the effort by sending anti-Mesmer pamphlets to the United States.

I studied the details of a dispute between two Italians, a painfully reclusive physician named Luigi Galvani and a brilliant but arrogant physicist named Alessandro Volta. Galvani had found in his experiments that if he touched one end of a metal rod to a leg muscle of a recently killed frog and the other end to its spinal cord, the leg would twitch. He interpreted the response as arising from natural electricity that flowed through the bar from the nerve to the muscle. Volta argued that Galvani's metal bar did not facilitate the flow of natural electricity but rather created electricity by chemical action at the moment the metal touched the frog's tissues. He proved the theory of chemical generation of electricity when he alternately stacked metal disks between moist cardboard, thereby inventing the battery.

Volta's invention had made him famous and from his bully pulpit he argued against the existence of animal electricity. But Galvani had conducted further experiments not involving the use of metal and proved that animal electricity existed. The truth was that *both* he and Volta were correct. Nevertheless, Volta continued denying the existence of animal electricity, and many years had passed before everyone recognized that nerves worked by means of electricity.

I perused the museum's materials that covered the century after Galvani, and I saw the impact of greed on the development of bioelectricity. A physician from Yale named Elisha Perkins had patented the medicinal use of a three-inch metal bar called a tractor. He had claimed that when the pointed end of the tractor touched painful or diseased areas of the body, the affliction was drawn out electrically. Perkins had made a fortune selling tractors to other doctors, but then someone began selling wooden tractors painted to look like metal; the physicians who used the wooden tractors, which were inherently incapable of conducting electricity, continued to report good results.

After the discovery of the laws of physics that governed electricity, new kinds of machines were developed and explosive growth ensued in the therapeutic use of electricity. Books had been published describing how electricity could be used to treat paralysis, angina, abnormal heartbeat, and other ailments. Alfred C. Garrett of Boston had been the first American physician to specialize in electrotherapeutics; his textbook included advice on how to use electricity to grow bone. In 1867 the first of ten editions of *A Practical Treatise on the Medical and Surgical Uses of Electricity* had been published, and soon thereafter journals appeared that were devoted to the study of electrotherapeutics.

Many eminent physicians had endorsed the medical uses of electricity, and by the turn of the twentieth century the rule, not the exception, had been that physicians treated patients with electricity to relieve pain, reduce inflammation, treat cancer, and combat a wide range of gynecological complaints. Electrotherapeutics had been so commonplace that the appearance of an electricity machine in a Norman Rockwell painting showing a visit to a doctor's office seemed entirely natural. But even before the painting had appeared on the cover of the *Saturday Evening Post,* economic and cultural changes had occurred that would doom the practice of electrotherapy.

The drug industry had developed soon after the Civil War. By writing prescriptions, rather than treating patients with electricity, a physician could see many more patients and at the same time avoid the necessity of buying expensive electrical equipment. Drug use had been unabashedly promoted in the medical journals, most of which were owned by the drug companies. When Wilhelm Roentgen discovered x-rays, which were a by-product of electrotherapeutic devices, the medical specialty of radiology began, and electrotherapists became the first radiologists.

The increasing reliance on drugs fed and fed off of the ascendancy of biochemistry. In 1910, an important study on medical education called the Flexner Report had dismissed as "unscientific" any medical practice that had no underlying biochemical theory. The report was aimed at electrotherapists, none of whom knew how electricity produced beneficial effects. In the decade that followed, the biochemical mode of explanation had become progressively more accepted as rational and complete, and teaching of electrotherapeutics waned.

The fatal blow may have been the antics of a physician named Dr. Albert Abrams. He had claimed every disease had a specific vibration frequency that could be detected in a drop of the patient's blood using a machine he invented and sold called the Pathoclast. Another machine, called an Oscilloclast, was used to administer what he claimed was the proper balancing frequency for each disease. Thousands of doctors had used the machines and numerous cures were reported. Many prominent people had become strong Abrams supporters, including the novelists Upton Sinclair and Arthur Conan Doyle. But when the physicist Robert Millikan examined the insides of an Oscilloclast he found that the wiring pattern was senseless, and concluded that the Oscilloclast was "the kind of a device a ten-year-old boy would build to fool an eight-year-old."

A few scientists had wandered in the electrical wilderness during the post-Abrams period, but with no impact that I could discern. The drug industry, the ascendancy of the perspectives of biochemistry and physics, changing fashions within medicine, and simple greed had worked together to exclude bioelectricity from the scientific mainstream. Then Dr. Becker began publishing, and the modern era of bioelectricity began. From my historiographic perspective bioelectricity looked more than ever like a rough diamond whose inner glow remained hidden by superficial scratches and a layer of volcanic pumice.

What is electricity? As a child I had asked my mother and she said, "It's a mystery, nobody knows." Someone once told me it could drip out of the wall sockets, particularly in hot weather, and cautioned me about standing too close to one. After my Uncle Angelo had returned home from World War II, he told me stories about admirals and generals who had sought his advice on how to beat the Japanese. I thought that he probably also knew a lot about electricity, so I asked him what it was. He said, "A long time ago, only the angels knew about electricity. They used it to send messages to each other. One day, an angel named Bobo decided to give electricity to people. But it was too strong for them and it burned down their houses, like when lightning hit the tree on Grandpop's farm. So Bobo put electricity into special machines that he gave to people, and when someone turned the handle, electricity came out."

In school the story changed, but not very much. The machine didn't make electricity – it made electromagnetic fields or EMFs. If the machine was connected to something by a wire, the EMF passed through the wire and was then called electricity. If there was no wire, the EMF could sometimes go through the air. In that case it didn't make electricity until it went into a wire.

Herman Schwan was one of the first people to talk openly about biological effects of EMFs, but the gist of what he had said was that they couldn't do anything to the cells in people. So when I had stood in front of the beams from the radar and the laser when I had worked for RCA, I no more thought about whether those EMFs could have caused changes in my cells than I thought about whether the color of my socks could have done so.

I had the same kind of blind spot even when I saw a grand-scale demonstration of the reality of EMFs in a physics course by a teacher who used a huge electromagnet. The students were each given an iron rod about a foot long and 1/2 inch in diameter and ushered into the vicinity of the electromagnet. When he turned it on its EMF instantly reached out and grabbed the bars, and twisted them with a force comparable to the strength of their muscles, even though nobody could directly feel the EMF.

Still the idea that the EMF could be affecting cells had never occurred to me. Then one day the thought popped into my head that everything Dr.

Becker had done, everything that Bassett or Brighton had done, everything that anybody who worked in bioelectricity had done, could be explained by EMFs alone, and that perhaps the EMF was the important thing and the bioelectricity was only an epiphenomenon.

I proposed an experiment to Dr. Becker, my first with animals, and he agreed. I bought some white rats and made an apparatus that held them spread-eagled. On either side of each of the hind limbs I placed two small brass plates, taking care that they did not actually touch the limb. When I applied a voltage to one pair, the region between them became filled with an EMF, as required by the laws of physics; I didn't apply a voltage to the other pair so that the bone in that leg would be the control for the experiment. The rats wiggled and squealed incessantly while they were confined; our secretary complained about the noise they made, which was a lot like the sound of a nail being dragged across a blackboard. While they were wiggling and squealing, the EMF was pouncing on their leg bones. I knew the rats had no sensation of its presence because they squealed the same whether or not the voltage was turned on, and they never reacted at the moment I turned the voltage on or off.

I gave the rats ten two-hour treatments and then killed them so that I could remove and examine the leg bones. I put gauze pads in the bottom of a jar, poured in a few ounces of ether, grabbed the rat by its tail and dropped it in the jar, put the lid on, and waited. When the rat stopped breathing I took it out of the jar, cut off the legs, and put them in formaldehyde. I dumped the gauze pads and rat feces from the jar into the trash can. Once, a few moments after I had emptied the jar, an orthopaedic resident who had been doing a research project in our lab cleaned his pipe and dumped the ashes into the can. The ether fumes exploded and filled the room with smoke, which quickly seeped into the hall. Andrew Bassett had been visiting Dr. Becker that day, and they both came running into the lab. The resident blamed me and I blamed him.

Our technician cut the legs into thin slices and dyed them so that different tissues exhibited different colors under a microscope. I hoped to see more growth in the slices from the limb that got an EMF, but all I saw was bewildering complexity. No two slices were the same, even if they had come from the same leg. I couldn't think of any way to turn what I saw into numbers, so I couldn't reach any reliable conclusion. Nevertheless I was as excited as I had been before the experiment, perhaps more so, because I thought that some of the complexity I saw through the microscope could have been caused by the EMF. Why not? What I needed to do now was learn how to do better animal experiments.

6
Knowing

An investigator named Lasalle published a series of papers that flatly contradicted Dr. Becker's work on limb regeneration. Lasalle had studied voltage change in injured and amputated salamander limbs and concluded that the changes had nothing to do with regeneration. Dr. Becker's project was critically dependent on his salamander studies, and I thought they had to be verified so that they could truly serve as a beachhead on a new continent. I wanted him to perform more experiments to decisively confirm he was right, and to delineate the errors Lasalle must have made. I said to Dr. Becker, in effect, "Let's repeat and expand your original work, put the results on an absolutely firm footing, and make our points conclusively." At that time I still believed something could be known "conclusively."

His attitude, however, was to ignore the problems on the coastline and advance into the interior. "Do you think the money will continue to come in if I keep doing the same experiments over and over again?" he said. I knew he wouldn't have repeated them even if he had received money for exactly that purpose because it wasn't in his nature to doubt what he had decided was true.

One day he decided he knew enough about regeneration to reasonably expect that he could grow a new limb on a rat, so he began trying. He amputated one of the forelimbs in each of a group of rats and attached a tiny electrical device whose signal mimicked what he had measured in the salamander after its forelimb had been amputated. It was hard to believe that the amazing result he hoped for could happen but, to an extent, it did. The rats started to grow new limbs. Not whole limbs, but tissue that looked like developing limb buds, as in a fetus. The amount of regeneration he achieved was more than anyone else had ever obtained in a mammal, and the response occurred in a matter of only a few days.

The results were published in *Nature* and then described in an article in the *Washington Post*, which Senator John Glenn entered into the *Congressional Record;* he said the work would "bring about exciting new scientific advances of great and unforeseen benefit." But Dr. Becker's work was not well received by other researchers, who mostly studied regeneration in salamanders with the aim of describing the details rather than learning how to trigger it in mammals, which common wisdom held to be impossible. An investigator at Purdue named Lionel Jaffe reacted with particular bitterness. He sent letters to the *Post,* the Veterans Administration, and the National Institutes of Health claiming that Dr. Becker had finagled the data to make the results look better than they really were. In a press release through Purdue's office of public relations, Jaffe called Dr. Becker's work "plain ordinary fraud."

Dr. Becker shocked me by asking for my advice about how to handle the matter; when I suggested he send a letter to the president of Purdue, Dr. Becker asked me to draft the letter. Shortly after he sent it he received a phone call from a contrite Jaffe who couldn't say "I'm sorry" often enough. As I had recommended, Dr. Becker asked Jaffe to apologize in writing, which he did. Dr. Becker was pleased with the outcome but disgusted that the controversy had occurred in the first place. "If he really believed that the regeneration didn't occur, why the hell didn't he just repeat the experiment?" he asked rhetorically. Well, it was pretty clear why: Jaffe wasn't going to waste his time doing experiments that he "knew" wouldn't work. He must have had a little green man who told him that.

Dr. Becker's green man was more enlightened, and he kept suggesting new ideas, like acupuncture. Dr. Becker invited an acupuncturist named Dr. Ho to give a demonstration to the hospital staff. Ho began by talking about how acupuncture points were connected by energy channels that conducted a flow of natural energy called chi. Then a nurse wheeled in a patient who had severe osteoarthritis. Ho rolled up his sleeves and motioned to the chief resident who gently picked up the patient's left leg, supported the knee from underneath, and pressed down on the ankle. After only a few degrees of motion the patient began to scream in pain, so the resident could proceed no further.

Ho spoke briefly to the patient while the nurse opened a sterile pack of acupuncture needles. Then he passed a needle several inches into the patient's left groin, without eliciting a reaction. Ho inserted needles at various points in the patient's leg; periodically he twirled one of them or moved it up and down a little. There was no anatomical logic for where he placed the needles; when he put them in the earlobes and in the web between the thumb and the first finger, one of the residents leaned over to me and said,

"When do we start dancing and burning incense?"

After about fifteen minutes Ho again motioned to the chief resident who sauntered over to the gurney, raised the patient's leg, supported the knee, and pressed down on the ankle. This time he was able to flex the knee through an almost normal range of motion, without producing pain. He repeated the procedure three times, each with the same result. Everyone clapped. The patient started crying for joy, and Dr. Becker smiled one of his rare smiles. The resident just stood there, scratching his head. Despite four years of high school, four years of college, four years of medical school, and five years as an orthopaedic resident, he had never encountered anything that could rationally explain what he had just seen.

The kind of research deemed desirable by those who had the power to make such choices changed in unpredictable ways, like haute couture, and acupuncture was no exception. It lacked a biochemical basis and therefore had been regarded as unfit for scientific study. Then President Nixon went to China and a window suddenly opened. The National Institutes of Health announced that grants would be provided for acupuncture studies, if the grantee promised to employ the methods of modern science. Dr. Becker received one of these rare grants and we used the money to measure the electrical properties of acupuncture points and meridians which, we found, differed from other points on the body. That work provided the first evidence of the kind considered acceptable in the West that acupuncture points and meridians actually existed. We published six articles on acupuncture, but when the time came to apply for a renewal of the grant the window had closed. The National Institutes of Heath terminated all its acupuncture grants, including ours.

One day soon after we got that bad news, Dr. Becker's green man must have said something to him like, "Bob, you ought to give some thought to the possibility that EMFs could sometimes hurt people." He had mentioned that idea several times but, as far as I knew, hadn't done anything about it. Then one of his neighbors developed cancer and asked Dr. Becker whether he thought the radio towers on the top of the hill where they lived could have been responsible. Dr. Becker learned that seven other neighbors had developed cancer, far more than would be expected in a rural area. When he plotted the addresses of the cancer cases on a contour map of the hill, he found a direct line of sight between each address and at least one of the towers, which he took to mean that the EMFs from the towers could have been involved in triggering the disease.

"When you turn on a television," he said, "a wave field is picked up by the antenna and amplified, and that's the picture that's displayed. If the wave can affect an electrical circuit in a television, why can't it be picked up by cells in the body and affect those cells?"

"But the TV is specially designed to pick up those signals," I said, "and there are no cells in the body that can do that."

"How do you know?"

"I just don't see how it could be possible," I said.

He was silent for a moment, and then said, "Just because you don't understand how it could happen doesn't mean it can't happen."

"The TV signal is probably absorbed in the skin," I said, "so the amount that penetrates further in is probably minuscule."

"Compared with what?" he responded immediately.

I had no answer. I didn't know how much energy penetrated into the body, or which organ might be susceptible, or at what level. Herman Schwan had done calculations he said proved EMFs couldn't affect cells. I didn't think he had proved anything, but I also didn't think that meant EMFs could alter cells.

Dr. Becker published the results of his cancer study in the *Journal of the New York State Medical Association*, and an article about the work appeared in a local newspaper. Some of his neighbors asked him to study something else because they were worried that the publicity would affect their property values. They might as well have saved their breath because nothing would stop him from doing what he believed in.

When the editor of MIT's *Technology Review* invited Dr. Becker to publish a paper about electromagnetic forces and life processes, he wrote: "I am concerned about the continuous exposure of the entire North American population to an electromagnetic environment in which is present the possibility of inducing currents or voltages comparable with those known to exist in biological control systems."

Meanwhile, one morning during breakfast I read the list of ingredients on a box of Cheerios and saw "BHT." I wondered what it was, so I wrote General Mills and learned that the letters stood for butylated hydroxytoluene, which was added to increase the product's shelf life by preventing bacterial growth. The reply said that BHT had no effect on people, and just passed through the body.

I had several conversations with a lawyer at the Food and Drug Administration about BHT. He told me that the law "deemed it safe" and would continue to do so unless someone presented "hard scientific evidence" to the contrary.

"What kind of evidence would be needed?" I asked.

"The law says that 'no additive shall be deemed to be safe if it is found … to induce cancer in man or animal.' So evidence that BHT causes cancer would suffice."

"What does 'deemed to be safe' mean?"

"Something that's safe as a matter of social policy, rather than scientific evidence." His reply was straightforward, but not necessarily approving.

"Do you mean that the scientific evidence doesn't matter?"

"No, just that if there is none, or if it's not definitive, the law fills in the gap and makes the decision."

Shortly after that conversation I found a report of an experiment in which rats exhibited no ill effects after having been fed BHT for a month. The investigators, who worked for the Hercules Powder Company, had concluded that the results proved that BHT was safe. Then I learned that Hercules manufactured BHT. Recalling Mr. Schmidt, I wondered whether the supervisor of those investigators had pressured them to change their results. I also wondered how the investigators had reasoned in inferring human safety from the study results on mice.

I bought some BHT, dissolved it in olive oil and, together with an anatomy professor at the medical school, injected it into the peritoneal cavity of mice. I hoped to find cancer.

Above a certain dose, the mice treated with BHT exhibited labored breathing, which we thought might indicate they had lung cancer. When we killed them and looked at their lungs we saw that they were congested and inflamed, but we didn't find any cancer. That wasn't good enough to get BHT banned, but I published the results anyway because they seemed to show that BHT could pass through the blood to the lungs and cause harm. I thought that would at least make people who ate Cheerios wonder whether BHT was safe. Then I got a letter from a scientist who asked, "Why did you suppose that the BHT you injected passed into the blood and was carried to the lungs? It may have diffused through the diaphragm and attacked the lungs directly." I had never thought of that. If that's what happened, then maybe my results were not a good reason to suspect the safety of BHT.

As I pondered what to do I got a phone call from Central Office asking why I had used Veterans Administration resources to study BHT. I told them that it was part of an effort to evaluate the possibility that BHT might make animals more sensitive to electricity, and that I wouldn't be doing any more such experiments. I felt I had done all I could.

Before my involvement with BHT I had never imagined that "safety" could be simply an assumption, or that the Food and Drug Administration sometimes didn't go beyond such assumptions. It was as if there were two kinds of truth, one based on evidence and a second based on "deeming."

The necessity for some kind of adversarial process to resolve questions about what was or was not true was usually the point in my classes in Evidence. My professor said, "The art of cross-examination is the greatest engine for finding truth ever invented. Only by direct confrontation can the holder of the greater truth be ascertained." I stared at him as he lectured, and then closed my eyes and thought about Dr. Franklin Hart.

"Dr. Hart, you testified that you measured the sound levels on my property, and found that they were 65 dB, is that correct?"

"Yes."

"Were these measurements taken at night?"

"Yes, between 1 and 2 a.m."

"Are you certain that the measurements were actually taken on my property, and not, for example, on property located much farther from the stone quarry?"

"I could clearly see that I was on your property when I made the measurements."

"How could you be so sure, considering that it was late at night, and that the land is covered with trees and underbrush?"

"It was moon bright that night, so it was easy for me to see where I was."

"I would like to show you a copy of the Farmer's Almanac. Would you tell the court and the jury what phase the moon was in the night that you claim you went onto my property to make the sound measurements?"

"Well ... according to this ... errrr ... this says it was a new moon."

"Then it really wasn't moon bright that night, was it?"

"No, I guess not."

"It wasn't possible for you to know where you were on the night you made those measurements, was it?"

"No, I guess not."

"By the way, Dr. Hart, are you being compensated for your testimony here today?"

"Yes. My rate for professional services is $150 an hour."

"How many hours' work have you devoted to this case, including testimony, preparation, measurements, and whatever other services you have provided?"

"Thus far, approximately 120 hours."

"Let's see, that comes to approximately $18,000, is that correct?"

"Yes."

"Does that include expenses?"

"No."

"Dr. Hart, you testified that you are a professor of acoustical science at North Carolina State University, is that correct?"

"Yes."

"In connection with your various duties, have you acquired a familiarity with the various textbooks dealing with acoustical science?"

"Yes, I am completely familiar with the literature in that area."

"Have you read the book entitled *Handbook of Acoustics*, by Malcolm J. Crocker?"

"Yes, I am quite familiar with it. In fact, I have cited it many times in my various publications."

"What about the text *Physiological Effects of Sound*, by Cornelius Clay? Have you read it?"

"Yes. I am familiar with that text."

"Dr. Hart, do you realize that you are testifying here today under oath?"

"Yes."

"And that you have sworn to tell the truth, the whole truth, and nothing but the truth?"

"Yes."

"Do you realize that this court has the power to punish for contempt any untruthful answers that you might give?"

"Yes."

"Bearing this in mind, Dr. Hart, I would like to ask you again whether you have read the text *Physiological Effects of Sound*."

"Well, I am pretty sure that I read it."

"Dr. Hart, would it surprise you to learn that there is no such text as

Physiological Effects of Sound and that no one named Cornelius Clay ever wrote a book dealing with acoustics?"

"Well ... I guess it's possible ... I might have made a mistake and confused the book with some other book. I..."

At this point Dr. Hart was sitting low in the witness chair, as if someone had removed his backbone. I looked at him scornfully and said, "That's all. We've heard enough. Witness dismissed."

Then I turned and walked toward the jury box and in a voice and manner that a preacher might have used to describe how the Holy Spirit enters the body of a convert, I said, "Ladies and gentlemen, scientists are only human beings. They are good and bad in the same way as other people. You cannot believe in the truth of what they say simply because they are scientists. You have to ask them how they know what they think they know. Sometimes they don't know at all. Sometimes, like Dr. Hart, they just lie."

It was a pleasant reverie.

7
EMFs

◆

I continued to think about whether electromagnetic fields (EMFs) could affect bone cells. "How could such a thing happen?" one antagonist asked me rhetorically, as if my having no answer vindicated his skepticism concerning whether it *could* happen. He was making the same mistake I had made when I talked with Dr. Becker about TV signals and cancer.

I attended a lecture by Richard Feynman, who had earned worldwide fame as a great teacher of physics. I was entranced from the moment he started talking. He was funny and self-effacing, and full of amazing insights that he revealed in little conversations with himself. "Why doesn't the earth fly off into space? Because the gravity of the sun holds it. Are there little ropes between the earth and the sun tying them together? Where is the machinery of this force? Nobody knows. It's a mystery." What an amazing thing for him to say! He had won a Nobel Prize, yet he admitted that he couldn't give a better answer than my mother.

He told a story about a child who throws a pebble into a pond and watches the water waves ripple outward. "When he tosses two pebbles in at the same time the waves begin at different points and then spread and begin interacting with one another. A little man who happens to be exactly at the place where the waves come together would bob up and down, but he wouldn't know how much of the bobbing was caused by one wave or the other. He wouldn't bob up and down twice, only once in response to the sum of the two disturbances at the instant they came together. It would be the same story regardless of how many stones the child threw. The little man would still bob only once in any particular instant, in response to the resultant of the interacting waves. He couldn't pick out just one wave and respond to it alone. Our eye is in the same kind of a position as the little man, only it's *much* smarter. Every second there are millions and millions of EMFs that are going through it from every direction. And from all of these EMFs it picks out just a few, and we say 'I can see.' Isn't that a remarkable and mysterious thing?"

I realized, for the first time, that even though the stories for some phenomena were better than the stories for other phenomena, the truth was that we really didn't know the "how" of anything.

I began searching for published reports about the effects of EMFs on animals or people, and I came across the work of a biology professor at Northwestern University named Frank Brown. He had put worms on an inclined board and observed whether they went to the left or the right after they crawled off. When he had placed a magnet on one side, more worms would turn in that direction.

"Why did you do those experiments?" I asked.

"I just wanted to see what would happen," he said. "In the wild they can't see or hear their food, and yet they find it. I thought they might use the earth's magnetic field as an aid."

"How has your research been received?" I asked.

"Not too well," he said. "I use a wooden board, two bricks at the back corners to elevate it, a magnet I bought at the dime store, a stopwatch, and my eyes to count. I don't have grand equipment, so I guess the work doesn't look scientific."

Even so, he was doing what interested him, and he seemed happy.

Brown had been the editor for the English edition of a book called *Electromagnetic Fields and Life*, written by a Russian physicist named Aleksander Presman. Brown told me that the topic had been studied intensely in Russia for a long time.

"Why do you think there have been so few reports of EMF effects published in the United States?" I asked.

"Probably because it just wasn't a fashionable subject here," he said.

Presman's book was marvelous; his references were mostly to articles in Russian so I couldn't read them, but the book still made sense to me. He said there were three kinds of electromagnetic phenomena. EMFs in the environment affected the metabolism and behavior of all living organisms. The environmental EMFs could be natural in origin, like the earth's magnetic field or electrical changes caused by the weather, and they could come from man-made sources like radio, radar, and powerlines. Second, there were natural EMFs within organisms that conferred life and mediated the processes of life such as growth and healing. Third, EMFs could function as a vehicle of communication between living organisms. Presman's book was almost poetry. I had never imagined that a physicist could achieve such a biological synthesis.

He must have recognized that there was resistance to the idea that EMFs were biologically active, because he wrote: "Physicists have concluded that weak EMFs are incapable of producing biological effects." But I guessed he too couldn't get a handle on exactly why they believed that, because he did not try to analyze their viewpoint. Instead he described the results of experiments that showed effects of different kinds of EMFs on animals and people. That was the sensible thing to do, I thought. If EMFs *could* affect cells in animals and people, then there had to be an equation that was the governing law, and there had to be a plausible story about how the EMF brought about the effect. Those things could be considered *later*, after everyone agreed there was something real to be studied.

In my original EMF experiment the yoked rats had struggled incessantly, so in my next experiment I decided to apply the EMF from outside the cages, to permit the rats free movement. I knew they had no conscious sensation of the field because they never visibly reacted when I turned it off or on.

To determine whether the field would make bones heal faster, I fractured one femur in each rat, casted each fractured limb, and put half the rats in cages that had an EMF and the others in cages that didn't. I had planned to recover the bones after three weeks and measure how much force it took to break them, expecting that more force would indicate the presence of more bone which would indicate more healing. Dr. Becker didn't like the experiment. He said that exposing the entire rat to an EMF might not be the same as treating only a limb. I didn't see how that would make any difference, and said so, at which point I got a lecture about how a systemically applied EMF might affect the brain and lead to serious consequences, maybe even opposite to those that I hoped to produce in the bones. He suggested I talk to Howard Friedman, the Chief of Psychology in the hospital.

Friedman told me about an experiment in which he and Dr. Becker had studied the relationship between the earth's magnetic field and admissions to psychiatric hospitals. They had compared changes in admissions rates with changes that occurred in the earth's field and found that when the field got stronger, the number of admissions rose. To test whether the association was causal, Friedman had exposed rabbits to EMFs and then examined their brains.

At first he got very excited, he said, because he found pathological changes in the brains of rabbits that had been exposed to EMFs. But then he found the same kind of pathology in the brains of the control rabbits. He was about ready to conclude that the EMFs had no relationship to the brain pathology when he realized that the brain changes were more frequent and more severe in the field-exposed group.

"I remembered reading about stress research," he said, "and that led to what I think is a good explanation of my results." He said that a virus lived in the brains of rabbits of the strain he used. The virus was normally kept under control by the rabbit's immune system, but when the rabbit was stressed the virus could multiply and cause damage. That was his explanation for why brain lesions occurred, and for why they were more common and more serious in the rabbits exposed to the EMFs. His story gave me a new perspective on the meaning of saying something "caused" disease.

Dr. Becker thought a similar thing might happen to my rats, and that I would wind up with sick animals from the stress caused by the EMF. I went ahead with the experiment anyway, and ran into terrific problems, although not the kind Dr. Becker had supposed. Sometimes the cast fell off because it was too loose; when that happened the broken limbs healed in grotesque positions. Other times the limb became necrotic because the cast was too tight. Sometimes urine and feces got under the cast, leading to infections. I didn't learn anything about EMFs, except that studying their effects on animals wasn't easy. "There ought to be a law against physicists doing experiments with animals," one of the people who took care of the rats said to me.

Someone who understood my goal and my limitations suggested, "Why not just weigh the rats and see if there's any difference after they live in the EMF for a while?" The basic idea, that if I put young rats in a field the skeleton might grow bigger so the rats would weigh more, sounded reasonable to me. I bought some rats, put half in cages that had an EMF, and the others in cages that didn't. Ten days after the experiment began, a graduate student who worked for me said, "Something is happening to the rats. Their eyes look swollen."

Dr. Becker thought the rats might have gotten an infection, but a veterinary ophthalmologist who examined them said the problem was not infection but rather "secondary glaucoma." After a month of exposure to the EMF, 17% of the rats had developed secondary glaucoma, whereas none of the control rats did. The condition occurred only in the right eye, which struck me as even more bizarre than the fact that the condition had occurred. Unfortunately, the average weights of the two groups were identical, so I had no evidence that EMFs affected bone growth. But I did have evidence that the EMF affected the eyes of the rats. That certainly wasn't what I had wanted, but that's what I got.

In principle, the rats could have had bad eyes when I bought them, and the EMF might not have had anything to do with why they got big. But I had assigned the rats randomly to the EMF or control group, and the odds that all those with bad eyes would wind up in the EMF group by chance were astronomical. Another possibility was that something I didn't know about was responsible. But what? The most likely scenario, I thought, was that the EMF made some atoms somewhere in some cell in the rat wiggle differently, and that set up a series of reactions governed by some unknown equation, and secondary glaucoma was the result.

"It's too bad the experiment didn't work out," someone said, but I thought to myself, "Well, in a sense, it did work out. EMFs didn't make the bones grow faster, but the EMFs did affect cells in the rat." What I had hoped would happen was just a matter of my personal desire. Whatever the EMF was going to do, it would do irrespective of what I wanted. The fact that the EMF did something other than what I wanted didn't make the result any less real. And if it was real, that was a new thing. The new thing was that EMFs could affect cells. Who knew where that could lead?

There were at least two different ways that I could look at my discovery. Maybe the EMF affected only atoms in cells that manifest glaucoma, or maybe it affected atoms in other cells leading to other effects that I didn't know about because I didn't study them, and the glaucoma developed as a consequence of the initial effects. I had no idea, then, which view was better.

I decided to do another experiment and measure something different. The anatomist who had helped me with the BHT experiment had an experimental model of cancer in which he could inject cancer cells into a mouse and they would multiply and kill the mouse in 2-3 weeks. Some of the cancer cells had defective chromosomes, and the number of such cells could increase or decrease depending on the kind of drug or other treatment the mouse received. We injected the cancer cells into mice, and when we analyzed the cells two weeks later 20% of those from mice that were treated with an EMF had defective chromosomes, compared with only 5% of the cells from mice that weren't exposed to the field.

I got excited. I bought more rats, exposed half of them to an EMF for three days, killed them, recovered the blood, extracted the serum, and measured the percentage of each of the four major groups of immunoproteins. I found that the exposed rats had a lower percentage of gamma globulin, which meant that their immune system had been altered.

Now I got really excited. EMF effects occurred everywhere I looked. On the other hand, I lacked expertise in doing biological experiments. Maybe all my results had no meaning. I felt like a soldier in a coastline bunker who peers out over the waves, watching for enemy ships. One foggy morning he sees what he thinks is a line of ships against the horizon, but no one shows much interest because it has never happened before, and he is wracked by indecision about whether to give the alarm.

While I was trying to decide whether to publish my results I learned that there would be a meeting at the New York Academy of Sciences on the role of electricity in biology, the first meeting this century on that topic. The organizers invited Dr. Becker to make the keynote address, and asked him whether he would like anyone else in his laboratory to present a paper. He recommended four of us, and we were all invited.

I looked forward to the meeting because I thought the people there would have interests similar to mine. I expected I would receive some help and advice, and that the whole experience would probably make it easier for me to decide whether EMF research was something worthwhile.

8
A Long Day

The New York Academy of Sciences meeting on the biological effects of electrical energy was held at the Barbizon-Plaza Hotel in Manhattan in September, 1973. Of the forty-five scheduled papers, five were from our lab.

At breakfast on the first day, some of the conference participants who sat at my table talked about the tennis match between Bobby Riggs and Billie Jean King to be held that evening. Someone said, "He acts as if it's impossible for a woman to beat him. That's why he'll win. It's not physical, it's psychological. He had Margaret Court beaten before their match ever started, and he'll do the same to Billie Jean King."

The first speakers described various kinds of electrical measurements on bone and teeth. Then someone from Germany showed pictures of how ink had flowed through bone specimens, and he somehow concluded that his results might explain why electrical current makes bone grow. Jonathan Black, who worked with Carl Brighton at the University of Pennsylvania, had something to say during each of the question-and-answer sessions that followed the talks. The measurement wasn't made correctly, or the wrong statistical test had been used, or an alternative explanation was better than the one that the speaker suggested. There was always something wrong.

When Black gave his talk he described in great detail what he had done. He even listed the initials of each "donor" of the bone specimens he had studied – as if he anticipated that somebody might stand up and ask him "What were the initials of the donor that gave those results?" Black concluded that he had presented "incontrovertible evidence of the existence of electrical signals from bone." But nobody had doubted that; it was as if he had discovered hot water.

Dr. Becker was allotted twice as much time as the other speakers. He described his conception of the big picture regarding bioelectricity and concluded: "In summary, I have described a primitive data transmission and control system located within the nervous system. This control system, I believe, antedated the nervous system and transmits its messages and control signals by means of analog-type direct currents in a solid-state matrix.

Inputs to the system consist of injury to the organism and environmental electromagnetic fields. Outputs from the system consist of the control of cellular growth by inducing specific electrical environments within the organism, and a general control over the functional level of the central nervous system."

The story was captivating, and Dr. Becker's oratorical style only added to the overall favorable impression he made on the audience. He spoke simply and directly, with surpassing confidence but no hint of arrogance. He was always at his absolute best when he spoke, and you just listened. The last thing he said was that he thought recognition and acceptance of what he called the "control system" would explain several of the problems connected with our understanding of "pain mechanisms, biological cycles, growth of control and healing, acupuncture, and the phenomenon of the response of living tissues to a variety of electromagnetic fields," and "other problem aspects of biology such as the question of the origin of life."

The audience stood and clapped, as if at the end of a Broadway play, and they continued to clap as he nodded, walked off the stage, and went down the center aisle to his seat. Cedric Minkin, the session chairman, began thumping on the microphone for order so that the next talk could begin. Bassett, who had walked to the podium in anticipation of being introduced, just stood there waiting. When the applause finally stopped Bassett said, "Ladies and gentlemen, we have just heard one of the most important talks in biology ever given."

But even at the pinnacle of his success, I could see there were problems with Dr. Becker's project. Many of his slides were from illustrations that had appeared in his papers five or ten years earlier. I felt that his success depended not only on a story of ever-expanding scope, but also on nailing down incontrovertible truths regarding each successive chapter in the story. But nothing was nailed down.

During Bassett's presentation I was surprised to learn that he was also using EMFs. I had spoken with him about them during the last time he had visited our lab – the day the resident had started a fire when he dumped his pipe ashes on the ether-soaked gauze pads in the trash can. I had told Bassett about my experiment with EMFs. He had no clear idea what EMFs were, so I had explained it as best I could. As I listened to his talk at the meeting all I could think was, "He stole my idea."

He described experiments in dogs aimed at accelerating bone healing. Each dog wore saddlebags containing circuitry that sent electricity to coils attached to the dog's hind legs, thereby producing an EMF. Bassett had cut the fibula in both limbs and then studied whether the one on the right, which got an EMF from the coils, healed faster than the one on the left, which got no EMFs. He didn't need casts because the tibia served as a natural internal support for the leg. In his first experiment he had used "1P" EMFs, which he said didn't work, so he switched to "10P" EMFs, which he said made the treated side stronger than the untreated side. At the end of the talk he briefly mentioned the use of the EMFs to treat children with bone problems.

In the afternoon a Canadian named Romero Sierra spoke first. He had a theory that birds could somehow detect low levels of microwaves because of their feathers. When he had exposed chickens to an EMF they had laid more eggs, but some of the chickens had died. He concluded, "The benefit of increased egg production may be completely negated by the increased mortality rate."

The tenor of the questions he received made it clear that the idea of microwaves killing chickens was disturbing, perhaps because of its implications for human exposure. Romero Sierra responded, "As man utilizes more of the electromagnetic frequency spectrum in the future for power transmission and communication, it is of paramount importance for him to scrutinize the risk beforehand, so that he can make well-informed judgments about the particular applications." I thought that was rather obvious.

Allan Frey, from Pennsylvania, reported that some people could actually hear the radar pulses emitted by military radars. He and Dr. Becker were contemporaries and had some kind of relationship that involved their common interest in electrical effects, but I didn't know the details. "Stay away from him," Dr. Becker said to me about Frey.

William Gensler, from the University of Arizona, described experiments in which he had measured the voltage produced by tomato plants. It changed in characteristic ways depending on whether or not he watered the plants, and he planned to develop a system that would tell a farmer when he should irrigate.

Dan Harrington, who got his Ph.D. working in our lab, used electrical current in an attempt to make skin incisions in rats heal faster. The healing process involved many different kinds of cells, so he couldn't say what cells were affected, if any, and his presentation was confusing. At the end,

Minkin observed sarcastically: "I think that fantastically complex cellular migrations and differentiations that occur in a wound-healing process, superimposed on the complex problems defining electrical parameters, present a problem that will keep us all busy for many years."

An investigator from the Karolinska Institute in Stockholm talked about whether the electrical signals from moist bone were due to piezoelectricity or streaming potentials. Streaming potentials depended on pH and there was always a pH at which the potential was zero. Piezoelectricity, however, didn't depend on pH, so his idea was to see whether the electrical signals from bone depended on pH. At a pH of 5.5 he got no signals, which led him to conclude that the bone signals were due to streaming potentials. But he had ignored the possibility that the piezoelectric signals were really there but couldn't be measured because they were shielded by a layer of water. I thought about pointing out that it was illogical to infer that something wasn't there when your measurement method wouldn't have allowed you to see it even if it were there, but I said nothing.

The next speaker was Lionel Jaffe, from Purdue, who only a year earlier had called Dr. Becker's work "plain ordinary fraud." Jaffe described a years-long project aimed at measuring tiny currents that he claimed flowed from cells when they were injured. Most people at the meeting had either measured voltage from tissue or applied current to it, but he was measuring current produced by tissue. There was no standard instrument for such measurements, so he designed and built one. When he put its tip near the surface of a cell, he got a reading. I couldn't decide if the instrument were genuine, or if it was like Dr. Abrams' Oscilloclast.

All of the speakers who measured voltages from cells had a different theory concerning what the voltages were doing. Clarence Cone, who worked at the Eastern Virginia Medical School, reported that voltages in dividing cells were lower than in nondividing cells, which he interpreted to mean that voltage controlled cell division. Someone asked him whether the change in voltage caused the cells to divide or whether cell division resulted in a change in voltage, but he said he didn't know.

At the end of the afternoon session there was a panel discussion of the question, "To what extent can electrical stimulation be used in the treatment of human disorders?" The prominent Dr. Bassett said he thought we were "standing on the threshold of a new era," in which solutions would be found to the problems of pain, regeneration, and cancer. "I would say that in 20 years almost as much electrotherapy as chemotherapy will be used in the medical community."

The rising Dr. Brighton emphasized that many different variables could potentially influence the result of treating bone with electric currents, and that the reasonable thing to do would be to systematically work out the nature and relative importance of the different variables. He said that's what he intended to do.

The heroic Dr. Becker warned of the problems in applying electromagnetic energy to patients, especially the "possible induction of malignant cellular changes," which he said might occur because the electromagnetic field either directly caused cancer or stimulated a pre-existing precancerous lesion.

My presentation was scheduled to be the first one in the evening session after the main dinner which took place at the Academy building on the east side of Central Park. I didn't enjoy the dinner because I was nervous, so I left early and went to the lecture hall. I placed the slides in the projector cassette myself and checked that they were all right-side up, and I made sure that the microphone and the projector remote-control worked properly.

When I was called on to give my talk, I told the audience about the rats with the bulging eyes and the mice with the abnormal chromosomes. When I finished and said "thank you" I felt relieved that the slide projector had worked, satisfied with my delivery, and excited that I had been able to describe my work to other investigators. But no one in the audience clapped, and no one asked any questions; that silence hadn't happened after any of the talks during the day. Some people cleared their throats, and I heard what sounded like muffled snickers. After what seemed like a very long time, the German who had done the experiment with ink and bone said that what I had presented wasn't "conclusive proof," and I agreed. Then Minkin said, "Your work is a very interesting presentation of phenomenology, but what importance does it really have?" He was what we called on Greenway Avenue a "wise-ass." He liked to poke people and watch them jump. I didn't like being poked, but I answered, "I am aware of no reports of phenomenologic effects of EMFs of the type I studied on mammalian systems." Everything was quiet again for a few seconds, and then someone said, "I suggest you read Presman's book." Another wise-ass. "I have read Presman's book," I said, "and my statement stands." I wanted to say more, but I didn't have the energy. It was as if I had won my match, but nobody was watching.

As Dr. Becker and I crossed E. 60th St. on the way back to our hotel, we were caught from behind by a hard-charging, chubby, uniformed Navy officer with a beet-red face. He ignored me and said to Dr. Becker, "I'm Paul Tyler. I am in charge of a program for the Navy to evaluate possible environmental impacts of an antenna that the Navy plans to build in Wisconsin. We are doing some studies and I would like you to be on a committee to review what we have found." At just that moment I heard someone behind me say that Billie Jean King had won.

Later that evening at the hotel bar I was approached by a man who said he worked for a columnist at the *Washington Post* named Jack Anderson. The man said he thought my talk was interesting and asked if what had happened to the chromosomes of the animals could happen to people. I told him I didn't know but that I thought it was possible. The reason I gave was the one I had heard Dr. Becker use many times when people questioned him about the relevancy of animal work for human beings: "Animals are just like people, only a little different."

"How could those kinds of effects happen?" he asked.

"Well, I guess there are two ways. Maybe the EMF affected the cells, and that's the reason they were different. Or maybe the EMF was a stress."

"What do you mean, stress?"

"I don't know exactly," I said, "but stress can make something that's already there even worse, that kind of an idea."

"Has anyone from the government talked to you about side-effects of EMFs?" he asked.

At first the question seemed odd, like asking me my wife's maiden name. Then some things that Dr. Becker had done over the years flooded my mind and almost crystallized into a pattern. Once in a while a stranger would come to our lab and spend an hour with Dr. Becker in his office. On those days his demeanor was particularly somber, and he always closed his office door, something he otherwise rarely did. "Who is he?" I once asked. "You don't need to know," he said. But another time he told me, "He works for the agency." I thought I knew what that meant, and a little while later I found out I was right. Dr. Becker usually asked our secretary to make his phone calls, and he would come on the line after the other party had answered. On rare occasions, however, he dialed himself. One day I looked in the phone book and saw that the Central Intelligence Agency had a listed telephone number, and that it was the same one which was pinned to the bulletin board in his office.

"No," I said again to the reporter, "why would you ask that?"

He didn't answer my question but said, "Have you ever heard of a man named Cecil Jacobson, at the George Washington Medical School?"

"No, who is he?"

"He's a researcher who studied the effects of EMFs on chromosomes."

"In rats?" I asked.

"No, in people."

The reporter told me a fantastic story, part of which he read from one of Anderson's columns about the Russians using EMFs for "brainwashing," and part of which he said that he had heard from Senate staffers during recent Senate Commerce Committee hearings dealing with health risks from microwave-oven EMFs. The gist of what he said was that the Russians had begun irradiating the American Embassy in Moscow with EMFs about ten years earlier, and that the Central Intelligence Agency had advised the State Department to look into the health records of the Embassy employees and determine whether there was any increase in health problems after the irradiation had begun. A doctor named Herbert Pollack had examined the records and concluded that there was no microwave-related health problem, but the State Department wasn't satisfied so it hired Jacobson to look at blood slides from the employees to see if he could find any abnormal chromosomes.

"Did he?" I asked.

"Yes," he replied, "but State canceled his contract and clamped a lid on the whole affair. Nobody's talking."

"What about Jacobson?" I asked.

"He won't talk to me. But before he lost the contract for the research on the blood he told some of the people he worked with at George Washington University that he had found genetic defects on the slides he was analyzing."

"But how could the government keep the affair secret when so many employees are involved? Surely one of them will talk," I said.

"They don't know anything about it. They were told they had to give a blood sample because a virus that caused Montezuma's revenge was going around. They thought the purpose of the blood test was to see if they were infected."

Anderson's article mentioned a secret research program, called "Pandora," dealing with the effects of EMFs on the brain, but the investigators doing the work wouldn't talk to the reporter.

"Who are the investigators?" I asked.

"One is Ross Adey, who works at the Veterans Administration in California. Do you know him?"

"No," I said.

"Another one is Don Justesen who works at the Veterans Administration in Missouri." "I don't know him either," I said. "What kind of work did they do?"

"It's classified so they won't give me any details," he said, and then he asked me, "Did any studies about the effects of EMFs on people go on in your lab?"

"No, not exactly."

"What do you mean, 'not exactly'?"

"Well, Dr. Becker did a study about EMFs from antennas and cancer, but it had nothing to do with brains."

Then we said goodbye, and that was the end of my long day.

9
Wagging Tails

♦

I thought the atoms in the eyes of the rats exposed to the EMF must have been compelled by an equation to begin jiggling in just the right way to produce the glaucoma I observed. To support my theory I did the experiment again, but only after an ophthalmologist certified that the rats had no preexisting eye defects. Unfortunately, after a month's EMF exposure there wasn't even one instance of glaucoma. The inconsistency between my two experiments seemed like a stain on the majesty of science, but inconsistency was what I experienced.

A paper was soon to be published in the *Annals of the New York Academy of Sciences* reporting that EMFs caused glaucoma in rats, and I needed a story for why I didn't get the same results when I repeated the experiment. I adopted Howard Friedman's rationale, and concluded that the negative results from my second study suggested that the glaucoma in my first study had been due to exacerbation of preexisting eye defects caused by a stress reaction to the field. The *Journal of Ophthalmology* rejected the manuscript; the editor said no one had ever proved that there was a link between stress and glaucoma.

I had no way of knowing at the time, but during the next thirty-five years thousands of EMF experiments would be performed, and the same pattern of positive and negative effects would be repeated endlessly. Only when I recognized the pattern did I realize that the primary meaning of the experiments was the *pattern*, not the *data*.

As I continued exposing animals to EMFs, looking for more evidence of their power over living things, my focus progressively broadened beyond the idea of using EMFs to make bone grow. Bassett and Brighton were far ahead of our lab in that area but, more fundamentally, I thought the EMF story probably involved much more than bone.

I was fascinated by the idea that fields somehow caused "stress." I didn't know much about stress, but I wasn't concerned because it had no necessary relation to *what* EMFs did; stress was only a story for *how* EMFs did something, so the story could change. Friedman could have been wrong about the role of stress, but that wouldn't have changed the fact that brain abnormalities in the rabbits he studied were more frequent and more pathological in the EMF group. The situation reminded me of the *Palsgraf* case. Maybe the concussion had knocked over the scales, or perhaps a frightened passenger had been responsible. However it happened, something had linked the explosion and the woman's injury. The most important thing was my discovery that EMFs could do *something;* how they did whatever they did was less important.

I remembered an experiment where the investigators added BHT to the food they fed their mice. After the pups had been weaned, they also were fed only BHT-laced food, and when they reached maturity the whole process was repeated, and then repeated again. In each of the three generations the number and growth rate of the mice were measured under the theory that any bad effects of BHT would be cumulative over the generations and reflected in the measurements. I thought that kind of experiment would be a good way to learn more about what EMFs could do to animals. Using batteries as the source of the voltage, as I had done in my previous experiments, would have been impractical because the experiment would last almost a year, so I used the voltage from an ordinary outlet. The results from the first generation were dramatic; fewer pups were born to the EMF-exposed mice, the percentage of pups that survived to weaning was smaller, they grew more slowly, and by the time they became adults they were half the size of the control mice.

Meanwhile, I studied law. The annual moot court competition during my senior year in law school was the case of a twelve-year-old boy named John Winship who had been arrested by the police in New York City for allegedly stealing $112 from a woman's pocketbook. A section of the New York Family Court Act provided that "a person over 7 and less than 16 years of age who does any act which, if done by an adult, would constitute a crime" may be judged to be a delinquent, and punished accordingly. A judge decided that Winship had stolen the money, and in making that decision he relied on another section of the Act which provided that "any determination that a juvenile did an act or acts must be based on a preponderance of the evidence." As a result, Winship was adjudged a delinquent and sent to a juvenile facility until his eighteenth birthday. He appealed the judgment on the ground that proof beyond a reasonable doubt should have been required.

I was assigned to a team that represented Winship. On the day I argued, the judge intoned, "We will hear oral arguments in Winship versus New York. Counsel, you may proceed." I rose from the counsel table, walked to the podium in the middle of the mock courtroom, said, "Thank you, your Honor, and may it please the Court," and began my argument.

"Winship should not be punished," I urged, "unless the evidence showed beyond a reasonable doubt the existence of every fact necessary to constitute the crime. The accused has an interest of immense importance at stake, his liberty. Society should not condemn someone for commission of a crime when there is reasonable doubt about his guilt. The reasonable-doubt standard is indispensable for fostering the respect and confidence of the community in the application of the criminal law. It is critical that the moral force of the criminal law not be diluted by a standard of proof that leaves people in doubt whether innocent people are being condemned."

The judge asked, "Do you claim that how convinced a mind needs to be to go beyond the preponderance standard and reach no-reasonable-doubt status can be described?"

I conceded that the labels used for different standards of proof were vague, but I argued that they communicated different notions concerning the degree of confidence that someone should have in the correctness of its factual conclusions.

He asked whether I thought there was a "public-policy justification" for using the reasonable-doubt standard for deciding what is or is not a fact in criminal cases, but only a preponderance standard in civil cases. When the issue had first arisen during my discussions with other law students, the answer seemed obvious. If we were going to send somebody to jail for committing a crime, we needed to be sure that he actually committed the crime. Then I began to see the big picture. Sometimes judges or juries will be wrong because human beings aren't perfect. The real question was why we had two standards for deciding we're not wrong. During our discussions someone made the point that there was actually a third, in-between standard in some first-amendment cases, "clear and compelling." If "beyond reasonable doubt" was 99.9% sure, and "preponderance" was 51%, "clear and compelling" was like 75%. Why were there different degrees of being certain that we knew what we thought we knew?

Well, when you thought about the consequences of error, you could see why. Juries can err in either of two ways. They can accept something as a fact when it really isn't, or they can refuse to accept something as a fact when it really is. In an ordinary civil case involving money, society doesn't care which mistake happens because either the plaintiff or the defendant would be injured unjustly, which are equally bad consequences. Society has no general preference for specifically protecting one party or the other. In a criminal case, however, the stake involves someone's freedom, which is more precious than money. The standard of proof affects the comparative frequency of the two types of erroneous outcomes, so the choice of the standard reflects society's judgment regarding the comparative social cost. As a society we believe that convicting an innocent man is far worse than letting a guilty man go free, so we go beyond the "preponderance" standard and adopt the "beyond reasonable doubt" standard in criminal cases. I told the judge that was the "public-policy justification" for the choice of the standard in criminal cases for "knowing" that something was a fact.

I got my law degree and was admitted to the bar. I planned to use my legal education only for something I really believed in – that much was clear. Putting Winship in jail for six years when a judge was only 51% sure that he actually stole the money would have been wrong, and I couldn't have argued the other side with passion and conviction. How or where I would use what I had learned about the law was far from clear. My friends in science knew nothing about law, and my friends in law knew nothing about science. I felt that I belonged somewhere at the interface between the two disciplines, but I didn't know in practical terms what that meant.

There was also a deeper problem, inchoate yet dimly perceivable. Science and law were two man-made systems for knowing what happened in the world, and each had its own method of knowing. Even in principle, whose method was supposed to be used at the interface?

Various construction projects around the house helped take my mind off the problem of what to do with the law degree. There were piles of limestone on our land, and I decided to use it to build a wall. I sorted the pieces by size and placed them so that they interlocked to provide mutual support; I wanted my wall to remain standing long after I was gone. I found that thinking about which stone ought to go where completely occupied me. Even when I tried to think about something else, I couldn't. Most of my neighbors thought I would never finish, but I did.

Lin had retired from teaching after our first son was born, and she remained retired as the next three kids came along. One day, just after she had finished picking blackberries that grew along the side of the road near our house, workmen from the county sprayed both sides of the road, leaving an oily white film on the vegetation. The white stuff was gone by the next day and everything looked normal, but a few days later the blackberries withered, like raisins; during the next two weeks everything turned brown except the grass.

I asked one of our neighbors, a biology teacher, if he knew anything about the spraying. "They killed all the wildflowers," he said, "dame's rocket, daylily, bellflower, bouncing bet, chicory, hawkweeds, everything."

"Have they done this before?" I asked.

"This is the third time in the last 5 years," he said.

I began to wonder what the white stuff was, and about how much we had eaten in our blackberry pies. I called the county transportation department and was told that the herbicides were a mixture of 2,4-D and 2,4,5-T, and that they were completely harmless.

At a garden store, labels for those products warned against eating food that had been sprayed or drinking milk from cows that grazed in treated areas. Then I learned that the 2,4,5-T was one of the herbicides used in Vietnam to defoliate the jungles, and that malformed infants had been born to women who were sprayed. I thought that the transportation department at least ought to warn the public, and said so to Joe Berndt, the head of the department. He told me the Environmental Protection Agency had said the herbicides were safe, and that spraying helped wildlife because when the vegetation died the animals moved farther away from the road and were less likely to get run over. I wanted to know why he didn't just mow the weeds, like most of the other counties in New York, but he said only that he knew better than I about taking care of roads.

I raised the spraying issue with my county legislator. Shortly thereafter I was asked to serve as chairman of an *ad hoc* committee to investigate the roadside spraying practices of the transportation department. I hadn't intended to get in that deep, but I felt I couldn't quit.

Berndt had begun working for the department the month he graduated from high school, and after 30 years had risen to the top job. I tried to tell him about the science of herbicides, but he regarded what I said as just a lot of mumbo-jumbo. All he needed to know was that the government had said the herbicides were safe. His view was that the government wouldn't have said that if it wasn't true. End of story. I thought he resented me for what he saw as my interference with him doing his job. That wasn't my intention. I respected him; he had worked hard every day of his life, just like my father. But in the area of safety of herbicides he was in over his head.

When I formally presented my final report I said, "We believe that maximum protection of the public health and welfare is preferable to minimum protection and later regrets," and I argued that the highways could be mowed for less than what Berndt had spent on herbicides, and that the fact they were legal didn't mean they were safe. When Berndt was asked to respond, he read a short statement: "The chemical companies spent $5 million and tested the chemicals for eight years. They were lab tested, field tested, and all known methods of testing toxicity and other harmful side effects were employed. For all practical purposes there is no danger when they are used according to label directions." He sat down but a moment later stood up and said: "We have sprayed the roads three times in the last five years and, except for Dr. Marino, we never had a complaint."

The council unanimously adopted my report and urged Berndt to refrain from using herbicides. Six weeks later, however, his department put out a bid request for four hundred gallons of 2,4-D, and six hundred gallons of 2,4,5-T, which was an increase of about 10%. The department bought the herbicides and sprayed them, but it never issued any warnings.

At work, I mated the mice to produce the second generation in my planned three-generation study. That same week Dr. Becker flew to Washington, DC, at the invitation of Captain Paul Tyler to attend a meeting.

"What happened?" I asked when he returned to work.

"The Navy wants to build the mother of all antennas."

"What kind of antenna?"

"A grid that will stretch over all of the counties in northern Wisconsin. It's designed to operate at extremely low frequencies. They didn't give details, but its purpose is to send radio waves through the atmosphere even if there had been a nuclear exchange with Russia and the atmosphere had been so screwed up that ordinary radio waves wouldn't propagate."

"Why would they want to communicate at that point?" I asked. "What could they say, 'Stay down forever?'"

"It's the other half of mutually assured destruction, I guess," he said. "They could tell them what to blow up."

Dr. Becker told me that the Navy had built a small test version of the antenna and was ready to build the real one, but had encountered several problems. For one thing, the technicians who operated the test facility had developed some medical problems. Also, the Secretary of Defense was Melvin Laird, who was from Wisconsin, and he didn't want the antenna built in his state.

"What was the purpose of the meeting?" I asked.

"To evaluate studies the Navy had funded to assess the environmental impact of the EMFs from the antenna," he said. As he did, he leaned forward in his chair, emptied the ashes from his pipe, refilled it with tobacco, and lit it with his lighter that never worked on the first try. That ritual usually signaled he was about to say something exceptionally important, and this instance was no exception. "At first," he said, "the Navy trotted out a bunch of physicists who did a lot of calculations they said proved that the fields were completely safe. But somebody had decided calculations weren't good enough, so the Navy had funded about twenty-five different biological studies. When information about the studies was presented at the meeting, to everybody's surprise most of them reported effects from the fields."

"What kind of effects?"

"All kinds, effects on rats and mice, chickens, fish, cells – stuff that nobody would have predicted. Effects that weren't supposed to happen."

"What did the committee recommend?"

"The only thing possible. That there needed to be more studies. Then, near the end of the meeting, someone pointed out that the antenna's fields were much weaker than those from high-voltage powerlines. Everybody got concerned about that, and we told Tyler we wanted to express our concern to somebody in the government. He said a Presidential committee was studying the effects of fields in the environment, and that he would report our concerns to the committee." To emphasize the importance of notifying someone in the government, Dr. Becker told me he had told Tyler about my research showing an effect on the growth rate of mice. At *that* moment I first realized there was a possible connection between my mouse study and powerlines, because the fields I was using had a frequency of 60 hertz, which is exactly the frequency of powerlines.

Years earlier, Dr. Becker had bought 130 acres of land near a little upstate New York community called Lowville. He had planned to live there when he retired, and in preparation for that he had built a beautiful nine-room house he called his "log cabin;" he went there as often as he could to hunt and fish, and be alone. He told me that the night he had flown home from the meeting, just after he settled into his favorite chair and began reading the latest copy of the Lowville newspaper, he saw a legal notice saying that a power company planned to build a high-voltage powerline that would come near and maybe cross his land.

The day after our conversation Dr. Becker sent a letter to the state public-health and environmental departments, the state commission that regulated powerlines, the power company that was named in the newspaper notice, and to our local power company. The letter warned that there might be a health risk from powerline EMFs, and recommended that further information be sought from Paul Tyler. The health department never responded to the letter. The others only thanked Dr. Becker for his interest, except for the state commission which called and set up a meeting with Dr. Becker.

On the day of the meeting, a lawyer for the commission arrived and Dr. Becker introduced me as someone on his staff who happened to be a lawyer, and said that I would also attend the meeting. Our visitor smiled, stuck out his hand, and said, "Hi, Dr. Marino. I'm Bob Simpson. I'm glad to meet you."

He had gone to work for the commission two years earlier, right out of law school. As things turned out, he influenced my professional life more than anyone I ever met, except for Dr. Becker.

The key point for Simpson was that a reputable scientist had said powerlines might be health hazards, and he wanted Dr. Becker to present that information in an ongoing hearing involving powerlines. Dr. Becker said he didn't know the details of the Navy studies, and that the man to talk to was Tyler. Simpson said he had spoken with Tyler and gotten copies of some Navy reports, but that the reports were useless without someone to present them in court, which Tyler had refused to do.

One didn't need to talk to Dr. Becker very long to see that he had a keen sense of public service, which Simpson exploited in an attempt to overcome Dr. Becker's reluctance to get involved in somebody else's fight.

"If you don't testify," Simpson said, "then the judge will have no information at all about health risks."

Dr. Becker asked Simpson where matters stood in the hearing. Simpson said that the only testimony on health risks was from Kanu Shah, an engineer who worked for one of the power companies. Shah had said that the lines would be safe.

"What did he base that on?" Dr. Becker asked.

"That he had seen cows under similar powerlines and they were contented."

I asked Simpson how Shah knew that the cows were contented. He pulled a transcript from his briefcase and pointed to a highlighted portion where Shah had said, "... because I saw that the cows were wagging their tails."

That did it for Dr. Becker. He took two puffs on his pipe, and said he would testify. Then, after a long pause he said, "I would also like Andy to testify." Simpson hemmed and hawed for a few minutes, but finally agreed, sensing, I think, that it was both of us or neither of us.

I asked how much Shah had been paid and Simpson said he thought $500 a day. Dr. Becker said we would not expect to be paid, but that our testimony should be in Syracuse so that we wouldn't need to drive two hundred miles to Albany.

"No problem," Simpson said.

My participation in the hearing seemed important to Dr. Becker, so I offered no objection. He presumably would say that EMFs were a risk or a potential risk, or something of that sort. I had no attitude about the subject, but I felt I would need to agree with Dr. Becker because I couldn't very well go on the witness stand and undercut his testimony. And if I did offer the same opinion as he, I would have the problem of explaining the basis of my opinion. My evidence consisted of the rats with glaucoma which I couldn't repeat, the mice with the abnormal chromosomes, the rats with altered blood proteins, and the first generation of mice that exhibited stunted growth. What did all that mean about the safety of powerlines?

I didn't know how I was supposed to act while I was testifying. Was I supposed to be like a lawyer who was representing Dr. Becker, trying to get the judge to accept what Dr. Becker believed? Was I supposed to organize the facts and arguments that showed that Dr. Becker was right and Shah was wrong, and then rebut each one of Shah's arguments? These questions worried me. I wondered what I had gotten myself into, and whether it would lead to anything good.

Part II

◆

Wandering: 1974-1980

*If you keep your mind on homecoming, and leave these unharmed,
you might all make your way back to Ithaca, after much suffering;
but if you do harm them, then I testify to the destruction
of your ship and your companions, but if you get yourself clear,
you will come home in bad case, with the loss of all your companions,
in someone else's ship, and find troubles in your household,
insolent men...*

(11: 110-116)

10
The Commission

The next time Simpson visited the lab he spoke with Dr. Becker for only a few minutes; the rest of the time he spent with me. Simpson told me the commission was considering approving two 765,000-volt powerlines, one running east-west to be built by a private company and one running north-south that the state planned to build. Only one powerline in the world operated at such a high voltage; it had recently been built in Ohio and caused a great public controversy, which Governor Hugh Carey was keen to avoid in New York.

"What does Dr. Becker want to say in his testimony?" Simpson asked.

I answered on the basis of my understanding of his sentiments. "Something like 'powerline EMFs could interfere with the body's electrical control system,' and 'further research is needed before the line is built.'"

"Is it a fact that powerline fields can affect the body?" he asked.

"I think so," I replied.

"Is the electrical control system and its susceptibility to EMFs generally accepted?"

That was a harder question. Generally accepted by whom? Physicians? I never met a physician who knew much about bioelectricity. In a sense, that meant bioelectricity wasn't generally accepted, but I supposed that wasn't what Simpson had in mind. Physicists? They usually didn't regard biology as "scientific," so they wouldn't be a fair jury. Perhaps biologists. The evidence could be presented, and they could vote. But I had no idea what their verdict would be.

I told Simpson that I didn't know.

He asked me whether Dr. Becker wanted to conclude that the commission shouldn't allow the lines to be built but then answered his own question. "Carey would go nuts – the north-south line is really his baby. Besides, Dr. Becker doesn't know how important the powerline is, so he can't just say, 'Don't build it.'"

I got a queasy feeling when Simpson asked me what I would say in my testimony. I wanted to support Dr. Becker because I supposed that was the reason he involved me in the first place. But I couldn't testify about the electrical control system he described because I had no clear idea about what it was. I felt I needed to get to "possible health hazard" directly, based on whatever reasons I could find.

The woman who had led the opposition in Ohio had written a book about her experiences. It contained a photograph of her standing under the powerline holding two fluorescent bulbs, both of which were lit by the powerline EMF "without benefit of cords or batteries." I didn't believe that could happen. On a moonless night, however, I held a fluorescent bulb under a powerline that passed near my house and the bulb glowed. If a field could reach out and energize a bulb, what could it do to a human being?

Simpson and I visited an industry research facility in Pittsfield, Massachusetts, where we saw a powerline operating at a million volts. The grass looked normal and a small patch of corn was growing under the center conductor, but both Simpson and I got headaches after about a half hour. The engineers had no sympathy for the people in Ohio who had protested the powerline. "They're just NIMBYs," one of them said – "not in my back yard." Two engineers told me about pains they had in their knees and other joints. The engineers suspected that at least some of the symptoms might be connected with working in the strong EMFs produced by the powerline, but if occasional joint pain was the price for doing a job that they enjoyed, they were willing to pay that price.

I learned that a power company had funded some biomedical studies at Johns Hopkins, so I spoke with an engineer named Tim Montpelier who worked on the project. He said he thought powerline EMFs weren't hazardous but that Russian engineers had a different opinion. A report written by his group had concluded "there were no significant changes of any kind" in the health status of the workers as a result of the EMFs, but a Russian report plainly said that prolonged exposure to powerline EMFs was hazardous to workers.

I found a published report about a laboratory study in which fields had altered the daily body rhythms of human subjects. Then I came across a report in which rats exposed to EMFs had developed bone tumors, and another report in which EMF exposure altered the triglyceride level in the blood of human subjects. The more I looked, the more such reports I found. EMFs apparently could cause broad and profound changes in living organisms. I began to feel a hopeful excitement, especially when I realized that no one seemed to be studying the phenomena systematically.

I remembered Herman Schwan's lecture at Penn concerning the effects of EMFs on cells, so I wrote and asked his opinion about my studies and those that I had found in the literature. He replied, "I am not aware of any reports or publications which demonstrate to my satisfaction harmful effects of weak electrical fields. ... I do not know how effects such as described by you can be explained." I thought he had danced around the seminal issue, which wasn't whether the effects were "harmful" or could be "explained," but rather whether they were real. My fear had been that Schwan would vitiate the importance of what I had found, but the opposite happened. I came away from the encounter feeling strengthened.

Despite my long-held belief that I really was smart, I recognized that I had never shown any compelling evidence of great scientific ability. On the other hand, everything new and worthwhile didn't need to begin with an Einstein, or even a Schwan. Something good could begin with me. There was a place not far from Syracuse where you could chip away at a rock-face searching for quartz crystals, called Herkimer diamonds. I went there once, and after only a little while I suddenly poked into a hollow about the size of a basketball and found a big jumble of well-formed diamonds. Others chipped for days and found nothing. More than being a good chipper was required – some luck was also necessary.

I wrote a draft of Dr. Becker's testimony, using a question-and-answer format as required by the rules of the hearing. The gist was that the body was electrical in nature and thus foreseeably susceptible to external fields, and that fields did affect the body – they caused stress, which predisposed toward disease. The conclusion was that involuntarily exposing the public to powerline fields was unethical, and potentially unsafe.

In my testimony I said that animal and human research showed EMFs could affect the body, and that such research was the basis on which science ordinarily assessed whether something might be a hazard to people. I concluded that the powerlines would be a health risk. I got to where I wanted to go by reasoning as did the Food and Drug Administration and General Mills. They knew of no biological effects due to BHT and therefore concluded it was safe; I had found effects of EMFs and therefore concluded they were unsafe.

Soon after the power companies received our testimonies, they fired Kanu Shah and hired other experts who wrote new testimony. The only expert I recognized was Herman Schwan. His testimony contained mathematical equations he called "biophysical principles;" he said they proved beyond doubt that the powerlines would be safe.

Another company expert was a botanist from the University of Rochester named Morton Miller. He described studies performed for the Navy by a private company. It had failed to find any EMF bioeffects, from which Miller had concluded the powerlines would not cause health problems. Dr. Becker could not recall being told about any of those negative studies during his meeting in Washington when the Navy had presented its EMF evidence. None of the positive studies he had been told about were mentioned in Miller's testimony.

On the first day of the hearing in Syracuse I took the witness stand and swore that what I had written was true; then the lawyers for the power companies began to cross-examine me.

"Do you agree with the biophysical principles that Dr. Schwan enunciated?" one of them asked me.

"I don't think his equations are principles," I replied.

"Are there any equations which would predict that biological effects would be expected to occur as a result of exposure of people to the EMFs of the proposed powerlines?"

"Equations don't predict that there will be or won't be. They are irrelevant. Attention should be focused on the experiments with animals and human beings to determine whether there are risks. That's BHT logic."

"What's BHT logic?" he asked.

"BHT is the preservative that General Mills puts on Cheerios. The company says BHT is safe because nothing happened to mice that were fed the chemical. That rationale would make sense only if the company was prepared to conclude that BHT was not safe if something did happen to the mice. The situation with EMFs is analogous, except that the animals and people that were fed EMFs exhibited changes."

There was a short article in the paper that evening about the hearing. It said my testimony was "irksome" to the power companies.

During the next several days the lawyers concentrated their questions on my research. I had been required to give them copies of all my data, and their statisticians parsed it, looking for errors. The table in front of one lawyer was covered with statistics textbooks, each of which had various paragraphs highlighted in yellow. He began by reading a sentence that defined "standard deviation" and asked, "Do you agree with that, yes or no?"

I didn't like "yes" because I had not read the book and didn't know the flavor of the meaning of the words in the definition. I didn't want to say "no" because the definition might have been the same as one I had used somewhere. I tried to explain why I didn't want to say yes or no, but he kept asking the same question over and over again, essentially demanding that I choose one or the other alternative. Then, a light went on in my brain and I told him, "I don't have opinions about sentences from books I haven't read."

I knew I had won when he put the books on the floor.

That evening I decided that if he established I had read a particular book, I would tell him, "I don't have opinions about sentences from books, period."

Frank Wallace was the lawyer who represented the state in its attempt to build the north-south line. He was a silver-haired authority figure to whom the other power-company lawyers routinely deferred. I had expected him to mount the stiffest challenge to my testimony, but his cross-examination was geared only toward embarrassing or irritating me.

"Did you consider the convenience of the judge when you requested that this hearing be moved to Syracuse?" he asked me.

"No."

"Did you recently give an interview to somebody at the *National Enquirer?*"

"Yes."

"Could you explain how you have time to give interviews to the *National Enquirer*, and yet you do not have time to journey to Albany for this hearing?"

Simpson objected to the question because it was "ridiculous" but Judge Thomas Matias overruled him.

"I would like to hear what the doctor has to say," Matias ruled, thereby giving me my first hint that he had a malignant mind.

"The doctor doesn't have anything to say," I told him.

Wallace continued: "I will read a quote. 'A top U.S. biophysicist, Dr. Andrew A. Marino of the Syracuse, New York Veterans Hospital, told the *Enquirer* he studied the Russian findings and "I agree with their conclusion completely. There is no doubt that powerline emissions can harm humans."' Is that article correctly quoting you, sir?"

"No."

"Did you ever write the *National Enquirer* and advise them that it did not accurately quote you?"

"No."

"Did you ever tell anybody that this is not an accurate quote?"

"No."

"But you are aware of the article, are you not?"

"Yes. It appeared right next to a picture of Jackie Onassis."

"Are you aware that the article described you as a 'top U.S. biophysicist?'"

"Yes."

"Do you agree with the article?"

"Well, I'm not going to question their characterization of me in that regard."

The lawyers for the east-west line were no improvement. One of them asked: "Do you know that a dog never wags his tail when it's sick?" Matias overruled Simpson's objection, so I answered, "No."

The lawyer continued: "Do you know that a cat manifests discontent when it wags its tail?"

"Objection."

"Overruled."

"No."

"Do you know that when a deer senses fear its tail becomes erect?"

"Objection."

"Overruled."

"No."

"Then how do you know that when a cow wags its tail under a powerline, it's not showing that it is content?"

"I don't know that."

"Then would you agree that when a cow is switching its tail under a powerline, that cow could be content?"

"Anything is possible."

As the hearing dragged on I had trouble sleeping, and I began having nightmares. In one, I'm running as fast as I can through a soupy fog. It parts as I move forward but I can't see anything to the left or the right. Just behind me, in the vacuum created by my motion, is a hideous creature holding a knife. Its arm is raised in preparation to strike. I slow down because it's hard to breathe, but the creature also slows down. When I ignore the pain and speed up, it matches me step for step.

In another nightmare I am seated, alone, on the stage in a big theater, and the audience, all men, are laughing at me, but I begin to perceive that the laughter reflects ridicule. As it continues I realize I can't move because my arms, legs, and head are rigidly strapped. I close my eyes and try to block out the laughter with thoughts of my children. We are walking in the woods through a bed of white trilliums, each trying to be the first to find a pink one. The laughter becomes louder. I begin to recognize individuals in the audience. They all have big barrel chests. Their jaws jut out. They all look like Mussolini!

After thirteen days of cross-examination I had reached the limit of my endurance, so I refused further questioning. The lawyers complained bitterly that the constitutional rights of the power companies to cross-examine me had been violated. Matias agreed and ordered me to return for additional examination, a legally impotent decision because he had no power to compel me to continue to gift my services to the commission after I had decided I would no longer do so. The commission then saved appearances by overruling him, saying that the companies had been given enough time, and that if there were any important questions they had wanted to ask, they should have asked them rather than use their time to ask "unproductive" questions.

I had major misgivings about how Dr. Becker would do on the witness stand because he was naïve about how cross-examination worked. I had hoped he would ask me for help, but he didn't. On the stand he answered questions as if he were in class teaching students rather than in a river full of piranha. When the power companies finally let him go, after four days, he said nothing to me about his experience except "Never again."

The day finally came for Herman Schwan to take the witness stand and face Simpson. Over Matias's objection but by order of the commission, the confrontation took place in Syracuse because that had been my condition for helping Simpson with the cross-examination, which I wrote. I sat beside Simpson in court and whispered in his ear whenever I saw a chance to undercut Schwan's testimony.

The whole experience was like feasting on unlimited meat and fine wine. The banquet began when Schwan said, without any uncertainty or doubt, that there was an equation from which he could predict that the levels of

EMFs from the powerlines were too weak to cause any biological effects, and therefore that the powerlines would be completely safe.

Simpson first asked Schwan whether he thought my research was valid, and he testified that the effects I had reported were artifacts due to poor design of my animal cages.

Simpson asked Schwan whether he accepted the research by Wever that showed EMFs could cause disturbances in human biorhythms. From Schwan's point of view, data couldn't show that fields affected people because such phenomena were forbidden by physics; so the thing to be explained was why the data *looked* convincing, which Schwan did by suggesting that Wever had rigged the results.

Simpson continued in the same vein for several hours, and Schwan dismissed study after study. He called some investigators incompetent, challenged the statistics of others, and hypothesized a litany of artifacts that he claimed vitiated the results. Finally, Schwan appealed to Matias for relief, and he ordered Simpson to stop asking such questions.

I realized that if Schwan could automatically deny some facts, he could automatically accept others, thereby showing the commission from another perspective that he was not a thinking man. On the spot, Simpson and I designed a line of cross-examination in which we presented Schwan with studies that had *not* found any biological effects due to EMFs which, principally, were those that Morton Miller had cited in his filed testimony. One by one, we asked Schwan whether he accepted those studies, and he said he did because their conclusions were correct. Then he admitted he had not read many of the studies and said, "Whenever an experiment was negative, I was not further interested in digging into the material." I could hardly believe my ears.

Simpson began a line of cross-examination designed to show the logical impossibility of determining "safety" in terms of concepts used by physicists.

"Doctor, do you know the mechanism underlying fracture-healing in bone or production of bone tumors?"

"No, no one does," he replied.

"Doctor, if the mechanism for these functions is not understood, how could anyone predict from a calculation of a given signal level in bone due to EMFs that there will be no effect on these functions?"

"Mr. Simpson, your question is utterly nonsensical. You are a very poor physicist," he fumed.

I had prepared a list of numerous biological processes that involved virtually every organ in the body. There were many things about each organ that were unknown, and it was our intention to ask Schwan how he could conclude that any particular calculated or measured level of signal in the body caused by the powerlines would be safe with respect to each process in each organ. This line of questioning continued for some hours and in each case Schwan answered, in effect, that he didn't know. Finally, he told Simpson, "That's it," and folded his arms. Simpson looked at me and smiled, then turned to Matias and said, "No more questions, Your Honor."

In two days on the witness stand, Schwan had gone from being the world's premier expert on the subject of the biological effects of EMFs, to someone who seemed biased and incoherent. To some extent, however, so had Dr. Becker when he was on the stand. It had been as easy to make a great man with real knowledge look like a fool as it had been to do so to a man who lacked the knowledge that he claimed. Both men lived for about 30 more years but, with one exception, never testified again.

When Morton Miller took the witness stand he described the Navy EMF research that had found no effects which, he said, meant that the powerlines would be safe. We pointed out that he had ignored the positive results and tried to force him to agree that, using his logic, they suggested powerlines might be a health risk. He fought us tenaciously, never conceding even the smallest point. In the end, we achieved the only goal possible with an obdurate witness, which was to fill the record with desperate protestations that could be distinguished from rational science if one cared to do so.

I was very pleased with the result of the cross-examination, and didn't think it could get any better, but it did. We had presented Miller with a study by investigators at the University of Wisconsin who had found that EMFs caused growth changes in cells. According to Miller, Schwan had said his calculations showed that the particular EMF used by the Wisconsin investigators was not relevant to powerlines and, believing this, Miller had lavishly praised their study. But there was a huge intellectual gap between Schwan and Miller, and as I sat listening to him I thought of a way to exploit it. During a break in the hearing, Simpson and I worked out a series of questions in which we would suggest to Miller that he must have misunderstood Schwan, because the EMF *did* apply to powerlines. As Simpson pursued this line of questioning, Miller began to soften. First he said that there was a "little problem" with the Wisconsin study, and that it had a "lack of appropriate controls." Then he said the results "may well be an artifact." Finally, he said, "Now I am criticizing the experiment, saying it was not a properly controlled type of experiment."

As Simpson stalled for time, I frantically paged through my notes to find Miller's exact words concerning the study, spoken when he believed it did not apply to powerlines. I found Miller's words and gave them to Simpson, who asked, "Dr. Miller, isn't it correct that you previously stated, 'The work is an outstanding study. It is a beautiful example of a well-controlled, well-analyzed experiment.' Did you not so state?" "Yes, I think that's correct," he replied.

Matias called a recess, and Miller got up slowly and walked along our side of the room. As he passed behind us he said, "I guess you guys got me there."

That night my grandmother cooked a home-made spaghetti dinner for our team, and we drank chianti, told jokes, and celebrated our victory. Simpson began talking to her about cooking, and she told him how to make kidney pie: "You've got to make sure that you boil the piss out of it." "Exactly what you did to Miller," someone said to Simpson.

I had always thought that the proposed design for the powerlines wouldn't be approved by the commission if it believed that doing so would jeopardize people's health. The commissioners were decent men and women – too political for my tastes, but nevertheless decent. Now that we had met the test and shown the powerline EMFs would be a problem, it seemed reasonable that the commission would widen the right-of-way or order the powerlines be built underground. But when I had shown that Berndt's herbicide practices bore no close scrutiny, the Onondaga County transportation department didn't change its policies. I worried that the commission might also pull back from the decision that so plainly seemed correct.

Even before Dr. Becker and I had become involved with the commission, the state-owned company that wanted to build the north-south powerline had purchased all the material necessary for its construction. As the hearing dragged on, the stored material sat like a brooding lion. The head of the commission was an avuncular figure named Alfred Kahn. He had been the fertile ground in which Dr. Becker's letter had taken root, leading to Simpson's approach to Dr. Becker. But one day the righteous Kahn was replaced by an opportunistic man named Edward Berlin, and soon thereafter construction of the powerline started. At a news conference Governor Carey said, "We need the jobs, we need the power," and Berlin dutifully authorized immediate construction of the north-south powerline. He said that if the commission ultimately decided the powerline caused health risks, it would be dismantled.

Tom Meadows was an environmental specialist who worked at the commission. His constituency was the plants and animals and archaeological sites in the state. In his efforts to protect them he was always overmatched against the proponents of the massive building projects that were the commission's regular fare. We each sensed that the other was a kindred spirit, and from time to time during the long hearing process he told me things about the inner working of the commission, confidences that even my good friend Simpson had not shared with me.

I asked Meadows why Governor Carey had approved the powerline in spite of the testimony that came out in the hearing. Meadows said, "Carey doesn't know what to believe. Kahn suspected there was a problem, that's why you were given a forum in the first place. Most people at the commission had expected you to collapse, and when you didn't, that presented a huge problem."

"Why?"

"Because Carey had already signed contracts with the Canadians to transmit their power to New York City, and the state had bought all the equipment and supplies necessary to build the line. Berlin is telling Carey that there is really no problem with EMFs. Carey doesn't trust Berlin, and he wants to do the right thing, but there are economic realities."

Meanwhile, out of the blue, during a phase of the hearing that was being held in Albany, Frank Wallace told Matias, on the record, "I have received information that indicates that either Dr. Marino was mistaken or that he committed perjury." The next day his client, the state power company, issued a press release saying that I might have committed perjury. Various newspapers throughout the state published editorials lamenting how easy it was for a perjurer to hold up construction of a powerline that would be an economic boon to the state. I repeatedly asked Matias and Berlin to investigate Wallace's charge, but they refused, preferring instead to let it fester.

Matias worked on drafting his decision. One evening, on the local news, a woman announced that he had issued it, and that he had concluded the powerlines would be safe. As she spoke, Matias's words "Marino lacks credibility" flashed on and off over her chest.

My copy of the decision arrived two days later. Matias had written that my experiments had not been conducted carefully enough for the results to be believable. He said I had a clear lack of expertise in doing experiments, and that I had presented no evidence showing that powerlines could be harmful. There was a long section entitled "Dr. Marino's Credibility," in which he expressed resentment that hearings were held in Syracuse to accommodate me while I had time for other things, like giving interviews to the press. He accused me of stubbornly refusing to change my experiments to follow the advice of the experts who worked for the power companies. He said that my testimony was emotional and unfair, that I lacked candor in reporting all important facts and data, and that I lied. He said that someone who would do all these things could not possibly be believed, and that consequently he didn't believe anything I said. He concluded there was no health problem with powerline EMFs, and no need for an independent research program.

Everything had gone so bad so quickly, it took my breath away. What a fool I had been to trust the commission. For several weeks I couldn't sleep. There was a gigantic weight on my chest that I had to push up when I breathed in, and when I breathed out it squashed me against the bed.

When the worst of my anguish had passed, I began calling people at the commission, trying to make sense of what had happened. Meadows told me, "Matias thought that cutting your balls off would please Carey and Wallace. Matias is far worse than you know."

"How could he not believe any of my testimony but yet believe Schwan and Miller?"

Meadows sighed a sigh of exasperation, as one might do talking with a young child.

"You're not looking at this situation the way he does. He doesn't see it as the pursuit of scientific truth, but rather as a matter of the practical interests of everyday life."

"What interests?" I asked.

"Pleasing your boss. Getting a job. When Matias retires from the commission he expects help from Wallace in getting a teaching job at the law school."

Meadows paused, and then said, "In addition to that, there was the deal you made with Simpson that your testimony would be taken in Syracuse. Matias had to accept that arrangement, and he wasn't happy about hauling his fat ass two hundred miles down the Thruway thirteen times just for you. For him, that alone would probably have been enough to attack you."

I asked my friend about the thing that stung me almost as much as Matias's decision. "Did he know about Frank Wallace's scheme to call me a perjurer?"

"Sure," he replied. "So did Berlin. They both had to agree not to investigate the charge. Otherwise the plan wouldn't have worked. The press release accusing you of perjury was written before Wallace invented the specifics of your crime."

"What did you mean earlier when you said Matias was 'far worse than you know?'" I asked, but my cooperative friend became less cooperative. "I can't tell you about that. But you can figure it out. Ask yourself how you think Matias, a lawyer, and far from the brightest bulb in the pack, could boil down 31,000 pages of scientific mumbo-jumbo in order to write a decision for the commission." How? I couldn't imagine how.

Shortly thereafter, at a Syracuse University basketball game I ran into a former girlfriend who worked at the commission. During our conversation she surprised me when she asked, "Have you seen Asher lately?"

"Who?" I said.

"Asher Sheppard."

I had met Sheppard early in my career, just after we got our Ph.D.'s. At first we had similar views about EMFs, but then he wrote a book that more or less dismissed EMFs as any kind of a threat, and I had lost touch with him. "No," I replied. "Have you?"

"Yes, he recently visited one of our hearing officers."

"Who?"

"A man named Thomas Matias."

The moment she placed Matias together with Sheppard, the wheels in my mind began to turn. I sent a freedom-of-information request to the commission, asking for any correspondence between Sheppard and Matias. I didn't know exactly what material the commission had, which allowed it to deny my request on the ground that it was too vague. I persisted, varying the legal formula in which I couched my request. Finally, one of my phrases keyed the lock and opened the door, and I received memos, letters, a contract, and even Sheppard's travel vouchers. I learned that Matias had secretly initiated a contract with Sheppard to write an analysis of my testimony, and then had incorporated the report verbatim into his decision. When the commissioners learned that Matias had hired a ghost-writer they were shocked, like Captain Renault when he learned that gambling was going on at Rick's Cafe.

Understanding the events that led to Matias's recommendation made me feel empowered, as when one learns the name and something about the strategy of an enemy, and the embarrassment he suffered after I publicly revealed what he had done was a matter of great satisfaction to me.

While I waited for the commission to issue the final decision, I decided to seek revenge against the silver-haired Frank Wallace for what he had done to me, so I sued him for defamation. At one point during the subsequent complex legal proceedings, different aspects of the case were in all three levels of the court of general jurisdiction in New York, as well as in the Court of Claims, where I also had to go because Wallace represented a state agency. The amount of money I eventually won provided some satisfaction, but not nearly so much as knowing that I had the power to fight back.

While all this had been happening, I was called by a producer for *60 Minutes* named Richard Clark, who proposed that I be interviewed by Mike Wallace. Clark and I spoke several times concerning the necessary arrangements. I also sent him a written report describing the background of the powerline EMF conflict so that Wallace could ask the pointed questions for which he was so famous. When he arrived in our lab, he chose a spot in front of the electromagnet for the interview, between the high-vacuum system and the electron spin resonance spectrometer, and said, "This looks scientific."

As we waited for the sound and camera equipment to be set up, he talked about his experiences during the Vietnamese War and joked with the people in his crew. Then he began, "Dr. Marino, how'd you get into all this controversy over low-frequency radiation and what it might do to us?"

"Because of some experiments we did involving mice and rats exposed to EMFs that are similar to those associated with powerlines," I replied.

"What do you believe could be the effect of, let's say, the 765-kilovolt powerline across New York state to the animals and the human beings that are going to be underneath it?"

"The evidence we have is laboratory experiments, of which now there are 60 or so, in which investigators have exposed various animal systems – primates, monkeys, rats, mice, all the way down to amoeba – to EMF strengths close to that of a powerline. And just a whole variety of effects have been observed. Stunted growth, body chemistry alterations, blood chemistry alterations, cardiovascular system alterations, the whole gamut of possible biological effects."

"You know what you're saying is pretty scary to a lot of people because we are all exposed to powerlines."

"I appreciate that," I replied.

"What you're saying is that we live in an atmosphere of electric smog and we do not know the effect of all of it upon the human animal."

"That is the bottom line."

"And what you're suggesting is that the power companies want to experiment with us."

"They're doing it. It happens every day."

"And you seem to be saying that the power companies and the government either do not know, do not understand, or do not care."

"I am saying that the power companies are well aware of the problem, and the first line of defense has been to deny it."

When the interview ended, Wallace leaned over to me and said, "This is going to shake people up." As the crew retrieved the cords and cables that they had snaked throughout the laboratory, one of the sound technicians told me he thought I was "going somewhere."

The opportunistic Edward Berlin departed the commission and went to work for a power company. It took Governor Carey and the remaining commissioners more than six months to clean up the mess that Matias had created and to accommodate the conflicting interests that had emerged in the hearing. The commission's final decision manifested the wisdom of Solomon but, unlike Solomon, it actually divided the baby. First, the commission reversed Matias's *ad hominem* analysis of me and said that my testimony suffered from only one ultimate infirmity: "It deals in possibilities and probabilities, not certitudes." Then, to no one's surprise, the commission officially approved the already-built north-south line. What followed, however, was a progressive series of bigger surprises. First, the commission ordered that the right-of-way be widened so that the EMFs at its edges were no greater than those of existing powerlines. Then the commission disapproved the east-west powerline, and it was never built. Finally, the commission ordered that the power companies in New York should be taxed to pay for new scientific studies that would resolve with certitude the question regarding whether powerlines were health hazards.

It was a happy day when I received the commission's decision. The ordeal of my involvement with the state was finally behind me, but I could only wonder about what the future held.

11
Monsters

◆

After my interview on *60 Minutes*, many people asked whether I thought the powerline in their backyard was a health risk, or could have been responsible for whatever health problem they had. All I could say was that I didn't know, because neither the power companies nor the government had ever seriously evaluated such possibilities.

I completed the multigeneration study I had begun before Dr. Becker's trip to Washington, DC. In all three generations, fewer pups were born in the group exposed to the EMF and they died at a higher rate; those that survived were stunted. It was as if the fields had turned the cages into mouse ghettos. Mindful of what had happened when I tried to repeat the experiment that had produced glaucoma in rats, I began the mouse experiment again to see if, this time, I could replicate my results. In other experiments, I measured corticosterone levels in the blood of rats exposed for a month to EMFs and found that the hormone was altered, indicating that the EMF caused stress, as Dr. Becker had supposed. I published my results and then began to do more experiments.

During the New York hearing, the Veterans Administration Central Office had received anonymous complaints that Dr. Becker and I were criticizing power companies. The cumulative effect of the calls changed the climate at the hospital and made it more difficult to do the EMF work. Then the National Institutes of Health canceled Dr. Becker's grant for bone research, and it turned down my grant request to study EMFs.

Even though we were under greater scrutiny and had fewer resources, I persisted with EMFs because I thought I might be on the cusp of something important. When I was a child, my father would take me to a shoe store that had a machine which looked like the big radio in our living room. When I slipped my feet into holes in the bottom of the machine, I could look down and see the bones in my feet. Later, everybody learned that x-rays, the basis of how the machine worked, were dangerous and should be used for important purposes, not to sell shoes. I thought that EMFs might similarly be underappreciated, and that powerlines would be built underground when due appreciation developed of the risks of overhead construction. At that time I did not understand the complexity of the situation.

The research program that had been ordered by the New York commission got started, but the state-owned power company that was building the north-south powerline gained control of the money. When I sent in a grant request, the head of the program told me I was "not competent to do the EMF studies you propose." Millions of dollars were spent for research, but in the end the program produced no worthwhile information.

I had traveled a long way since the idea that fields could alter the chemistry of the body had first occurred to me, but I couldn't see where answers that could decisively resolve the safety issue would come from, or who cared enough to pay for the necessary research. I felt as if I were in a foreign country on a moonless night and I didn't know where the good people were.

I spoke again with Tim Montpelier, the engineer who earlier had given me much useful information about EMFs. He told me about recent experiments in which he had found harmful effects on baboons, but he said that the study sponsor, the Electric Power Research Institute, had canceled his contract. Its head was Chauncey Starr. I called and told him, "Our laboratory discovered the EMF problem. I know a lot about it. I can find some answers, but I need money." I felt like a suppliant seeking a gift, but I thought I deserved it because of my previous work. I also asked him to respect science enough not to withdraw the money in mid-project, as he had done to Montpelier. Starr answered me as if I were a fool to ask him for money, but he assembled a herd of friendly investigators that he funded lavishly.

The fattest pig in Starr's herd was Richard Phillips. Eventually he published numerous studies saying that powerline-type fields did not cause stress in mice, did not affect the heart or the prostate gland in rats, and did not alter the metabolism of pigs. He spoke at conferences in Portland, Ottawa, Finland, Palo Alto, Kiev, Tashkent, Minneapolis, and Toronto saying, in effect, "We looked as hard as we could, but couldn't find any evidence of health risks."

The herd included Robert Banks, who held seminars for power-company employees where Richard Phillips and Morton Miller gave their versions of the scientific evidence, lawyers described the legal concerns, and public-relations experts advised how to explain the industry position to the press.

Dwight Spieler, an expert risk assessor from Carnegie Mellon University, produced a booklet about EMFs, hundreds of thousands of which were distributed nationally. Its well-crafted illustrations and direct prose projected a sincere interest to communicate the truth. The booklet said, effectively, "You are right to be concerned about EMFs from powerlines. It is natural to wonder whether they can do anything. But there are risks everywhere, even with the food we eat and the air we breathe. You should maintain a rational perspective. Experts who investigated powerline EMFs found that the risk was no more than that of getting a sunburn from a flashlight. The problem, therefore, is not risk but rather baseless fear of a risk. You are more likely to become ill as a consequence of the aggravation you cause yourselves when you worry about EMFs than you are from the EMFs themselves."

The message of the booklet spread like a plague and was reinforced in many ways, as only the powerful are able to do. In an article in *Science*, Starr wrote that research he funded placed "emphasis on objectivity," and on "intellectual integrity," but the reality was completely otherwise. His contractors received large sums of money and in return provided information that was good for the Electric Power Research Institute, but for nobody else.

I learned that several federal agencies had formed a committee to study health hazards from powerlines, which excited me because I thought the government would be more respectful of science than Starr. I called the committee chairman, an engineer from the Department of Energy named Robert Flugum, to learn whether he had an honest interest, and whether he had money to spend. "There is a problem," I told him, "because it seems clear that EMFs like those from powerlines can affect animals. I think that means EMFs can probably affect people. There are many questions to be answered, and I need money to do experiments." Flugum responded, "Large powerlines have operated all around the world for many years, and we have no evidence whatever of health problems." Then he continued down the same dark path: "Other investigators dispute your evidence that EMFs can affect animals – Richard Phillips, for example and Morton Miller. Herman Schwan agrees with them."

I had never spoken with Richard Phillips and so it was impossible for me to understand how he could look so hard and always find nothing worth seeing. Morton Miller and Herman Schwan, on the other hand, I understood well. But there seemed no safe way to explain them to Flugum. One day while Miller was testifying in the powerline hearing he said he had been offered a great deal of money to do EMF research.

"Why would you want to do that kind of research, believing as you do that EMFs have no effect?" he was asked.

"Because he who has the gold makes the rules," Miller replied.

That was the story I wanted to tell Flugum, but I did not do so for fear that it was his gold Miller had received. Finally Flugum said, "Although we do not expect to find any problem, from an abundance of caution and in the interests of verification and sound scientific corroboration we are funding some EMF research, including studies to replicate some of your work. We will consider any proposals you care to make."

I thought that if only I could talk to Flugum face to face, I could make him see that the problem was real and enduring, and that only the government could adequately consider everyone's interests. On several occasions he invited me to attend the meetings of his committee in Washington, but the invitations always came only a few days before the meeting and never provided for my travel expenses.

For a long time I hoped for a contract from the Department of Energy, and while I waited I did whatever I could to convince Flugum I would do good science with an honest intent. But I was unsuccessful. Flugum gave contracts to Phillips, Schwan, Miller, and others, some in the herd of the Electric Power Research Institute and some not. While I was still on bended knee Flugum struck the blow that convinced me of the futility of my effort. A company he had hired to analyze the New York hearing produced a report saying that the opinions of Schwan and Miller had been more credible than mine. The Department spread the report around the country, and it largely accomplished its purpose of making the defeat of the power companies look like victory.

In the meantime, I completed the second multigeneration mouse experiment, and was amazed to find this time that the mice in all three generations exposed to the powerline EMF grew *larger*. I saw in these results a refutation of the view that nature was a robot whose future behavior was absolutely foreordained by its past. On the contrary, nature was quite capable of novelty. I began to suspect that the very idea of "replication" needed revision. As it turned out, such a revision is critical to understanding EMF health risks, but even today too few scientists have taken that step.

Richard Phillips came to my laboratory and asked questions about my multigeneration study. I told him everything about the results of the first experiment, but I said nothing about the results of the second. He happily told me of his plans to repeat my experiment, and to do so free of defects he imagined existed and accounted for what he thought were false-positive results.

I visited Phillips's laboratory, which was in Richland, Washington. He worked in a large factory for making research. On one floor his company did cardiovascular and endocrinological studies; on another they grew various kinds of cells in incubators. There was a floor where they studied whether their customers' product affected the growth of animals. The cheapest studies used mice but, if the customer could afford it, the company also used pigs. The Electric Power Research Institute was the only customer for pigs, but many others bought mice studies.

Phillips had magnificent resources; his exposure facility for the multigeneration study was made of stainless steel and plexiglas, and he had elaborate equipment for generating the fields and monitoring subtle changes in any environmental factor that might conceivably affect the results. Technicians recorded the data in the form of a code, the key to which was known by only Phillips and his chief lieutenant, so only they knew which measurement was from a mouse that had been exposed to an EMF and which was from a mouse that lived in the field-free area. Phillips showed me great hospitality but his attitude toward my work was ominous; "Business is business," he said.

As we were about to depart from the research facility, a technician referred to "the other exposure facility." Phillips had no intention of showing me that place, but I pressed my request to see it. "I showed you everything I had when you visited me. Besides, you must admit you're making a good living because of me," I said.

I got what I asked for. He took me down a hall and opened a door into a second, *identical* exposure facility, which he was using to simultaneously replicate the multigeneration experiment. He had everything he needed to accomplish my complete destruction, as one steps on a bug and then drags his foot so that no semblance of the original form remains.

When I knew Phillips's experiments must be over, I called Flugum to learn the results. "The experiments are not over," he said, and he left me to puzzle about the reason for the delay.

Three months later Flugum told me, "We have received the results of the multigeneration experiments and they were entirely negative." He was pitiless in his condemnation of me, saying that I had baselessly alarmed many people, and caused the Department of Energy to waste much time and effort pursuing a chimera. I believed his millions had not produced good science but only science that he wanted. However, I had no evidence.

An apparent opportunity to obtain some details of Phillips' studies arose when I was contacted by a lawyer for a farmer whose land had been taken by the Department of Energy for construction of a powerline. Because it crackled and hummed and gave shocks when he touched his farm equipment, the farmer had taken his case to court, and an expert had testified that the powerline was completely safe and had none of the consequences complained of by the farmer. "Who was their expert?" I asked the farmer's lawyer. "A man named Robert Flugum," he said.

I agreed to testify in opposition to Flugum, hoping that, as the conflict developed, he would be required to give up the richly detailed information he had received from his investigators – as I had been required to disclose all my data when I testified in the New York hearing. But the judge in charge of the farmer's case denied me any role in the case. "Dr. Marino's claim that powerline EMFs are health hazards is the subject of dispute within the scientific community and has not generally been accepted by scientists with expertise in the study of EMFs. He therefore may not testify." That was my first experience with the 70-year-old *Frye* Rule of Evidence which some judges used to disqualify witnesses who sought to present new scientific information.

I decided to directly seek the information I wanted, under the Freedom-of-Information law. I requested a copy of all data that had been provided to Flugum by Phillips. But the wily Flugum responded by facetiously asking Max Cleland, the head of the Veterans Administration, whether I had been authorized to evaluate the Department's program in powerline research. Cleland wrote to Wayne Sarius, the head of the hospital where I worked, asking in effect, "Who is Marino, and what is he doing?" Sarius ordered me to stop sending letters to the Department of Energy, but he failed to consider the dedication of Dr. Becker.

Our hospital had been going through one of its periodic fiscal crises, and articles were appearing daily in the Syracuse newspaper describing our inability to provide basic medical services at our hospital. Nevertheless,

Sarius had decided to remodel his offices using rosewood paneling costing $800 a sheet. The extravagance had scandalized the hospital carpenters, and they provided Dr. Becker with documented information about the project's expenses. I never learned exactly how Dr. Becker put it to Sarius, but it must have been something like, "Back off, or this information will be in the newspaper tomorrow." Sarius was then quick to accept my explanation that I didn't claim to represent the Veterans Administration, which is what he reported to his superiors, and they reported to the Department of Energy.

Flugum turned down my request. I greatly expanded the scope of the information I sought, an effort that required many pages of detailed specifications, and appealed to James Schlesinger, the Secretary of Energy. I told Schlesinger I thought that the information I had requested would prove Flugum was biased, that he had awarded research contracts on bases other than peer-review or competitive bidding, that his practice was to withhold scientific information that ought to be made available generally, that his herd of investigators consisted solely of dedicated nay-sayers, and that he attempted to intimidate me from inquiring into his activities.

Flugum's supervisor, Bennett Miller, sent me a letter defending Flugum and turning down my request. "All proposals," he said, "were evaluated by expert panels or through peer-review procedures, with official procurement regulations and all contract actions followed rigorously." So I wrote to Vice President Walter Mondale, telling him my story and asking for his help. A short time later I received five cartons of material from the Department of Energy.

The web of deceit disclosed in the information was more nefarious than I had imagined. I saw that Flugum was on the board of advisors of the Electric Power Research Institute, and that Chauncey Starr, the head of the Institute, was on the board of directors of the scientific society that published *Science*, likely accounting for his ready access to that famous journal where he had extolled the Institute's "objectivity" and "integrity." I saw that Flugum, in correspondence with the National Institutes of Health, had vetoed my presence on the American team of scientists that took part in exchange visits of U.S. and Russian scientists aimed at resolving the contradictory positions adopted by the respective power industries concerning possible health risks from powerline EMFs. Instead, Flugum had designated Richard Phillips, Herman Schwan, and Morton Miller to represent and defend U.S. science. I saw that Flugum had invited his friends to apply for contracts, which he had then approved and funded.

Other documents told the story of Richard Phillips's two multigeneration experiments. They had been conducted with about a month's offset between them. When the coded results of the first experiment had been analyzed, he had been horrified to learn that the exposed mice, both male and female, were significantly smaller than normal, just as I had observed, suggesting a pervasive effect of the EMFs. Upon learning of that result, Flugum gave Phillips more money to extend the experiment to a fourth generation, in the hope that the results from the first three generations would straighten out. But the effects also occurred in the fourth generation. That ended the first experiment. Phillips expressed the hope that the results of the second experiment would be better, but neither he nor Flugum yet knew the worst. When Phillips broke the code for the second multigeneration experiment he saw that the male and female mice exposed to the EMF were significantly *larger* than their controls, opposite to the results of his first experiment, thus exactly duplicating the pattern of experimental results that I had found.

The depth of the anguish and confusion that Flugum and Phillips felt was plain to see in their correspondence, which neither had ever expected would be seen by others. The implication of the experiments, that powerline EMFs could affect the growth of animals, was directly opposite to what Flugum had consistently assured his superiors in the Department of Energy, thus calling into question the practicality of the Department's wish to build overhead powerlines of unlimited size.

Flugum was only a bureaucrat; for him, science was just a managerial tool. Phillips was different. His concept of science was far too simplistic, but at least he thought about science at the conceptual level. For him, science was a straightforward matter. A thing was or was not, with no uncertainty or ambiguity. Throughout his career he had regarded the presence of either as proof of the absence of science, so he had never imagined that his two experiments, which were as much alike as a human being could achieve, might truly have had law-governed but opposite results. Indeed, such a phenomenon had never been clearly described during the long history of science. From his perspective, chance was the only possible explanation for his "inconsistent" results. If Phillips had had the soul of a true scientist, someone who discerns and describes nature for what it is – what the green man sees rather than what the will chooses – he might have earned the respect and admiration reserved for those who go into the laboratory and make a marvelous discovery that heralds a shift in paradigm. But he lacked the necessary intellect, and so he averaged the results of his two experiments and concluded that EMFs had *no effect* on the mice. As if a man who was frozen at one end and burned at the other had not sustained injuries because the average temperature had been normal.

But Phillips had done more than squander an opportunity for glory – he had descended into the abyss of science where he joined the likes of Mesmer, Perkins, and Abrams. Whenever Phillips had found a positive effect in his other studies, he repeated the experiment as often as necessary until he got a negative result, and then reported the negative result as the correct one. In one case he had found that rats exposed to EMFs developed inflamed prostate glands. He repeated the experiment and got the same result, but the third time he did the experiment there was no effect on the prostate gland. In his report he advised Flugum to alter the data from the first two studies to "minimize the risk of abuse by others," because "it could be construed by certain parties to support their claims that EMFs produce adverse biological effects."

In other cases, the design of his experiments made it impossible to see anything about nature, like a man staring at the sun. The worst instance of maldesign was his use of mouse cages only two inches high and rat cages only four inches high, each half the minimum height specified by federal rules for stress-free housing of the animals. Using these cages, he failed to find evidence that powerline EMFs were stressors, but that conclusion had been foreordained because both the EMF and control animals were already stressed as a result of living in cages in which the animals could neither turn nor rear up. The additional effect of the field, although real, was tiny by comparison and therefore almost impossible to detect.

Phillips had published corticosterone studies that he said refuted my published results regarding stress. Because of the bounty the Vice President had provided, I saw that there was much more to the story. Phillips had first performed a corticosterone study in which he exposed rats to powerline EMFs for thirty days, and then analyzed the hormone levels in the blood using a measurement technique identical to mine. When he found the average level in the exposed animals was only 70% of the normal level, indicating that the EMF was a stressor, he analyzed the blood a second time, using a different method, and still found that the levels in the exposed rats were less than normal. He repeated the experiment with twice as many rats, using both measurement methods, and found the same general results: 78% using the first measurement method, and 51% using the second method. He then repeated the experiment a third time, and a fourth time, and in each experiment the average value from the exposed rats was lower. At this point he wrote Flugum that "The data appears to be consistent with similar findings by Marino."

Phillips then repeated the study a fifth and a sixth time, and in these experiments he reported to Flugum that there were no effects due to the EMFs. When Phillips published a paper in *Bioelectromagnetics* describing his research on the effect of EMFs on corticosterone, the only results he disclosed were those from the two negative experiments.

There can't be a democracy if people don't vote. There can't be a literature if people don't write. There can't be a science if scientists don't tell the truth. So I resolved to tell the truth, as it appeared to me. At the annual meeting of the Bioelectromagnetics Society I delivered what Don Justesen, the president of the society and a friend of Phillips, called my "philippic." I had been required to disclose to Justesen what I planned to say about Phillips. Since Phillips knew the nature of my criticisms of him and would speak after me at the meeting, I supposed that he and Justesen counted the situation as advantageous for them, and therefore gave me a pulpit.

Proceeding rapidly from point to point so that I would finish within the 15 minutes I had been allotted, I told the audience why the studies Phillips had done had no merit. When the yellow light went on I talked even faster, as in a reverie. I pounded on the lectern, gestured excitedly, and concluded that misconduct and errors of omission and commission pervaded the work and completely destroyed its value. There were some people who heard what I said and understood that science produced by people like Phillips was not science at all. How many such people, I couldn't tell. Few, I supposed, since the meeting was paid for largely by the Department of Energy and the Electric Power Research Institute. When it came time for Phillips to speak, his reply was cut short because lightning knocked out the electricity, and the auditorium went silent and black.

The next day I spoke with Phillips. He told me he had never expected I could cause him such problems. Then he predicted that I would never get money for the research I wanted to pursue, and that even if I did, whatever

I found and reported would be contested so that its meaning would never be clear, and that even if my publications gained acceptance, there would be no practical consequences because the world would take no notice.

12
Risk
◆

At a meeting of the Bioelectromagnetics Society I heard a speech by Dwight Spieler, the expert on risk assessment who had evaluated the risks of powerline EMFs and written a booklet on the subject. He said that a risk assessor persuades the public about the implications of scientific studies, just as a physician guides his patient by telling them what will or will not be harmful. In a confident tone he said that EMF risks were minimal, and that the only thing to fear was fear itself. He mentioned that his booklet had been successful, as evidenced by the fact that 100,000 copies had been printed, and because the booklet had been instrumental in the decisions of several state boards and commissions that they shouldn't worry about risks from powerlines. When he announced that the prestigious National Science Foundation had given him a grant to print more booklets, he received a respectable round of applause. He seemed confident he was correct, and I had to admit that he really sounded as if he knew what he was talking about. But he never really explained how he was able to conclude that powerlines were safe.

During the question and answer session someone asked if he could also talk about what was good for health. He replied that the principles of risk assessment also applied to that issue, but that his clients were interested only in what might be bad. I asked him what kind of persuasion he used, and he got a big laugh from the audience when he replied, "You're a lawyer, Dr. Marino, so you ought to know. It's the kind of persuasion employed in courts."

As I was leaving the auditorium I bumped into a good friend of Asher Sheppard's, a man named Hans Tauschen whom I had seen many times since I first met him when he visited my laboratory many years earlier. We were joined by Tim Montpelier, the engineer who worked at Johns Hopkins, and then by Spieler, to whom I was introduced by Montpelier. Spieler and I shook hands and exchanged a few pleasantries, and I decided to try to get to the bottom of how he knew what he claimed to know. He was good at writing booklets and giving speeches but, so far as I knew, he had never really been pushed to explain how he had learned about health risks of powerlines. So I began asking him questions with the intention of either getting short direct answers, or interrupting him if I thought his replies became too long-winded or irrelevant.

"Would you agree with me," I asked, "that there is a difference between knowing that something is the case, and *believing* that it is so?"

"You are right," he said. "Knowledge and belief are obviously not the same thing. But..."

"For example," I interjected, "a lawyer doesn't necessarily know what the truth is. He simply represents his client."

"If you say so," he replied.

"When you wrote your booklet, did you hope to create knowledge or belief in the minds of the readers?"

"Belief," he said. "A booklet can't lead people to knowledge about a complicated matter like risks from EMFs. You see when I first..."

"Can a speech do that?" I said.

"No. How could..."

"Would you try to persuade someone that, for example, a particular method for measuring corticosterone in the blood was more reliable than another, or that an animal experiment should have been designed in one particular way? Do you expect to persuade people about these kinds of matters?"

"Technical issues are decided by specialists, and I accept their consensus. Risk assessment deals with what is right and wrong, which is a question for a generalist not a specialist. For example, the decision to send men to the moon was made by someone who had no special technical knowledge about building spaceships. That decision-maker was obviously convinced that he understood the risks of the project, so he could assess whether they were outweighed by the expected benefits."

"Risk assessment seems to be incredibly important," I said. "Everybody needs it all the time."

"True!" he said. "The fruits of all other sciences are available to the public only because of risk assessment. For example, take the risks of pesticide residues in food. If an expert on the metabolic pathways of organophosphates began to explain the structure of the chemicals, and their mechanism of interaction with mitochondria and ribosomes in the cell, most people would not understand what he said."

"Their eyes would glass over," I said.

"Right, but an expert in risk assessment could persuade the people about what was right regarding the risks of pesticides."

"So that's why you say that when you offer your opinion about the risks of a chemical, or a technology such as powerlines, you do not teach your audience facts like, for example, a professor in a classroom. Instead, you create a conviction in your audience about what is right or wrong."

"Yes! Then the best course of action becomes clear."

"What do you mean?" I asked.

"Well, take powerlines for example. A major question is whether the powerline should be built underground or suspended overhead from towers. Overhead lines cost less and are therefore the logical choice, but there was a perception that they produced health risks. When I was called upon, I synthesized and organized the available information and provided an assessment of the putative risks. The risk of getting cancer from powerlines is no more than that of getting a sunburn from a flashlight, and I can make this point more persuasively than engineers who talk in scientific jargon about how EMFs interact with the body. That's why I think you were quite correct when you said earlier that risk assessment is incredibly important."

"Risk assessors seem to have great power over the minds of decision-makers," I said. "It is imperative that anyone who assesses health risks must always do so with an eye toward what is best for society."

"Science should always be used to promote the public good," he replied. "I remember…"

"Then you agree that an expert shouldn't say that powerlines were safe if he didn't know they were safe?" I asked.

"Yes," he replied, "he must think what he says is true."

"If another risk assessor told the public that powerlines were safe and you knew that were not the case, would you have an obligation to speak the truth?"

"Well…yes," he said. "Otherwise, public support for science would be threatened. People must not think that scientists are like car salesmen."

"Is that because car salesmen don't necessarily speak the truth," I asked, "but rather say what will help them close the sale?"

"Not all car salesmen," he said.

"Earlier you said that risk assessment involves persuading an audience about right and wrong, rather than providing knowledge."

"Yes."

"From where does an expert in risk assessment gain his knowledge about right or wrong?" I asked.

"From studying the scientific evidence and the credibility of its proponents," Spieler replied.

"Suppose the president of a power company called on a risk assessor who was like Robert Flugum of the Department of Energy?"

"What do you mean 'like' Robert Flugum? In what way?"

"His agency's mission is to expand the powerline system of the country," I said. "He is therefore predisposed to see things as good or bad depending on whether they help or hinder what his agency is trying to do. Would you agree?"

"If he had a different mindset, he wouldn't be well suited to his job."

"Do you think his mindset would influence his advice to the president of the power company?"

"It could happen."

"So risk assessment, if it is always to be done correctly, seems to require unusually brave experts," I said.

"What is your point?" he said.

"Just this. If someone like Flugum gave advice, how would the company president, or anybody else, know that the advice wasn't distorted by ulterior motives, since an expert in persuasion would be likely to successfully cover up any such ulterior motivation?"

"Honesty is the cornerstone of science," he said. "Every scientist should be honest, and I think the overwhelming majority are honest."

"Even so," I said, "Flugum might decide that furthering the mission of his agency was the more important consideration, and consequently offer his persuasive but concededly ignorant advice from this perspective. Isn't that possible?"

"It's possible."

"This is not the worst problem with risk assessment," I said. ""I can think of at least two others."

"What else are you prepared to criticize?" he said. "You might as well get it over with. If I don't think you make any sense, I'll simply ignore what you have to say."

"That seems fair," I said. "Suppose the risk assessor was someone like Herman Schwan, who believes in the primacy of equations, and in his ability to understand how they explain and constrain what happens in the world?"

"All right," he said, "suppose he is."

"Someone who thought the Messiah was coming would be unlikely to accept any sort of evidence that he was mistaken."

"Yes."

"And to someone who believed that the priest's words turn bread and wine into Christ's body and blood, the results of chemical tests on the bread and wine after the words had been spoken probably would not matter."

"Yes."

"And if someone thought he would be surrounded in heaven by many virgins, there would be little point in arguing with him."

"Yes."

"So, if a man truly believed something, it would be pointless to argue against his belief."

"Yes," he said. "Belief comes from the heart, not the mind."

"If a believer like Herman Schwan had your ability to persuade a lay audience, do you agree he would be a powerful force in the direction of misleading the public, even though he might have no specific intention to do so, and probably should be credited as having the absolute opposite intention?"

"I suppose that someone like Herman Schwan is enough of a scientist as to be able to separate his beliefs from the facts. I think he deserves the benefit of the doubt on this matter. But you said there were two problems with risk assessment. What is the second?" he asked.

"Just this," I said. "Suppose the risk assessor was someone like Richard Phillips who operates a researchery that produces whatever information suits his client's interests. Wouldn't such a man be expected to make unjust use of his power to persuade?"

"Credit or fault always lies with the individual," he said, "not the science. If a surgeon, for example, were to use his skill to kill rather than cure, one would not disparage the practice of surgery but only the surgeon himself. The situation with risk assessment is exactly the same. A scientist should always make proper use of his science, but if he does not do so, you must not, in my opinion, attack the science itself."

"When you have the time," I said, "we should have a discussion about what 'science' is."

"Why does this issue arise in your mind now?" he asked.

"Because I think that science is something that explains and describes in natural terms what has happened or can happen in the world."

"I certainly agree with that," he said.

"But what you are saying now is not in tune with what you said earlier when you described risk assessment in terms of being able to persuade, and as resembling the activity of a lawyer."

"Are you suggesting that risk assessment is not a science?" he asked.

"It doesn't look that way to me," I replied.

"What do you think it is?" he said.

"A knack for persuading an audience," I said.

"So you don't think risk assessment is a good thing?" he said.

"I'm explaining what I mean. Why not let me finish."

"All right. Tell me."

"It's like the ability of Bob Hope or Billy Graham. They had a certain knack for making an audience feel good. There are no rational principles that can unfailingly say what will be funny or inspiring. If there were, we would have many more geniuses like them."

"Are you saying that risk assessment is identical to what Bob Hope or Billy Graham did?" he asked.

"No. Only that they are all the same kind of activity," I replied.

"What activity?"

"The occupation of a smart and enterprising person who is naturally skilled in dealing with people, and is therefore able to deceive people concerning the matters about which he speaks. I would like to say more exactly what kind of persuasion I think risk assessment is."

"Please do."

"I agree with you that it resembles the activity of a lawyer in a courtroom, but it is a counterfeit version of that activity."

"Why is it a counterfeit of what lawyers do?"

"When the lawyer seeks to persuade a jury, his goal is to achieve justice for his client under the law. The law defines the standard of conduct and provides for an advocate to present an opposing construction of the facts, because facts do not speak for themselves. The law also provides for cross-examination so that each side can test the other's case. In contrast, the goal of the expert in risk assessment is to manipulate the opinions of his audience so they see the world through his eyes. He adheres to no standard, has no adversary, and no one who calls him to account when he uses combinations of words and phrases that please but mislead his audience."

Our conversation went on for a few more minutes, but didn't advance any further. Then Montpelier spoke for the first time and began making the point that I was wrong about risk assessment, that it really was a science. He said: "Prestigious organizations like the Electric Power Research Institute employ risk assessment. To say that what it is doing is not real science seems ridiculous."

I didn't think that advanced our conversation, because it amounted to saying I was wrong just because some people disagreed with me. So I told Montpelier how I felt and asked him to say something that I could analyze, which he tried to do by making the point that what he called "powerful" organizations routinely employed risk assessment. His main example was the Electric Power Research Institute. Our conversation then focused on what "powerful" meant.

"By 'powerful,'" I said, "do you mean something good to have?"

"Certainly," he replied.

"Then in my opinion the Electric Power Research institute is not powerful, so you haven't made any argument that risk assessment is a science."

"What? They can give or withhold money for research so that everything comes out best for them. I don't necessarily like it, but I recognize the situation for what it is."

"You are talking about two different things," I said. "The Electric Power Research Institute is not powerful because it always does only what seems best to them."

"Isn't that what it means to be powerful?" he asked.

"No," I said, "as you have conceded."

"I have conceded?" Montpelier said. "I maintain the opposite."

"That's not correct. You said that power was something good to have. The Electric Power Research Institute can produce false information but to do so is bad, not good. So its efforts to persuade falsely do not indicate that it is powerful, unless you think power is the ability to hurt oneself because it seems good when, in fact, it is not good."

"Of course I don't think that," he said. "But I know Chauncey Starr, who is the head of the EMF program at the Electric Power Research Institute. He is quite happy. How in the world can you say that he is hurting himself? That's crazy!"

"He would hardly think it good," I said, "if he understood that his actions were hurting himself and his family because they increased the hazards that his loved ones must endure, which is what happens when he uses risk assessors to persuade people that EMF pollution is harmless. What could he possibly regard as less good or more bad than a cancer in one of his grandchildren caused by EMFs in the environment that he put there and then passed off as something completely safe."

"That's absurd," Montpelier said.

"Why? Say something that isn't offensive, but rather advances this conversation."

"Well, I would like to understand what you're talking about."

"Then I will explain my point again, but even more simply. To be 'powerful' means to get what is actually good. It is weakness, not power, if what you bring about is actually evil, even if you are doing what pleases you."

"By your logic, Starr should be miserable, but I think he is happy."

"Is it your opinion that Starr could be so wicked and still be happy?"

"I think so," Montpelier replied, "at least if he is not caught."

When he made that remark, he, Spieler, and I all laughed. Then I asked them whether we could all at least agree that someone who intentionally injures an innocent person commits an unjust act. They both nodded in agreement. However Tauschen's face screwed up in disgust and flushed deep red in anger, and he said explosively, "Dr. Marino, are you serious or are you joking?"

To outward appearances, Tauschen and I had been much alike at the inception of our careers. The reality, however, was that we had started with different concepts of morality, and that difference determined our careers and our present views of right and wrong. On every occasion, whatever his client said and however he described things, Tauschen never disagreed, but rather shifted back and forth to always provide support. The experiments Tauschen did for Chauncey Starr and for Robert Flugum were meaningless, which was exactly what they wished. Yet, at Gordon Conferences and other meetings attended only by scientists and those with money to spend on science, with my own ears I heard Tauschen say that EMFs could be environmental hazards, and that more research was needed. His position on the science of EMFs varied so greatly that, no matter what he said, I could show that Tauschen would not agree.

"Your rhetoric has run wild," he continued. "You claim to practice morality, but you actually drag us through tiresome popular fallacies about the purpose of science and the consequences of the actions of strong men like Starr. Science always has been the special advantage of the best men, and weaklings who cannot stand up for themselves must accept their fate. The weak invent rules to prevent the strong from dominating them and to try to shame the strong by saying that dominating others is unjust. That's why you call the advantage Starr enjoys over the mass of people an 'injustice.' However, nature says that the stronger should have the advantage over the weaker, and the more able over the less."

"I want to be clear on one point," I said. "Do you think that Starr can build powerlines in such a way that maximizes his profits, irrespective of the misery he causes? Does he have a greater right to his profits than the public has to its health?"

"Of course," he said. "He is doing as he pleases and getting what he wants, which is what it means to be happy. How could anybody be happy when he must do what others tell him to do, rather than what he himself wants to do? Anyone in his right mind spends his life going after what he wants. Those who have courage and intelligence are able to get what they want. Those who fail conceal their impotence by claiming that wanting greater profits is shameful. It is their own cowardice that leads them to praise justice, which is no more than an impediment to those with natural gifts and talents that enable them to accomplish great things. One ought to have strong desires, and then proceed to gratify them."

"Like Chauncey Starr?" I asked.

"Like Chauncey Starr," he repeated, with emphasis.

After I returned home from the meeting I thought about the conversations I had there. When I had arrived at the meeting I had no intention to discuss the practice of risk assessment and try to determine what it actually was, although from time to time I had thought about that question because it had always been something that was difficult to categorize, whether it was physics, biology, psychology, medicine, or something else. I hadn't ever thought that risk assessment might not be any of those things, nor any other science. Spieler's cooperation had allowed me to quickly get to the bottom line, which was that a professional risk assessor was more like a drug dealer who could make people feel good than a scientist who had some reliable knowledge about the world. Had Spieler given long speeches or gone off on some tangent, the conversation would have never amounted to anything. But he had not thrown any curveballs that I could remember, so the conversation remained on track. That was the first time that I had clearly seen the power of a dialogue between two persons who both stayed with the point of the discussion and avoided giving speeches.

What Tauschen had said troubled and excited me the most. His views carried me back to what I had read in college about Nietzsche, who could radiate an insight like a shaft of light that appears just before sunrise. But his views about morality were so black that my Jesuit teachers had refused to even discuss them in class. After college I had put aside thinking deeply about moral questions and focused instead on science, which seemed utterly unconnected to morality. That was a childish perspective, and I realized that I had Tauschen to thank for my education in that matter, although of course I don't think he had any serious intention to help make me a wiser man.

Chauncey Starr, who had great power in the world, shared Tauschen's view that ordinary people were no more than impotent cowards. Spieler provided Starr a veneer of science over his aggression.

Although many questions and confusions remained, I was happy in regard to one particular point. The worst thing in the world to me was not knowing or understanding the way things were or how they worked, the feeling of helplessness like a cork bobbing in the ocean. After I had reflected on what happened at the Bioelectromagnetics meeting, however, I realized the value of Father Wallner's advice on my last day in college: "Always do the right thing. No one can tell you what it is. You must figure it out for yourself. You now have all the tools to do that."

13
Dangers

The Navy chose to build its antenna in Michigan because the soil there was well suited to orient the antennae EMFs toward the ionosphere, from where they would be reflected down to the submarines. The Navy's Paul Tyler asked Philip Handler, the president of the prestigious National Academy of Sciences, to appoint a committee to evaluate whether the EMFs would harm the people in the state, which he did, with Herman Schwan as its chairman. It deliberated over the same evidence that had alarmed the committee on which Dr. Becker had served, after which Herman Schwan announced, "There is no clear evidence of harmful effects." I asked him, "What is clear evidence of a harmful effect?" I figured that was something necessary to know so that if "clear evidence" ever appeared, he would recognize it, but he basically said, "I will know it when I see it."

After the Navy had built and begun operating a small version of the antenna, stories about medical problems among the men who operated it began to circulate, and some people in Michigan became alarmed. Tyler, brandishing Schwan's report that bore the imprimatur of the prestigious National Academy of Sciences, appeared in various public forums where he reassured the audience that the antenna would be safe. "Prestigious" was joined so frequently with "National Academy of Sciences" in news accounts of the meetings that it appeared the term was part of the name of the organization.

Some of the people whom Tyler had not converted to his opinion asked Dr. Becker and me to appear at their meetings and speak on their behalf. We both believed there had been no fair discussion about whatever consequences the large antenna's EMFs might produce, but an involvement in that problem was not imaginable for us. We had earned criticism and suspicion from the Veterans Administration for our work with the commission in New York, and for my persistent freedom-of-information requests to the Department of Energy. In many ways, both subtle and direct, we had been warned by VA officials at Central Office in Washington, DC, that further involvement in public controversies would trigger the destruction of our laboratory.

I saw that one course which might allow us to avoid danger but still help the people who had sought our help would be to encourage them to engage Paul Tyler in a dialogue that could illuminate their concerns about the EMFs. "Schwan's report is against you," I told them, "but the report of Dr. Becker's committee is in your favor. You should press Tyler to explain the differences in the judgments of two committees that evaluated the same laboratory evidence." I had hardly finished when I heard, "What report? What committee was Dr. Becker on?"

Paul Tyler would not speak to me, but his next-in-command told me that the report of the committee on which Dr. Becker served was for official use only, and would not be released. Near the end of our conversation, Tyler's lieutenant said, "Our nation needs the antenna. We are at war. People die in war." Not disagreeing, I said, "Your interest is different from that of the people in Michigan. They can't conduct their own laboratory studies so they must rely on the science you produce. Surely they have a right to see it." But he continued to insist that the report must remain private to the Navy.

While I pondered what I would do, someone on the staff of Wisconsin senator Gaylord Nelson called me, and I told him about Dr. Becker's committee. He asked me for a copy of the report, which I sent to him after receiving Dr. Becker's permission and a promise from the senator that he would not disclose the source of the report. I scrutinized the report before I mailed it to insure it didn't contain an inconspicuous code that might indicate to Tyler which committee member's copy had been disclosed. Senator Nelson released the report during a speech on the floor of the Senate in which he criticized the Navy for its secrecy about the antenna's possible hazards.

Publicity concerning the antenna increased and local governments in Michigan, worried they might be hosts to the antenna, passed resolutions urging the Navy to take steps to ensure that their constituents wouldn't be placed at risk. One result of their concern was the appointment by Philip Handler of still another committee which, incredibly, included Herman Schwan and Morton Miller. The head of the committee, a biologist from Harvard named Woodland Hastings, asked Dr. Becker and me to testify before his committee. Dr. Becker, however, had had his fill of conflict stemming from trying to be honest about science, so he declined Hastings' request. I, however, was only at the beginning of understanding what I could bear.

The idea of being a supplicant before a panel of judges that included

Schwan and Miller was disgusting to me. I thought that anyone who would even ask me to do that must be either cynical or ignorant. I called Hastings, to test my suspicions about him, and as we spoke it became clear to me that he knew nothing of the opinion they had given under oath that EMFs from powerlines were completely safe. So I told him what they had sworn to, and explicitly recited the reason that their presence on the committee would corrupt its deliberations. "Powerline EMFs are almost a million times stronger than those from the antenna. What do you expect their opinion will be regarding the antenna?" Between moans and groans from the pain that shot through his back as he rolled from side to side in his bed at the Harvard infirmary, I proposed that he appoint me and Dr. Becker to his committee. "Then I could confront Schwan and Miller," I said. A shocked and somber Hastings told me he would ask Handler to appoint Dr. Becker and me to the committee, but Handler refused.

I had the sense that pursuit of the biology of EMFs might lead to new insight into how living things worked, thereby benefiting everyone. In this regard I believed in Dr. Becker's project and wanted nothing more than for it to succeed. I saw EMF biology as an emerging flower, full of promise but with a future value that could barely be perceived. I soon learned, however, that the men on the National Academy of Sciences committee saw only a weed, or at best a negligible bloom that deserved scant consideration. They had been placed on the committee, it seemed clear to me, precisely because they held that view.

I decided to submit a written statement that openly expressed my discontent at Handler's decision to provide the public with opinions about EMFs that came from a committee burdened with bias, ignorance, and divided loyalties. I did not really expect that the statement would induce the committee to alter its course. My goal was to publicly establish that there was disagreement on the safety issue. I believed that truth emerged through conflict, not consensus.

In my statement I did not directly oppose the antenna because I recalled the words of the Navy officer who told me the U.S. needed it. But I also thought that those asked to bear the burden should accept it for reasons that seemed worthwhile to them, and not have their lives stolen by some mysterious force that leaves only unexplained death.

Still, it was the Veterans Administration we worked for, not the Environmental Protection Agency or Health and Human Services. I sought the advice of a wise woman who earlier had helped me to understand the people in Central Office. She read my draft statement, sat back in her chair, and after thinking about what she would say, told me, "Philip Handler was a good biochemist, but he has come to believe that science is the solution to all mankind's problems. When he became president of the Academy he began to pursue his agenda ruthlessly. He is very dangerous, and you would do well to avoid conflict with him." As she spoke she held up her fists and shook the one on the right, glancing at it as she spoke. Then she looked at the one on the left and said, "You must be careful for another reason. Many retired Navy people work in Central Office. They will not take your statement kindly, as it seems to attack values they hold dear."

I did not believe that such a famous and powerful man as Philip Handler would take notice of me personally; instead, my attention was fully focused on the Veterans Administration, and the possibility that my public involvement might lead to the closing of our laboratory. I made what changes I could to emphasize a need for honest research to uncover the mysteries of EMFs, unfettered by implications such knowledge might have for the goals of any organization. By emphasizing pure science I hoped to avoid antagonizing the Navy sympathizers who worked in Central Office.

I handed the revised statement to Dr. Becker and asked if he thought it might jeopardize our laboratory. The gist of what he said was that we had two possible destinies. If we did not speak out, the laboratory would live as long as he chose to practice medicine at the hospital and we pursued science that had no likelihood to give offense. If we sent the statement, however, we would invite the enmity of Central Office which, he reminded me, controlled our money. He talked in this vein for a few moments, but I knew where he would wind up. He cared about ordinary people and about the need for a free science. It disturbed him deeply that the Navy had swept the results of his committee under the rug, and that the decision of both the first and second committees of the National Academy of Sciences had been rigged. He was a veteran and a conservative Republican, no anti-military bleeding-heart liberal, but his ultimate loyalty was to the truth. He slowly took out the old-fashioned fountain pen he always used and signed the statement.

Soon we had troubles. Marguerite Hays called from Central Office and said there would be no more money for us and that our laboratory must close. The clever witch did not arrange this result directly. Rather, she used the innocent processes of science to strangle us by choosing Lionel Jaffe, of Purdue, as the scientific reviewer of our research, knowing he would say only bad things about our work. In his review, he wrote that Dr. Becker's work was plain ordinary fraud, and that the work of anyone who collaborated with him must also be worthless. Thus armed, Hays ignored all of Dr. Becker's achievements and awards, and ordered that our laboratory should be put to sleep, like an old dog.

While Dr. Becker searched desperately for some source of money at Central Office that was not controlled by Hays, we received another severe blow when the journal *Science* attacked us in an article entitled "Critics Attack National Academy's Review Group." The author described our views as one might describe the opinions of young seminarians who had disagreed with the Pope on a matter of faith and morals. In what looked to be even-handed journalistic treatment but Dr. Becker and I knew was not so, the article quoted officials of the National Academy of Sciences who stoutly defended the integrity and balance of the committee, and the method by which it was chosen. I worried about the effect the article would have on my career, and on my ability to obtain grants when the day came that I must emerge from Dr. Becker's protection, but I was helpless to do anything.

In due course, the National Academy of Sciences committee decided there was no clear evidence the Navy antenna would cause harmful effects. How they reached their conclusion they never explained; it was as if the knowledge had been delivered to them by Hermes. Hastings told the press that the committee had given the antenna a "clean bill of health," but that description was a travesty; the committee was far from a physician who regards his patients' interests as paramount, and besides, it wasn't the antenna's health that was at issue.

When Jimmy Carter had campaigned for the Presidency, he promised the people in Wisconsin and Michigan that the antenna would not be built if the majority opposed it. After he was elected, however, the Secretary of Defense announced that it would be built notwithstanding the results of local referenda, which were strongly against the antenna, because the nation needed it and the prestigious National Academy of Sciences said it would be safe.

It appeared that our involvement with the antenna had run its course, but the issue of the antenna's EMFs caught the attention of *60 Minutes*, and Dan Rather came to our laboratory to interview Dr. Becker. Rather's manner was low-key and disarming, and they both smiled frequently as they spoke. Rather said, "Dr. Becker, you seem to have the Navy pretty upset. What's going on here?"

It was as if Rather had just arrived from another planet and knew nothing about the risks from EMFs, so Dr. Becker began at the beginning. There was an electrical control system in the body, external EMFs could influence it and produce stress, stress could cause disease, the Navy downplayed the risks, and the Academy committee whitewashed the whole issue.

Rather brought up the report that Senator Nelson had released and asked, "Is it true that the Navy repressed the report for better than two years?"

"Let's just say the Navy did not disseminate the report widely," Dr. Becker replied. Near the end of the interview Rather said, "Is what you are trying to say that we are playing with a stacked deck?"

"I think so, yes," Dr. Becker said.

The interview enraged Philip Handler. In a letter to the president of CBS, which he also sent to many newspapers, he said the committee was surely no stacked deck, the antenna was certainly safe, and that claims to the contrary were laughable. He said the interview was a twin disservice to science because it lowered respect for science by criticizing its greatest citadel, and it inflamed public opinion about nonexistent health risks. CBS offered no response, leaving it up to the public to decide whether Dr. Becker's interview or Handler's letter was closer to the truth.

Dr. Becker also made no response to Handler, concentrating instead on attempting to find a path around the death sentence Marguerite Hays had issued. One bright day we received the wonderful news that administrative control of our laboratory had been moved to a division in Central Office headed by a man who liked Dr. Becker and respected his work, and who would continue to fund it. Marguerite Hays could only watch as we sailed past her open mouth.

I returned to my studies of EMFs, and grew stronger and more confident in my general understanding of the biological significance of EMFs, although I sometimes sank to near despair because there were so many aspects that I understood so poorly. I designed better experiments, though, and got some remarkable results. In one case, my plan was to withdraw blood from each mouse prior to field exposure and then a second time after exposure, and to compare the two results. I was concerned, however, that the stress experienced by the animals during the withdrawal of blood would interfere with my ability to register the effect of the field. After seeing measurable stress from the withdrawing of blood, I redesigned the experiments so that the effect of the measurement procedure could be separated from the influence of the EMF. Surprisingly, the changes in the blood of animals that went from no field to field exposure were similar to those in animals that went from field exposure to no field. As far as I knew, such a symmetry in the ability of the animals to respond to an alteration in their environment had never been reported previously.

The editor of the journal that accepted the paper for publication called and asked what topic I would choose if I were invited to write an editorial. I told him that I would say that science was not necessarily a truth-seeking machine. Sometimes it was like a whore that could be mounted by any special interest with the money to pay her price. The journal published my manuscript, but the invitation to write the editorial never came.

Bassett and Brighton had each established companies and had begun developing commercially viable treatment systems for growing bone. Dr. Becker, however, devoted his energy to his work on regeneration. He was interested in growing an entire limb, which he believed was possible because the body was electrical as well as biochemical in nature. His studies attracted the attention of Max Cleland, the head of the Veterans Administration, who praised Dr. Becker's research in a letter to the *Washington Post*, saying that Dr. Becker was "a world-renowned expert in the study of limb regeneration and his pioneering work offers hope of major breakthroughs."

The story interested Alan Cranston, the senator from California, and he secured passage of a bill that specifically provided money in the VA budget for Dr. Becker's regeneration research. But Cleland, who was probably busy with more important matters, gave administrative control of the appropriated money to Marguerite Hays, and she squandered it in pursuit of exactly the opposite goal that the senator had intended. She spent much of the money hosting an elaborate conference in Bermuda to which she invited scientists who thought that growing new body parts on human beings was impossible. Not surprisingly, speaker after speaker dismissed Dr. Becker's conception of the possibility of regeneration as childish and naïve. The great negativity generated at the conference was reported extensively, and the resulting publicity drove a stake into the heart of Dr. Becker's vision.

In my mind I returned again and again to Philip Handler, who had controlled the work product of the Academy committee like a fisherman with a fish on a hook, corrupting science by bending analysis of the antenna's EMFs to his will. I discovered that the corruption I had seen was nothing new for Handler or his prestigious National Academy of Sciences. No longer a sleepy honorific society as it had been when it was first chartered by Congress, it had become an engine for making choices, powered by a thousand employees, each of whom held his job at Handler's pleasure while he answered to no one on earth. His ambition was to make the National Academy of Sciences the Supreme Court of scientific advice, and to make its advice the supreme form of human knowledge, two goals that he seemingly had accomplished. But it was not a court where both sides of a contentious issue could be heard, but rather a court that sometimes furthered the interests of special outsiders with whom Handler identified. He championed nuclear power, a strong military posture, and the healing power of pharmaceuticals, and was hostile to environmental constraints, concern with social problems, and the general drift of the times which, he wrote, subverted the youth. He ran the National Academy of Sciences as if it were an oracle. Whenever Congress or an executive agency came to learn what course was best, Handler would inhale the intoxicating air of science and his divinely inspired utterances would be interpreted by his committees. My contempt for him grew. I was like a log on a fire covered with ashes on the outside but glowing cherry red inside.

One day, a writer from *Saturday Review* named Susan Scheifelbein came to interview us about EMFs and our theories on the importance of stress. She was small and thin, with a timorous voice; her body movements were guarded, as if her bones were made of eggshells. She explored the subject of EMFs with us, and in the process I forgot the advice of the wise woman in Central Office who warned me not to cross Philip Handler. Scheifelbein interviewed others and then wrote her story, which included this passage:

> Dr. Becker, who had spent his career studying how to use EMFs for healing bone and inducing regeneration of limbs, was asked by the Navy to be on a committee that would evaluate possible hazards from an antenna that had been proposed for Wisconsin. The committee reviewed animal and human studies and decided it was urgent they be continued, and that warnings be given of the possible significance to the public posed by EMFs from powerlines. But the committee report was not released, there were no

warnings, and the studies were not continued. Senator Nelson released the report, saying that the Navy had intentionally hidden it. The National Academy of Sciences appointed another committee. Marino told the chairman that committee members Miller and Schwan had been paid by power companies to testify that EMFs from powerlines were safe. The chairman promised to put Becker and Marino on the committee, but the Academy refused. They sent a protest letter to the Academy, which got very upset, but nothing changed and the committee gave the antenna a clean bill of health. Even so, Becker and Marino may be right. Hans Selye was the man who discovered the effects of stress, and showed that the condition elicits hormonal responses in man. Selye told *Saturday Review* that EMFs can cause stress, and that Becker and Marino were the experts on EMF-induced stress.

Shortly after the article was published I received a phone call from Scheifelbein. She was so distraught that at first it was hard to understand what she was saying. Then I heard, "I just got off the telephone with Philip Handler, and he is very angry with the article."

"What exactly is his problem?" I asked.

"He said that my article attacked the objectivity of the Academy, and he demanded that I publish a retraction."

Before I could even ask what he wanted retracted, and why, she said Handler had told her, "I'm going to use every penny we have in the Academy to break you, and break the *Saturday Review*."

Scheifelbein was no more than fresh meat to a lion like Handler, and the only reasonable thing she could do was stay away from his teeth. So I urged her to neither say nor do anything that would allow him to claim she had retracted her story, and to refuse any more of his phone calls.

She did as I advised, and when Handler realized he could not influence her directly, he appealed to Norman Cousins, a former editor of *Saturday Review*, asking him to prevail upon her to apologize for questioning the process by which the National Academy of Sciences produced its report about the antenna. Cousins did as he was asked, telling Scheifelbein that, essentially, her story claimed the existence of a conspiracy among the Academy, the Navy, and the power companies, which he said was ridiculous on its face. He also said Handler had assured him there was no health threat from powerlines. Scheifelbein, however, resisted the pressure and her editor stood behind her.

Handler then turned his attention to me. He sent a letter to the editor of *Saturday Review* together with a manuscript that he demanded be published or, he threatened, the Academy would sue the magazine. In the manuscript he said my research with rats and mice was worthless, that I had personally been discredited by my peers, that EMFs from powerlines and from the antenna were safe, that the committee was formed and functioned with only the greatest probity, and that Scheifelbein's article was "slanderous." The editor refused to publish Handler's manuscript, so I waited for him to start the lawsuit, hoping it would come. I had battle in my heart, believing it far better to confront him directly than have him above me where he could yap abominably from his lofty perch.

Despite my fervent hope, the lawsuit never came. Handler got cancer, and it ultimately killed him. In all my life I never knew a man who did more harm to the idea of science than Philip Handler.

14
Zaret

◆

From time to time Dr. Becker had talked to me about the work of an ophthalmologist named Milton Zaret; they shared an interest in the health risks of EMFs although they had never worked together. A book about Zaret appeared in which, according to a review in the *NY Times,* he had claimed that microwaves caused cataracts and that the supporting evidence had been covered up by the Defense Department and the Central Intelligence Agency. The *Times* had interviewed the Navy's Paul Tyler who denied there was any health problem or cover-up and said, "I'm a doctor whose specialty is preventative medicine, and I obviously have to be very concerned about the health of our men on shipboard. If I thought we needed money for research on microwave effects, I'd ask Congress for it."

When I asked an ophthalmologist whether microwaves could cause cataracts, he told me, "The Navy did many studies about that on servicemen, but couldn't find any connection."

Soon after the book appeared Zaret visited our lab. I spoke with him long enough to see that he had much experience studying EMF bioeffects, so I asked him for an interview regarding that work, and he invited me for a visit at his home in Scarsdale, New York.

We sat at his dining-room table and talked from mid-morning until late evening, stopping only to eat a dinner that his wife had prepared. He smiled easily, but not deeply, and spoke with a quiet dignity as he told his dark story all the way through to its sad ending. He showed me reports he had written for the Defense Department, and told me about conversations he had had with Tyler and others.

Early in his career he had been consulted often by government officials regarding EMFs, and had been respected both for his knowledge and his loyalty. Then the same officials cast him out like a leper, and the soft-spoken man with the hunched shoulders genuinely seemed unable to understand the reasons for the reversal of his fortunes. After our discussion I talked with many of the people he had mentioned, and I read his reports and those of his antagonists. Ultimately, I came to see that his career could not have had any ending other than the one that occurred.

As a teenager Zaret had pleaded with his mother for permission to join the Navy. She refused at first, and relented only after he threatened to enlist in the Canadian Armed Forces. He told me how he had been greeted when he reported for duty at Pearl Harbor. "The first thing the captain said was, 'Here's your copy of the Navy's rules and regulations. I want you to read it and remember that every one was written to be broken, but with foresight.' That was the best advice I ever got, because the war wouldn't have been won if we hadn't been breaking regulations." Zaret had been on duty the day the Japanese attacked.

He completed three tours of combat in the Pacific as an engineering officer, and after the war went to medical school and then trained in ophthalmology. Not contented with only a clinical practice, he began doing research on the nature of the ocular hazard posed by lasers, which was a pressing ophthalmological problem of that time. In studies for the Defense Department he showed that the intense red beam could instantly burn the retina of rabbits. That work led to the first warning of the dangers of lasers.

When a question arose concerning whether servicemen exposed to microwave EMFs from radars and communication equipment were at increased risk for developing cataracts, the Defense Department again sought Zaret's help. He began a study in which he was to perform a particular ophthalmological examination on several thousand servicemen who had been identified by the military as working in a job that involved exposure to microwaves.

As the study neared its end, without Zaret having found even one cataract, several electronics companies asked him to examine some of their employees, civilian microwave engineers who had been diagnosed with cataracts by other ophthalmologists. The engineers had a practice of looking through a peephole in a wave guide to observe the operation of the device that generated the microwaves. Zaret found that the cataract had always occurred in the eye that the engineer had used to look through the peephole. In these inadvertent experiments, the microwave energy had signed its name at exactly the same place in each man, the tissue in the front of the eye immediately behind the lens.

"It is seldom possible to establish a cause-and-effect relationship with scientific certainty in human pathology," Zaret told me. "In these cases I was sure beyond any reasonable medical doubt that the repeated exposure the engineers got by looking through the peephole was what had caused the cataract."

Corroboration of his hypothesis had not materialized in his ongoing clinical study. However, the kind of ophthalmological examination needed to examine the tissue behind the lens had not been one of the planned study examinations, so he would not have found evidence of an EMF-induced cataract of the type he saw in the engineers even if it had been present. But enrollment in the study was not quite over, which gave him the opportunity to perform the necessary examination on the last of the servicemen who entered the study. When he did, he observed a roughening and thickening of the tissue behind the lens that could have been the inception of the pathological process that results in a cataract.

In his final report he said that he had not found any cataracts in the study cohort, which Tyler interpreted to mean that microwaves posed no ocular hazard. But Zaret had not systematically examined the tissue behind the lens, and he had told Tyler that the health implications of the study could have been far different if he had done so. He also told Tyler, "I have seen cataracts in my private patients that seem clearly linked to repeated exposures to low-level microwaves." Tyler, however, discounted Zaret's findings in his private patients and in the handful of servicemen whom he had examined near the end of the study.

A belief had developed in the military that any damage done by microwaves would involve the burning of tissue in the eyeball, as with lasers. Because of the way Tyler had construed Zaret's results, military physicians continued to tell sailors who were bathed continuously in microwaves from the hundreds of shipboard antennas and soldiers who operated weapons and communications systems that they were not harming themselves so long as they did not feel heat in their eyeballs. At the same time, however, the Defense Department funded additional studies by Zaret on occupationally exposed servicemen to try to resolve the issue of whether cataracts could develop in the absence of microwave-induced heat; in this study he performed all of the necessary examinations.

As he carried out this research, private patients including veterans and active-duty servicemen continued to seek him out. Over the next several years he found cataracts in many men who had been occupationally exposed to EMFs but had never looked through a peephole, and could not recall having felt heat in their eyeballs. Some of the men had worked with radar, radio antennas, or microwave ovens, others with walkie-talkies or cathode-ray tubes. In all these cases, young, healthy servicemen had suffered damage to the tissue in the front of the eye behind the lens caused by something that was unseen and unfelt, and could only have been microwaves at intensities too low to cause heat.

Zaret finally decided that it was medically necessary to change the level of microwave EMFs that the Defense Department regarded as safe for servicemen, because that level had been predicated on a thermal-effects-only mechanism of action of microwaves, which he was now certain was wrong. He was told, however, that the proof needed was a direct showing in animals that cataracts could be caused by exposures to nonthermal levels of microwaves. The Defense Department asked Zaret to perform the study, and he agreed.

The Navy owned a small uninhabited island in Pearl Harbor, and it gave him an exclusive five-year lease to use the island for his experiments. He acquired hundreds of old- and new-world monkeys and began his work. Meanwhile, the use of microwaves in weapons and communications systems, industrial devices, and home appliances continued to expand, unrestrained by the possibility of side-effects, ocular or otherwise.

One day, four well-dressed men from the Central Intelligence Agency came to visit Zaret. As they sat around his dining room table eating a dinner his wife had prepared, the man who would soon become Zaret's case officer told him that the country needed his help. He gave Zaret translations of Soviet scientific articles purporting to show that low levels of microwaves could affect the brain, and asked: "Is it possible that microwaves could be used to create something like a Manchurian Candidate?" Zaret told him it wasn't obvious how it could be done, but that the possibility couldn't be completely ignored.

Thereafter Zaret performed EMF research on rats for the Central Intelligence Agency, mirroring work done by Czechoslovakian scientists, and he observed effects on the nervous system that were similar to those that they had reported. These paired experiments allowed the language in which Soviet scientists described the strength of the microwave EMFs used in their studies to be translated into the different scientific nomenclature used in the United States, like a microwave Rosetta Stone.

Soon after that, Zaret's case officer brought him to the Pentagon where he met a man named Samuel Koslov, who Zaret later learned was the chief advisor to the President on matters involving EMF hazards. Koslov told Zaret something that was known to only a handful of people in the U.S. government. The Soviets were irradiating the American embassy in Moscow with low-level microwave EMFs that were beamed from antennas concealed in buildings on the opposite side of the boulevard from the embassy building (just as I had been told by Jack Anderson's assistant, in New York in 1973). The CIA had analyzed the microwaves and determined their precise characteristics which, to the surprise of U.S. government officials, matched those in published Soviet experiments involving behavioral effects in rats. This had suggested to Koslov that the microwaves could be altering the biochemistry inside the brains of embassy personnel, thereby causing them to make erratic decisions. He gave Zaret the money necessary to replicate some of the Soviet research, and sent him to international meetings to mingle with Soviet scientists and gather information that might help solve the problem of the purpose of the Moscow EMF beam. He was trained by the CIA to recognize his counterparts working for Soviet intelligence, and to avoid their stratagems for compromising Americans.

He learned only a few details about the research of the other scientists who also labored in the black for Koslov. What Zaret did learn only added to his concerns that EMFs could affect the nervous system, and perhaps cause other diseases besides cataracts. Meanwhile the civilian and military embassy staff and their families continued to work and live in the embassy building in complete ignorance that they were being bombarded by a microwave beam.

After the ambassador and several others who worked at the embassy developed cancer, Koslov began a secret study in which the blood of the embassy employees and their families was examined for evidence of cancer. The study was directed by a State Department physician named Herbert Pollack, who told the people that the blood tests were "routine."

Zaret had a great ache in his heart. What kind of an employer subjects his workers to danger without providing any warning? What kind of a general fails to take what steps he can to protect the safety of his men? And what kind of doctor conducts secret research on his patients? Nevertheless, Zaret continued to examine the eyes of civilian microwave workers and servicemen whose duties put them in harm's way. He also continued to study the eyes of the monkeys on Laulaunni Island; he exposed them to EMFs and then looked for the roughening and thickening of the tissue in the front of the eye behind the lens that preceded formation of EMF-induced cataracts, which was what the Defense Department had demanded as evidence of the hazards of microwaves. As he worked, true to the duty that he felt he owed the government, he said nothing publicly concerning what he knew or suspected about microwaves.

By a process that he never explained to Zaret, Koslov ultimately came to the conclusion that the Moscow microwave beam was not an effective mind-control weapon. He ordered Zaret and the others to discontinue their research and, far worse for science, he destroyed all the data and records that had been produced by each of the investigators who had performed the secret studies. Because of this, Zaret's work, and that of Ross Adey at the Veterans Administration Hospital in Loma Linda, Don Justesen at the Veterans Administration Hospital in Kansas City, and many others, was lost. Years later Koslov told Congress, in effect, that he had destroyed the material because he didn't have any room to store it.

Finally a time came when the forces inside Zaret resolved themselves, and he saw his duty in a different light. At a scientific meeting he described the early signs that occur in the tissue behind the lens of the eye of someone who is repeatedly exposed to microwaves. At another meeting he detailed several dozen cases in which the degenerative process in the eye had not been interrupted by withdrawal from the hazardous environment, leading to formation of a cataract. At still another meeting he told his audience that the microwave cataract was a preventable environmental disease. In a publication, he said explicitly that the Defense Department standard for microwave exposure was not clinically credible.

Although there was no ostensible response by the government to what Zaret had said about microwaves, the gears had started to turn. Secretary of Defense Harold Brown met with his advisors and received estimates of the cost in money, injuries, and lowered military efficiency of changing the microwave standard as Zaret had advised, and estimates of the costs for caring for the injuries that would occur if the EMF standards were not changed. Tyler and others argued that a more stringent standard, which would require that radars be used at lower power levels, would threaten the Navy's ability to carry out its mission and would "kill ships," by which he meant make them more vulnerable to the enemy. They pleaded with the Secretary to retain the standard, which is what he decided to do. I know that this is true, or mostly true, because I also heard the story from someone who was at the meeting who told someone else who told me, and because the story explains the pounding that poor Zaret took in the aftermath of his public call for a change in the official microwave standard.

He first knew that he had troubles when Tyler issued an order canceling Zaret's lease on the island in Pearl Harbor and terminating the monkey experiments; when Zaret appealed, the Navy chose Koslov to review Tyler's decision. Zaret's mentor and fellow cold war warrior told the Secretary of the Navy that Zaret was not trustworthy, and that neither his scientific nor clinical research was reliable. The Secretary sided with Koslov and authorized Tyler to take control of the island and destroy the monkey colony. That guaranteed there would never be evidence sufficient in the opinion of the Defense Department to require a change in its microwave standard.

There followed a drumbeat of articles in scientific journals that directly attacked Zaret and supported the microwave standard that his work threatened. The public face of the effort was Colonel Budd Appleton, the head of the Ophthalmology Department at Walter Reed Hospital, who authored several studies that appeared to vindicate the military's myth about microwaves. Anyone who read his studies carefully could see that they were not proper science. However, there was really no motivation for anyone to parse his work; the military didn't want to jeopardize its mission, industry saw only the profit microwaves could yield, not the pain, and the public blindly trusted the experts whose opinions appeared in the press.

Appleton and Tyler were Zaret's chief antagonists, but there were also others who took part in the mission to destroy him. Don Justesen was among the most vicious and effective. I attended a meeting in Richland, Washington, that was held mostly for the edification of contractors who made their living by performing EMF research for the government. Justesen gave an after-dinner speech in which he cruelly exploited a debate he had had with Zaret on a radio program in Kansas City. Justesen spoke to the Richland audience for a few minutes about microwaves, and then asked in a mocking tone, "When did the microwave cover-up begin?" When the laughing died down, he held a tape recorder near the microphone and Zaret's voice said, "It goes back to the project that was set up by the Defense Department to investigate the microwave irradiation of our embassy in Moscow." More laughter.

Justesen then asked, again mockingly, "Are the government and industry allied in keeping from the American public the possible hazards and dangers from microwaves?" and turned on the tape recorder for Zaret to reply, "The answer to that question is yes." Justesen smiled broadly as the audience booed Zaret.

Next Justesen asked, "What does microwave radiation do?" Zaret replied, "The danger lies in the repeated exposure at low levels of microwaves. Each exposure in itself might not be too meaningful. But repeated often enough, it could produce severe damage." More laughter and boos.

Justesen asked, "Do you think the answer to the microwave problem is government action of some kind?" Then he pulled the electronic string on his puppet one more time, making Zaret reply "Yes."

During the time it took for the Defense Department to pound down Zaret, no one in the so-called community of scientists or brotherhood of ophthalmologists spoke against his mistreatment, and the idea that ordinary microwaves could be toxic to the eye or cause other diseases became unfacted. Appleton and those in league with him drowned out Zaret with their chronic bellowing.

Zaret began testifying in civil cases on behalf of civilian workers whose eyes were injured by microwaves produced by microwave ovens. Despite opposing testimony given on behalf of the oven manufacturers by Appleton, who appeared in court in full uniform complete with medals on his breast, Zaret's clients prevailed in the first of these cases.

"How do you decide whether somebody has a cataract that was caused by microwaves?" I asked him.

"First I examine their eyes and look for signs of the cataract. I ask them about the kind of microwave exposure they had, and how long they had it. If I can see the cataract, and if the patient had the exposure, then sometimes I can say that the cataract was probably caused by microwaves," he replied.

"Why couldn't it have been caused by something else?"

"It could. No doctor can ever be certain. But look at the facts. The patient has a cataract, and he has experienced a lot of microwave exposure. Now, something caused the cataract. Something made it happen. Something that, if you took it away, you would take away the cataract. I know that microwaves can cause cataracts. It's possible something else that we don't know about caused the patient's cataract, but that's unlikely because the other things we know can cause cataracts weren't experienced by the patient. So, it's more likely that microwaves were responsible, because we know they can cause a cataract and were extensively present in the case."

But just as Zaret's disarmingly direct logic had begun to make an impact in the courts, he was checkmated. The endgame began when two women visited his office to have their eyes examined. They had worked in a restaurant where they opened and closed the doors of microwave ovens several hundred times a day, but had quit after being diagnosed with cataracts by their ophthalmologist. The women had commenced suit against the oven manufacturer, claiming that they got cataracts from the microwaves that had leaked out of the ovens, and they asked Zaret to testify on their behalf. He examined their eyes but remained undecided about the presence of the signature cataract, the opacity in the tissue behind the lens, so he told them to return in six months for another examination. He heard nothing more about the two women until someone showed him a copy of an affidavit in which they swore that they had been instigated to sue by Zaret, and that he had lied to them and misled them regarding their medical condition. They asked the judge to dismiss their lawsuit because it had no merit.

Zaret had little understanding of why the women returned his efforts on their behalf with such vile lies, but I could see their motivation plain enough. They couldn't sue their employer because the law prohibited lawsuits by employees, so they had sued the oven's manufacturer. It would have been a coup for the women if Zaret had agreed to testify for them, and a disaster for the company. The litigation would have generated adverse publicity and a flood of future suits in which the essential element of proof in each would have been disarmingly direct testimony by the world expert in microwave cataracts that the plaintiff's cataract had probably been caused by the microwaves that leaked from the oven. So the lawyers on both sides made an unholy bargain in which the women bore false witness against Zaret and, in return, the company gave them the money they craved.

The affidavit soon leapt into Tyler's hands and he used it against Zaret like a knife, not as a surgeon who cuts into flesh to preserve life but as a butcher who intends to kill. Tyler circulated copies of the affidavit to any potential litigant he came across, effectively saying, "Look, see what an evil man Zaret is. You will receive honest advice from Colonel Appleton." In the face of this concerted defamation, microwave patients stopped going to see Zaret, and so the old soldier's last tour of duty came to an end.

Despite everything he had been through, Zaret seemed at peace with himself. My most vivid memory of my day in Scarsdale was his serenity and gentle smile when he said goodbye.

Tyler and Koslov had devoted their careers to protecting the nation against external threats of every sort – real, potential, and imagined. They would have used any fact that aided in the completion of their mission, and opposed any fact that interfered. Perhaps fine men to follow in matters of war, but surely not in matters of science.

15
Bone EMFs

The use of one form or another of electricity to make bone grow was studied by many investigators, some primarily motivated by a desire to advance science and others by the possibility of commercial applications. The most prominent investigators were known as the 3 B's – Bassett, Becker, and Brighton. As it turned out, only Andrew Bassett's team succeeded commercially, and he became one of the richest orthopaedic surgeons in history. His intrigues during that process ultimately led to lawsuits in which I took part as an expert witness. I saw that he had abused science, but that he had done more good than harm to his patients.

Bassett and Dr. Becker had grown bone in dogs using electricity delivered from wires, and both had recognized that the phenomenon might be a treatment for patients who suffered from nonunions, the name given by orthopaedic surgeons to the condition that results when a fractured bone fails to heal. Carl Brighton had proved that the treatment worked in patients but Bassett was pessimistic about its practicality. He felt that few patients would accept wires passing into their bones for the months necessary to effect a cure, and would therefore opt for surgical treatment. Brighton, however, worked with single-minded determination to develop a wire-based system for the nonunion market. Bassett resented Brighton's refusal to acknowledge publicly who had discovered the principle that electricity could make bone grow.

"He will fail, and that's what he deserves," I heard Bassett tell Dr. Becker as they discussed Brighton's likelihood of commercial success.

Dr. Becker saw some merit in Brighton's effort, and thought it well-suited to someone who lacked vision but was persistent. Dr. Becker's interest was regeneration. More than once the thought had occurred to me that a practical success in a narrow area like bone nonunions might have been more beneficial in the long run, to us and to science, than an exclusive focus on the lofty goal of regeneration of limbs and nerves. Dr. Becker, however, did not see things that way.

One day Bassett read an account of the use of EMFs to stimulate muscles and nerves. When the field had been applied, the subject's muscles trembled uncontrollably and he said that he felt as if his skin were being pricked with pins. Bassett wondered if EMFs could be made weak enough so that they produced no sensation but still strong enough to make bone cells react as when they were treated with electricity delivered through wires. For a while Bassett had taken no concrete steps to test his idea because he did not know how to build the equipment needed to make EMFs. Then a chance encounter on an airplane with a scientist who worked for a company that manufactured batteries led to an invitation for Bassett to give a seminar on his bone research. The company liked what it heard and offered to provide Bassett with equipment to produce any kind of EMF he desired.

Years earlier he had made electrical measurements on bars of human bone prepared from specimens he got in the operating room. He had clamped the bar at one end and routed wires from its top and bottom surfaces to a voltmeter. When he pushed down on the free end of the bar the voltmeter recorded a voltage spike, and when he released the bar he got a spike in the opposite direction. The spikes were almost never symmetric. Sometimes the first was larger, sometimes the second. He hadn't understood that the pattern displayed on the voltmeter depended on exactly how he bent and released the bone, and on the electrical characteristics of his voltmeter. Instead, he had interpreted the inequality between the spikes as an inherent property of bone. He told the company that he wanted an asymmetric EMF like the one he had measured, which he said he had discovered but God had invented.

A company technician designed a circuit to produce the asymmetric EMF and the company applied for a patent, listing its technician as the inventor. The company told Bassett that the patent application covered only the design of the circuit that created 1P, which was the company's name for his asymmetric EMF. In actuality, the application also covered the method of using asymmetric EMF to grow bone. Shortly thereafter the Johnson & Johnson Corporation, for which Bassett was an orthopaedic consultant, signed an agreement with the battery company to jointly support his research, and he began using "1P" on patients.

He had first applied this EMF to a nine-year-old girl who had been in a wheelchair her entire life because of a congenital nonunion. After a month's treatment she could walk, and x-rays showed the shadowy white image of bone in what had been a black gap in the little girl's tibia. Another success followed, and soon many desperate parents of children who had never walked began going to the mountaintop on West 168th Street in Manhattan to see the doctor who had developed a treatment that could make their children normal. But God's EMF did not grow bone in most of these children. The disappointed parents wrote their final check to Columbia and went elsewhere looking for their miracle.

Bassett had little reason to continue his novel therapy in the face of such capricious results, but the scientist whom he had met on the airplane, Arthur Pilla, theorized that 1P did not work consistently simply because it was too weak. He suggested that it be made stronger, but not so strong the patients felt that their skin was being pricked with pins. Bassett used the new asymmetric signal, designated "10P" because it was about ten times stronger than God's first EMF, and patient after patient got better. He had mentioned this work when I saw him at the meeting at the Barbizon Plaza Hotel, but not in a way that I caught the sense of excitement and hope that he had actually felt at that time.

Johnson & Johnson and the battery company were less certain than Bassett regarding whether the EMF had grown the bone; they thought it might have healed naturally, following good orthopaedic care. The companies insisted on an animal experiment to show unambiguously that 10P could make bone grow, so Bassett and Pilla designed and began an experiment on dogs.

Bassett cut the dog's fibula in half and pinned both parts to the tibia to stabilize the transected fibula. Pilla provided coils to produce the 10P EMF and Bassett attached them to the conjoined bones; the coils were powered by several pounds of batteries housed in a bag strapped to the dog's back. For purposes of comparison, Bassett did the same surgery to the other hind limb but did not connect the coils to the batteries. The original plan was to kill the dogs after a month and measure the strength of the fibulas. Stronger bones in the dogs that received EMFs, they reasoned, would show that the field could make bone grow.

There were problems from the start of the experiment. The dogs attacked the hardware, often destroying it and injuring themselves in the process. They developed infections because, unavoidably, they urinated and defecated in their wounds. From court documents that Bassett and Pilla had been forced to produce during litigation that occurred ten years later, I could plainly see their agony. They needed good results to maintain funding for the clinical studies, but they faced a looming failure in the dog experiment that, from their viewpoint, was irrelevant because they had already shown that their method worked in people. In this situation, Bassett and Pilla succumbed to the Siren song every scientist hears but is sworn to resist. Exactly how it happened I couldn't tell, but I think it was like this. One day they tested the bones from a dog that had become infected and found that the EMF side was weaker, so they eliminated the data, perhaps saying to themselves, "One can't obtain valid data from a sick animal." Then there was a case of infection where the EMF side was stronger, but they did not eliminate that data precisely because it supported the story they wanted to tell. After that it got progressively easier to ignore results that didn't support the right story. In the end, vindication of 10P was the only possible conclusion of the experiment, like inferring that all the men in some city had cheerful dispositions after first expelling all the melancholics.

According to the data that Bassett and Pilla elected to include in their final report, the 10P-treated bones were stronger; that meant, they wrote in *Science*, that the treated bones had healed faster. Bassett and Pilla should have been punished for what they did, but instead they were widely acclaimed. People believed their research was good because it had been done at Columbia and published in *Science*. The experiment they had described could never really be done because dogs treated as were theirs could never yield reliable data, so no one would ever be able to disprove what they said. Nevertheless, things didn't work out the way they had hoped because officials of the two companies that had paid for the research didn't believe what they read in *Science*. They had frequently visited the lab at Columbia, so they had seen the conditions in the kennels and heard from the workers who cared for the beagles. Questions were asked that Bassett and Pilla didn't answer satisfactorily. Doubts led to more doubts and, in the end, both companies withdrew from all involvement with Bassett's EMF research.

Undaunted, Bassett and Pilla made secret plans to continue their work. Pilla quit the battery company and Bassett hired him at Columbia. Unknown to the school's administration, the two men started a company that began making 10P units, which Bassett used to treat patients.

Pilla applied for a patent for the method of using 10P to make bone grow, listing himself as the sole inventor to thwart a claimed financial interest by Columbia that they anticipated would arise if Bassett's role were recognized. But the partners had forgotten about the battery company's patent application involving 1P. To the dismay of the two men, the patent office granted a patent to the company but denied Pilla's application because it was too much like the 1P patent.

The resourceful Pilla learned that he might get a patent if he claimed an EMF that was entirely different from 1P. So he crafted an EMF of such extraordinary complexity that it could not have been arrived at by any possible variation of God's signal. Their company built equipment to generate the new EMF, and Bassett soon discovered that it also grew bone in his patients. This time Pilla got his patent, based on an affidavit from Bassett who swore that the EMF was better for treating patients than 1P or 10P.

After Dr. Becker borrowed a 10P unit from Bassett and used it successfully on his patients, he did not question the clinical efficacy of 10P. But he believed the *routine* use of EMFs for treating patients with nonunions was too risky because the body's ability to respond to EMFs was an inextricable part of what living organisms were, and that Bassett was blindly tinkering with primordial forces. On several occasions Dr. Becker said to me, "How do they know they're not going to start a cancer?" He pointed to the controversies about the health risks from the Navy antenna and from powerlines, which he felt were ample evidence to suspect that bone EMFs might do harm as well as good. He frequently urged Bassett to restrict the use of EMFs to patients who had failed surgical treatment, and not to treat children at all because their cells grew so rapidly that there was an increased chance of initiating cancer. Bassett disagreed, and accused Dr. Becker of not understanding biophysics; the bone EMF was so precise, he claimed, that only bone cells could respond to it.

Bassett and Pilla's business fortunes improved dramatically after someone who had been dating Bassett's daughter told a friend about Bassett's research. The friend told someone else, and ultimately a group of venture capitalists acquired a controlling interest in the company in return for their large cash investment, business skill, and Wall Street connections. Thus matured and richer, the company built a factory and began large-scale manufacturing of equipment that produced Pilla's EMF, and he and Bassett worked hard to popularize it.

The company operated through Bassett's office at Columbia. The business plan provided that he would not hold an official job with the company "in the interests of preserving the sanctity of the University positions and titles that are so important to the corporation." Thousands of letters went out on Columbia letterhead in which Professor Bassett recommended use of equipment made by a New Jersey company, but his connection with it was left unstated. Like Dr. Becker, many orthopaedic surgeons found that the equipment could cure patients.

Bassett and Pilla also gave equipment to many laboratory investigators who then published papers dealing with effects of the EMF on the metabolism of bone cells. The papers popularized the notion that, in some mysterious way nobody but Pilla knew anything about, his EMF had the power to affect the basic machinery of cells. As similar reports continued to appear, Pilla came to really believe that the extraordinary complexity of his EMF was responsible for the metabolic changes that occurred in the cells. His theory was that his EMF had been successful because it was a code that only bone cells could decipher, and that it told them, "Build bone." Then he decided there must be other codes for other cells. Perhaps a signal for cancer cells that said, "Stop growing," and another that said to severed spinal cords, "Attach." Unfettered by any insight into the complexity of cells, he convinced himself he understood them, and he worked diligently to capitalize on his vision.

Meanwhile, Brighton continued his efforts to develop a marketable device for treating bone nonunions that was based on electricity delivered by means of wires. He performed numerous animal experiments and always interpreted the results in terms of an effect of electricity on oxygen levels, thereby reducing the mode of action of electricity to the prosaic level of a drug rather than something closer to the nature of what life was, as in Dr. Becker's conception. Brighton's company did clinical studies and, like Bassett, found that four out of every five patients got better.

Brighton started a scientific society devoted to bioelectricity. At its annual meetings he and Bassett argued about the relative merits of their respective systems, and about the scientific principles that they imagined could explain their observations. Dr. Becker rarely attended. He was largely uninterested in their science and socially uncomfortable with them, a cowboy among the ivy-leaguers. He also did human studies, the nature of which sharply distinguished him from the other two B's. His research was progressively innovative, labor intensive, and focused on the seriously ill.

He began his human studies after an investigator in our laboratory named Joe Spadaro had discovered that atoms of silver could kill bacteria, like an antibiotic. Spadaro had placed pieces of silver wire in concentrated bacterial cultures and observed that the bacteria in the zone adjacent to the wire had died because, he soon realized, the silver atoms that had dissolved from the wire were bacteriocidal. He then discovered that he could accelerate the rate of dissolution and drive the silver atoms further from the wire by passing electricity through it, thereby producing the zone of inhibition more quickly and extending it further.

Dr. Becker's sickest patients were those with infected nonunions in which the bacteria had burrowed into the nooks and crannies of the bone and destroyed the cells. The standard treatment was to surgically remove the dead bone, but bacteria that had escaped the debridement would often recolonize live bone leading to more surgery and ultimately to an open gaping wound that constantly drained pus. Amputation was the only alternative for these patients.

Dr. Becker placed a silver wire in the bone hole of one of these patients and routed the wire to a small black box that housed the necessary circuitry. At first he made the silver wire electrically positive to drive off the silver that would kill the bacteria. Then he switched the polarity to negative to produce the current that would stimulate bone cells. Every day he adjusted the position and polarity of the wires, and the nurses cleaned the wound and changed the battery in what they called "Becker's box." Healing granulations soon appeared, and the wound began to close. Finally the day came when the hole was gone, and for the first time in years the patient could walk without draining pus. Dr. Becker treated other patients and most of them also healed.

My work with bone went off in a different direction. When Pilla had first published the description of his EMF and claimed that it was highly specific for triggering bone cells, I was amazed by what appeared to be a great discovery. But the more I studied his publications the less sense his theories and assertions made to me. Dr. Becker did not pretend to understand the equations Pilla used to explain himself, but just trusted the fast-talking Pilla who, it turned out, had graduated from St. Joseph's University two years before I had begun studying there. He spoke French and had earned his Ph.D. at the Sorbonne. Dr. Becker was more relaxed whenever Pilla visited us, sometimes even smiled, and always listened carefully to whatever Pilla had to say. For a while I worried that Dr. Becker intended to replace me in the lab with Pilla.

I decided to find out for myself whether there was anything special about Pilla's signal. He had given me equipment that generated his signal, and when I applied it to cells growing in plastic dishes I found that their metabolism differed from that of unexposed cells, as Pilla had said. But when I modified the equipment so that the EMF was less complex than the original signal, the cells reacted metabolically as they did to Pilla's signal. Every reduction in complexity had the same consequence, and when I reached the EMF whose form was irreducibly simple – a sinusoid – the cells still responded exactly as when I had used his so-called special signal. In other experiments I also found that cells reacted similarly to EMFs and current delivered from wires. As far as the cells were concerned, all forms of electrical energy were more or less the same.

When I read some old journals stored in the basement of the medical library, I saw that the effects of stimuli on bone cells were even more nonspecific than I had imagined. I found descriptions of experiments in which iron rods had been inserted into the legs of dogs. The iron decomposed over time, releasing particles that caused a massive production of new bone. Other experiments suggested that virtually any material that decomposed in the body could elicit a response from the bone cells, unlike platinum or glass, which did not decompose and did not elicit a response. Still other experiments showed that acid, chronic pressure, and various chemicals could all stimulate bone growth. I also learned that the experiment Bassett and Dr. Becker had done in 1961 in which they grew bone in dogs using electricity had been done in 1858.

Much became clear to me after I had performed my experiments using Pilla's equipment and read the reports in the old journals. Bone is an irritable tissue that grows in response to any stimulus that is not too small to be recognized nor too large as to cause bone's immediate destruction. The dog experiment Dr. Becker had done with Bassett had forked the path to the enchanted land Dr. Becker had envisioned in his purest moments at the inception of his career. One path led to the use of EMFs as a tool to grow bone, perhaps more convenient or salutary in particular circumstances than another tool, but still just a tool. And as any hammer could drive a nail, any EMF within wide bounds could grow bone. Bassett and then Brighton took that path. I joined Dr. Becker on the other path, and after a while he and I came to a fork. He continued toward his dream of regeneration, and I turned to the study of the side-effects of EMFs – the harm humans do to when they ignore their essentially electrical nature and subject themselves to electromagnetic fields.

From my point of view, matters regarding electricity and bone came to a head after Brighton organized an international scientific meeting on that topic and invited me to give talks about why I thought electricity made bone grow, and about any risks I perceived in such treatments. I did as he asked, and also took the opportunity to directly tell him, Pilla, and Bassett what was in my heart about their work.

First was Brighton. He had erred the same way in each experiment, always claiming that his results had only one meaning, that electricity affected oxygen levels which in turn stimulated bone cells. In reality, however, his endless studies never explained why or how the electricity he delivered to bones of his patients through the steel electrodes he drilled in through the skin had the power to heal. I said that to him, perhaps awkwardly because he always seemed to be thinking about something else, so he was a difficult man to engage in conversation. Not unexpectedly, my words had no observable impact and elicited no substantive reply.

When I spoke to Pilla I said, more or less: "Your signal is not special in any discernible way. Many keys can turn the same lock. Even if you believed the opposite, it still seems to me far better to avoid speaking so forcefully about the reasons for the clinical success of your EMF until you find convincing evidence for your theory. Otherwise you could ruin the whole field of bioelectricity."

The gist of his answer was, "I'm not like Bob Becker. It's not my goal to be memorialized after my death as a great scientist. All that really counts is how well one provides for his family. I don't expect a second chance at the kind of success I can now foresee, and so my plan is to capitalize on my opportunity. That is the path I'll take."

To Bassett I said, "EMFs cause rats to make hormones, thereby showing that the field is a stressor. EMFs also affect the brain, and what it controls. The possibility that your EMF has side-effects of these types should be conceded when you use it to treat patients, and balanced against its potential benefits."

He said: "All our evidence indicates that our signal interacts with the body only because of its special properties. You should not lump all electromagnetic phenomena into the same category and denigrate the lot. You have a one-sided view."

When the talks and disputations were finished we all went to dinner.

The restaurant rotated slowly many floors above Atlanta, and the activity that we saw in the streets far below changed continuously in unpredictable ways for unknowable reasons.

Soon after the meeting I learned that Bassett and Pilla's company had applied for a license to sell their device as a treatment for nonunions. I warned the Food and Drug Administration that the risk of EMFs remained unevaluated, and that although Pilla's theory promised much, it delivered nothing. But the official in charge of the application issued the license; he knew Bassett from the time he had chaired the device section's advisory panel.

The businessmen who ran the company made plans to take it public. At that delicate moment, when all the investors were so close to Fortune they could see her smiling face, the undisciplined Pilla announced publicly that his technology could do more than grow bone, it could cure cancer. The underwriters of the public offering thereupon withdrew it, fearing charges that Pilla's claim had been made solely to drive up the price of the stock. After a few months, however, the fears of the underwriters receded and the stock offering proceeded. It was fully subscribed within a few days, making multimillionaires of Bassett and Pilla.

Brighton's company also received a license from the Food and Drug Administration. But the method of wires drilled into the bone was not accepted by patients, and his company was squashed in the marketplace by Bassett and Pilla's company. The defeat slowed Brighton for a while, but then he developed a system using wires that did not require them to be drilled into the bone but only to be glued on the skin. He hoped that the electricity passing from one side to the other of a limb with a nonunion would find the dormant bone cells and awaken them. Although there was no good reason for that to happen, it did, and four of every five patients he treated got better. Even so, Bassett's prediction that patients would not tolerate wires came true a second time, and he again defeated Brighton in the marketplace.

As though in an effort to escape his fate, Brighton went over to Bassett's ideological camp and adopted the philosophy of EMFs, which was his third system for treating nonunions. Brighton hoped to exploit the simplest of all EMFs, which both the battery company and Pilla had failed to claim as their own. But this time Brighton couldn't generate enthusiasm among businessmen for the invention of a therapy that had already been invented and was being profitably practiced by the most successful small business in the United States. So he finally retreated from the bone wars. He had achieved the chairmanship of the Department of Orthopaedic Surgery at the University of Pennsylvania, and was respected for his many contributions to science. He educated many men who became first-rank orthopaedic surgeons, and his research added respectability to the study of bioelectrical phenomena, which had been discredited in preceding generations. But it seemed there was something in his blood for which he had to be punished, so he was denied what he most wanted, commercial success with his electrical devices.

Bassett's company sailed along on a calm sea beneath a bright sun until, one day, the battery company that owned God's EMF appeared and claimed that Pilla's signal infringed its patent. It asked the court for three times the profits that Bassett and Pilla's company had earned, as provided by law. At first, their company defended against the charge of infringement by claiming the validity of Pilla's patent. But the company had to change its tune after it learned of an agreement Pilla had signed with his former employer in which he had promised that any invention he made within six months after quitting would be owned by the battery company, and the evidence showed that Pilla had created his extraordinarily complex EMF within the period of time that he owed the obligation. Eventually the lawsuit was settled, although I never knew how much Bassett and Pilla's company paid the battery company. Nevertheless they remained rich because the settlement didn't affect the value of their company's stock. However, the lawsuit did impact Columbia's opinion of Bassett; when it learned it had been cuckolded, his path to the chairmanship of the Department of Orthopaedic Surgery was forever blocked. He retired and spent his time giving speeches aimed at rehabilitating his tarnished image.

Pilla continued trying to cure cancer, but failed. It eventually became clear to almost everybody who knew anything about EMFs that his theory was pure bombast, and the attention focused on him evaporated. The best I can say for him is that he provided well for his family.

Dr. Becker continued with his silver treatment, but few other physicians cared to lavish the efforts required for its success. The market for a device for infected nonunions was too small to attract the interest of a big company, and the necessity of confronting the Food and Drug Administration was too daunting for him to even consider.

Bassett and Pilla's company sought to identify orthopaedic diseases in addition to nonunions for which its EMF was a cure. The patients who submitted to the clinical trials had no clear idea that the possibility of success was no more than a hope, undisciplined by science and guided only by guesses. They had placed their trust in the men in the white coats and in the Food and Drug Administration, which knew much about political intrigue but little about the science of EMFs. There were no reasons to expect success in the clinical trials, contrary to what was believed by the patients who participated, and not surprisingly all the trials failed.

In the end, the fate of EMFs in orthopaedics was like that of unfortunate Elpenor, who lay at the foot of Circe's castle, neither alive nor dead.

16
Turning Point

━━━◆━━━

I saw strange and provocative photographs in various medical journals; the images were called "auras" or "energy fields," and they appeared as numerous thin colorful lines that radiated from the subjects, which included people, animals, and plants. Some auras were said to have been caused by drugs, others by sexual arousal, or disease. The most fantastic photographs were of leaves where the top had been cut away but its energy field could still be seen, more faint than that of the part of the leaf that remained but equally as complex.

Some commentators said that the auras captured the nature of life while others regarded them as nothing more than trick photography, but nobody really knew, one way or the other. If auras did reveal the energy of life and I could prove it, I thought I would become famous.

An aura camera was a sheet of photographic film on a metal plate connected to a high voltage. A part of the body or a plant leaf placed on the film would give off light that was too weak for the eye but could be captured by the film. I built an apparatus that operated at 100,000 volts and began recording the complex, beautiful images, no two of which were ever exactly alike.

My goal was to reproduce what had come to be called the "phantom leaf" effect, which I considered to be the clearest evidence that auras really were biological energy fields. But whenever I removed part of a leaf and photographed what remained, I saw only the aura of the remaining leaf, never that of the missing part. One day, just after I had placed a leaf on a sheet of film in preparation for a photograph, I was called away. When I returned I saw that the leaf had dried so I replaced it with a freshly picked leaf, which I photographed. When the film was developed I saw the image of the second leaf, as expected, but also a weak image of the first leaf.

I soon realized that a leaf placed on photographic film spontaneously deposited numerous tiny water drops, each of which gave off auroral light when the high voltage was applied, and that the pattern of the drops was determined by the distribution of pores in the leaf. Consequently the image on the film resembled the leaf. After that realization, I could create a "phantom leaf" by placing a leaf on the film for a few minutes, picking it up and cutting off a piece, and then returning what remained to its original position. A photograph would then reveal an aura of the leaf and a weaker "phantom" of the discarded part, thanks to the effect of the voltage on the pattern of droplets on the film.

I embalmed leaves in formaldehyde and found that the auras were identical to those I took when the leaves were fresh. I found the same result with animals; the aura of a salamander was the same after it was dead as when it was alive. Everything pointed to the conclusion that the auras had nothing to do with biological energy fields, and that's what I said in a manuscript that I sent to *Science*. The editor, however, declined to publish it; he said, "Photographing auras is one of the many crackpot methods generated by a lack of understanding of natural phenomena and a bent toward mysticism. Publication of the paper may unintentionally, because of *Science's* prestige, give a certain 'aura' of scientific importance to the 'technique.' Ignoring the auras is the avenue to be followed."

Soon thereafter I spoke with an investigator at Drexel University in Philadelphia who told me he had finished work on a contract with the Defense Department to explore possible military applications of auras. It turned out his research had been identical to mine and he had reached the same conclusion as I. He told me of his plan to send a manuscript to *Science*. When I explained why there was no chance *Science* would publish it, he said: "I received $300,000 from the Defense Department; that amount of funding should state a case for the scientific importance of the phenomenon."

"But what about the editor, who thinks that nothing about auras is worthy to appear in *Science?*" I asked.

"We know somebody there," he said. Shortly thereafter his article appeared in *Science*.

The incident disoriented me for a while. I had no idea that favoritism of that kind occurred at *Science*. Every time I thought I understood how science worked, something happened that made me realize it was more complex than I had imagined. Still, I had learned what I had set out to learn, and my regret at being denied publication in *Science* was mitigated by the negative character of my experimental result, because there was little glory in telling the world that the wonderful thing I sought was not real.

Dr. Becker was deeply disturbed by the journal's intellectual dishonesty.

He sent a letter to the editor complaining about his unethical behavior. The editor dismissed the complaint, calling it "slanderous."

I had continued to do laboratory studies of EMFs, but the funds that Dr. Becker generated were becoming progressively scarcer. Then something happened that led me to begin a kind of research that required little money and no laboratory, and yet was capable of producing results of great public interest. The story began when I was asked for help by a physician named Stephen Perry, who practiced in the Midlands section of England. He had noticed an increase in the number of his patients who suffered from clinical depression, some of whom had committed suicide. When he mapped the addresses of his patients, he found that they often lived near high-voltage powerlines.

I knew that disease was sometimes studied by analyzing questionnaires, lung cancer for example. Investigators had asked each subject, "Do you smoke?" and "Do you have cancer?" and to a 95% certainty the percentage who had cancer was higher in the group that smoked. In the language of statistics, that meant smoking and cancer were "associated," and the kind of "association" that leaped into most minds was cause-and-effect.

Looking for "associations" struck me as exactly how ordinary people dealt with the world. When two events were conjoined, the habit of the mind was to regard the earlier as a cause of the latter. Clearly "association" didn't always entail causality – no two things were ever more associated than day and night, yet no one believes that one causes the other. Even so, the association between smoking and cancer was accepted by most people as causal, and it occurred to me that the questionnaire method might also allow me to find a link between EMFs and disease – for example, depression.

The information we needed for a questionnaire study of powerlines and suicide in the English Midlands was available from public records. We formed a comparison group consisting of people who had no known mental illness, and then measured the EMF at the front door of each subject's home. From a statistical analysis of the data I found that people who had committed suicide were more likely to have lived where the EMF was higher. The kind of association I suspected was cause-and-effect.

After we published the results, reporters from television and the print media showed a lot of interest in the health hazards of powerlines. The link between EMFs and suicide that the questionnaire method revealed had resonated with ordinary people. Equally important, the study had been cheap and easy to perform.

For my next questionnaire study, I planned to plot the addresses of everyone in New York state who had died from cancer or heart disease and examine whether an unusually high number of them had lived near powerlines. There were so many homes that I couldn't actually measure the EMF at each one, but I could calculate the fields if the power companies gave me pertinent details about the operating characteristics of the powerlines. The companies, however, refused to do so, and the Public Service Commission turned down my request to order them to give me the information I needed. "It's not politically possible," I was told.

By that time, however, my taste for the method had soured because of the education I had received from a man who had devoted his whole career to questionnaire studies. "When you pick your control addresses," he said to me, "they could have differed from the suicide addresses in many different ways besides the EMF, and any one of these differences could have been responsible for the suicides. The best you can say is that the EMF *could* have been connected with the suicides. The best anyone who does a questionnaire study can *ever* say is that the factor he singled out to study *could* be connected with the disease he chose to study. It is illogical to assert anything stronger." Instantly I recognized what he meant, like saying Joe was a fish simply because both he and fish could swim.

While all this was happening, a psychiatrist named Murray Cowen was measuring electrical signals in human subjects, much as I had done in goldfish and Dr. Becker had done in salamanders. Cowen was one of about a half dozen staff physicians at the Veterans Administration Hospital in Syracuse who were protégés of Dr. Becker in the sense that they tried to follow his lead and study bioelectrical phenomena in the context of their specialties. Cowen made his measurements using electrodes attached to the forehead and back of the head. He believed that the signal was rich in meaning about psychological and physiological events in the brain, if only he could decipher the signal's code.

One day he measured my signal. When he projected a triangular arrangement of black lines on a white background, my signal changed slightly as I formed the percept, and then resumed gently changing with time, guided by some mysterious force that prevented it from ever repeating itself. When he projected a more complicated image my signal woke up for a few moments, as my brain did whatever calculations brains do when they piece together parts of a stimulus and present the mind with a picture, and then settled somnolently.

Cowen found similar results in other clinically normal subjects, but the response patterns of psychotic patients differed markedly. He hoped his method of measuring the signals might put diagnosis of some diseases on a firmer footing, and that application of signals to patients might be therapeutic. He had sought the money he needed for his research from the National Institutes of Health, but with bitterly disappointing results. "A committee of radiologists reviews all requests to study bioelectrical phenomena, and time after time they turned me down. All they care about is x-rays."

I suggested that a turnover in the membership of the committee might result in an improvement in its attitude toward EMF studies, but he said, "They always appoint people who think as they do. Nothing changes. The gene stock is frozen." As I was leaving his lab he said, "When you apply for a grant to study EMFs, you will see what I mean."

Soon after that I sent an application to the National Institutes of Health in which I asked for money to study whether powerline EMFs would affect the growth of mice or produce signs of stress in them, but, as Cowen had predicted, the radiology study section refused to fund my research. The section chairman told me that the National Institutes of Health funded only basic science, and that the study of disease caused by EMFs was applied research and should be done by power companies.

I called the Environmental Protection Agency, hoping that the people there would be interested in the kind of research I proposed, and that they understood the futility of relying on the power companies to assess the issue honestly. I was told to reapply to the National Institutes of Health, and that money for my research would be passed by the Agency through the Institutes to me. When I reapplied, the radiologists expressed surprise at my chutzpah and had nothing good to say about my proposed work. Nevertheless I received the pass-through money as I had been told would happen.

I performed the mice experiments and found that the EMF affected their growth and produced signs of stress, as I had hypothesized. That added evidence to my indictment of powerline EMFs as a malevolent force that could also cause stress in people, sending some of them to an early grave without leaving any fingerprints. Other investigators also published results showing that powerline EMFs could cause stress. This led to articles in news magazines and other publications which added to the rising public profile of the EMF issue. Even my 86-year-old grandmother had heard about it. "Andrew, I read an article about electricity causing cancer," she remarked in a tone that suggested she was proud of me, "like you said." I believed I was making progress in my battle with the power companies, especially after John Mattill, editor of MIT's *Technology Review*, asked me to write an article about EMF health hazards; it had been his invitation to Dr. Becker in 1972 to publish an article that had led to the first public warning that EMFs could be a health risk.

Then, everything went bad very quickly. Karen Ray, who had worked for the hard-hearted Chauncey Starr at the Electric Power Research Institute, left that job and went to work for Mattill. Soon thereafter, while I thought Mattill was reviewing my corrections to the proof of the article that he had asked me to write, he called and told me that he had withdrawn it from publication. Then, in another unexpected development, Barry Commoner, the editor of *Environment*, published an article by Morton Miller that disclaimed all risks from powerlines. "That one slipped through," is what the chastened Commoner said to me about the article written by the man whose guiding principle of science was "He who has the gold makes the rules." Commoner eventually published the article that I had written for Mattill, but the damage was done because the premier journal in the field of environmental science had pronounced high-voltage powerlines safe.

When I applied for renewal of my grant, the National Institutes of Health told me that not only would I not receive any money, no one who proposed to study the propensity of EMFs to cause disease would *ever* receive federal money. That's what happened. It would be fifteen years before political changes resulted in a novel law that specifically required the National Institutes of Health to determine whether powerlines caused cancer.

Meanwhile, the State University of New York Press asked Dr. Becker to write a book on EMFs. He invited me to be a coauthor and I quickly accepted. I saw the book as a way to organize the area of EMF biology, and to describe his seminal contributions so that he would get the credit he deserved. I proposed a two-part structure in which we would first discuss the role of natural bioelectricity, and then describe the side-effects of the man-made EMFs in the environment. That was the way he had always approached the subject and, as I expected, he readily agreed.

For each part, I suggested we first present our experimental evidence and explain its meaning, and then analyze the studies and arguments of those who reached different conclusions and show why they were wrong. Dr. Becker emphatically disagreed with this plan. He said his work in the early 1960s about electrons moving in nerves, electrical properties of bone, and bioelectric fields controlling growth had all been published in good journals and had never been disproved, so he saw no reason to do anything other than recount those results in the book. I thought that approach was weak because the reader would rightly wonder what had occurred during the ensuing two decades, and perhaps decide that the original work had been stillborn. The situation regarding EMF side-effects made an even more compelling case for including the work of others. Although he had been the first to sound the alarm, many investigators had joined that chorus and there were cogent reasons, scientific and otherwise, why their contributions needed to be discussed, not the least of which was that of avoiding the appearance of us being a two-man band.

Weeks went by and it began to look as if the book project was dead. Then I made another suggestion. "Suppose you write the part about natural bioelectricity the way you like, and sign your name to those chapters. I'll write the side-effects part and sign it, and you will be first author on the book." "That would be appropriate," he said.

I found many articles on the biological effects of EMFs. Some dealt with powerlines, others with microwaves. Almost all of the microwave studies were from the Soviet Union. As I sifted through the evidence I noticed that the type of biological effect didn't depend on the frequency of the EMF. One report described a change in heart-rate in rats caused by a low-frequency field like that from powerlines, and another report described the same results but caused by a high-frequency field like that used in radar. I found similar pairs of reports describing effects on neurotransmitters in mice, brain electrical activity in rabbits, and on many other biological endpoints. The time came when I realized that was a universal pattern. There were no uniquely important frequencies, signature biological changes, or tissues that were especially vulnerable to EMFs. Everything that could be measured was potentially susceptible to EMFs of essentially any frequency. EMFs, like changes in temperature, were somatic factors capable of producing changes in *any* biological endpoint.

I wrote separate chapters on the effects of EMFs on the nervous system, the endocrine system, the cardiovascular and hematological systems, metabolism, growth, healing, and mutagenesis. The chapters all said essentially the same thing: "EMFs affect the body."

I had to decide how to treat the overwhelmingly negative research sponsored by the Electric Power Research Institute, Department of Energy, and the companies whose products produced EMFs. Had I set out to confront each report, the book would not have been credible because few would believe so much expensive research was worthless. Instead, I imagined a pointillist painting of a white rose. It was defined by a small number of white points that told where the rose was and an infinite number of dark points that told where it wasn't. The misleading science could be ignored because it conveyed essentially no information, like the infinity of dark points.

Sometimes a negative study had the character of a white point. Dick Phillips's multigeneration mouse study was such a case. He had built two EMF exposure systems that were as alike as human beings could make them. Yet when he performed the same experiment in each apparatus, the EMF-exposed mice were smaller than their controls in one experiment but larger in the replicate. He called the results contradictory and extracted the meaning that he desired by averaging the results of the two experiments and concluding that the EMF had no effect on the mice. But the results weren't contradictory, they were *opposite*.

Suppose an unrecognized factor had modulated the effect of the EMFs. Perhaps the temperature in the two different exposure rooms differed by a tiny amount, and it was *that* difference which determined the direction of the effect of EMFs on growth. The idea came to me after I heard a talk entitled "Does the Flap of a Butterfly's Wings in Brazil Set Off a Tornado in Texas?" According to the speaker, Edward Lorenz, the weather was so complex that even an infinitesimally small temperature change could lead to large, unpredictable consequences. I thought that suggestion, which later came to be called the theory of deterministic chaos, made more sense than averaging away two positive results from independent experiments, so I adopted a de facto chaos theory as an alternative to what Phillips had done.

Where and how the body detected EMFs were mysteries. Nevertheless, in every instance in which EMFs impacted a human being, it had to be true that there was first a time when the EMF and the body were spatially and temporally distinct. Then the EMF moved closer and closer to some electron somewhere in the person's body. Finally, the EMF embraced the electron. When that occurred its motion suddenly changed; that change altered the motion of the atom to which the electron belonged, which in turn altered the motion of the next atom, and so forth. In this way, the EMF was transduced into a signal in the body, as manifested by changed motion of the body's atoms. Although I knew nothing about the details of the process, I knew that it *must* occur *somewhere* in the body.

To provide a story about why the process leads to disease, I developed Dr. Becker's idea about stress. A person confronts many different stressors, each of which taxes the body's adaptive capacity or, by another name, the body's ability to resist disease. When this resistance is overcome, the result is some form of disease. I argued that the kind of disease produced by EMFs depended not only on the field but also on other factors in the environment, and on one's unique physiology. EMF exposure led to disease because it was a stressor that extracted a physiological tax. The image I had in mind was that of inflating a balloon, which corresponded to adding more stress. Popping the balloon corresponded to getting sick.

When I had begun my study of EMFs I would have been embarrassed by such an explanation – a myth instead of a mechanism! But I had learned that unreasoning insistence on knowledge of a mechanism was the perspective of someone who did not understand the complexity of living things or the nature of biological knowledge, and therefore thought that "health risk" came about by means of determinable gears and pulleys, and that it had a firm, objective meaning.

There was no biological truth of that type. The answer to whether environmental EMFs were health risks depends on one's perspective, as well as on the results of experiments. I don't mean to say there is a fixed something regarding which there can be different perspectives, as suggested by that visual metaphor. There is *only* experience, which is just an interpretation of something. The "truth" about health risks of EMFs is what emerges when one looks at the evidence from within one's own limitations and sense of values. There is no way to get to the bottom of the matter because there is no bottom.

In an important sense I was at the end of my study of EMFs because I realized that the reliability of the knowledge about what EMFs can do to people depended *not solely* on the intrinsic "truth" of that knowledge, but on the power of the proponent of the perspective. One could hope to assess which opinion was more warranted by the evidence, but not which was "true."

Dr. Becker approved each of the eight chapters I wrote but neither asked for nor received any comment from me regarding his three chapters. We both directly challenged the concept of scientific knowledge we had been taught, but in quite different ways. My chapters mostly dealt with the works of others, and were illustrated with tables and figures containing data. His chapters focused almost exclusively on his work and were illustrated using drawings he made.

The time was fast approaching for us to part company. Me to some uncertain future where I would have to stand on my own, without his leadership, protection, or support. He to some melancholy limbo to pursue his art and brood over the ways of the world. I wondered where I would go, and what my future would be. I hoped it would be something good.

17
Southbound

The head of the Veterans Administration was Max Cleland. He had lost most of his limbs in Vietnam so I had hoped he would take an active interest in Dr. Becker's work on limb regeneration, and insure that it was well-funded and protected from his enemies. But Cleland paid no special attention to what Dr. Becker was trying to accomplish, so Marguerite Hays had no trouble dissipating the funds that had been obtained for his use by Senator Cranston. Even worse, only months before the renewal application for the main grant that funded our laboratory was due at Central Office, Cleland appointed Hays as the head of all research at the Veterans Administration. Because of that, the path around her that Dr. Becker had previously found for funding our lab no longer existed. One of the first things she did after her appointment was visit our laboratory to personally evaluate our science.

She was a diminutive woman with skin that was yellow-white, like the hair of my rats. Her lips were abnormally thin, as if she had chewed most of the flesh off, and her top jaw protruded forward, which made her bucktoothed, gleaming upper teeth the prominent feature. She had been accompanied from Washington by an obese woman with a crewcut. The two frequently exchanged glances and touched one another.

Hays listened as Dr. Becker presented results of our research and described the experiments that we hoped to do. When he finished she said of his work on bone stimulation, "Others are far more advanced," which wasn't true because nobody else was even studying infected nonunions. Regarding limb regeneration she said, "You saw at the Bermuda conference how little other investigators think of your work," which was true because she had invited only those who despised it. When I presented my research on EMF effects she said: "Years ago you were given permission to appear for a single day in a public forum, and you expanded that far beyond one day, resulting in controversy and conflict with other federal agencies."

At lunch I had no appetite because I saw the looming end of our laboratory. Amazingly, most of my co-workers seemed to enjoy the meal, as if they were at a summer picnic. They engaged in small talk with Hays, who responded with some graciousness. Even Dr. Becker spoke to her with civility, perhaps not as he would to a friend but neither as to an enemy. I did not speak to her. But I watched her eat. When she opened her mouth it seemed all teeth. She took small bites. As she chewed, the strange conformation of her mouth made visible the action of her tongue as it stirred the bread and meat and saliva to create the paste that she swallowed. The biting and chewing and swallowing happened rapidly, as if she needed to keep up with the others, but the unseemly speed propelled bits of meat and bread and lettuce out of her mouth in small streams of spittle more or less continuously.

After lunch Dr. Becker resumed talking about his work with silver, saying that his method was effective and that even better results could be expected. Hays, however, dismissed the work as "inefficient," and said that the effort and time required for silver therapy was too great, and that amputation of the limbs would conserve scarce government resources. It was hopeless. We were like a doomed crew on a ship that has been sunk in an uneven battle.

Soon after Hays returned to Washington, DC, she ordered our laboratory to close. No one in the lab except Dr. Becker and I had federal job-tenure rights, so they were all just given final notice by the hospital personnel department. Dr. Becker was offered a job as an admitting physician, a position normally filled by foreigners unable to obtain a medical license in the United States. I was offered a job as a trainee, in preparation to become an assistant hospital director, which was the only available position at the hospital for which I was technically qualified by virtue of my education.

Dr. Becker had no more energy to fight. One evening, he simply turned off the lights and walked out of his office, ending his twenty-two-year effort. He left behind his library and files, and the plaque he had received honoring him as the foremost scientific investigator in the Veterans Administration. He also left his portrait of George Washington and the motto that had hung on his office wall as long as I could remember that said, "The inmates have taken over the asylum." The headline in the newspaper said: "International Authority in Regeneration Quits Battle for Funds." He went to live in his beautiful house in Lowville, New York, where he could hunt and fish in peace. He was only 56 years old.

Workmen soon began putting our laboratory equipment in boxes for shipment elsewhere. No one wanted the electron spin resonance spectrometer so it was junked, except for a few parts that I saved as souvenirs. The expensive microwave cavities that were the heart of the spectrometer were placed in a box labeled "assorted," along with screwdrivers, resistors, capacitors, electrical tape, and two rolls of wire, and shipped to a Veterans Administration hospital in New York City. Our high-vacuum system was sent to Virginia and our spectrograph and densitometer went to a laboratory in Washington, DC, along with the muffle furnace that I had used to heat pieces of bone, either to drive off the water or to burn off the protein. Platinum crucibles that had held the samples in the furnace were labeled "tin cups" by the workmen and, along with our voltmeters, ammeters, electrometers, and oscilloscopes, were sent to other laboratories. Our electromagnet went to a scrap dealer for extraction of its valuable copper. The dumpster behind the hospital was the final repository for the beautiful Wheatstone bridge that I had used to measure the dielectric constant of bone, and for the apparatus that I had built to apply EMFs to mice and rats. Investigators at the hospital picked through our remaining equipment.

One day, as our laboratory was rapidly disappearing, leaving me with the huge problem of trying to determine what kind of a career working in Dr. Becker's lab for seventeen years had fitted me for, I was approached by a visitor. "I'm Frank Anders," he said in a southern drawl. "I'd like to talk to you about Andy Bassett." Anders identified himself as an orthopaedic resident at Louisiana State University Medical School in Shreveport, Louisiana, and then told me an interesting story.

He had been working on a project for the National Aeronautics and Space Administration that was aimed at preventing the loss of calcium from the bones of the astronauts. Bassett had claimed that he could solve the problem by means of a special EMF, and NASA had awarded him a contract to test his theory on rats. But the space agency had judged the interim study results to be suspicious. In one instance when data from a crucial experiment was due, Bassett sent only a letter that began, "Disaster has struck," and went on to say that a power failure at Columbia the night before the test animals were to be sacrificed had ruined the experiment. Anders was visiting EMF labs to find out whether anyone thought Bassett could ever succeed.

We talked all day about Bassett, whose honesty I held in low regard, although nowhere near as low as I would years later after I learned more about his activities. As Anders and I spoke, we warmed to each other. That night, at dinner, he said, "Andy, why don't you come and work in Louisiana?"

"For who?" I asked.

"Jim Albright," he said, "the chairman of orthopaedics at the medical school."

"Frank," I said, "I have the reputation of being sort of...controversial. What would Albright think about that?" Frank just laughed and told me a story that, I saw later, captured the ineffable character of the remarkable man who was to become a dominant influence on the rest of my life. "Albright decided that the government's monetary policy was unconstitutional and paper money was worthless, so he asked the state to pay him in gold."

While I waited for the job interview that Frank promised to arrange, I learned that Philip Handler had died of cancer. He was eulogized in *Science* as a warm, tender man who never blurred the distinction between scientific and political questions.

The invitation for an interview finally came. The temperature was -10°F when I took off in Syracuse, and 72°F when I landed in Shreveport and met Jim Albright. He had a low-key, self-effacing manner, entirely unlike any other surgeon I knew. In the evening I had dinner at his house. His wife, Merrilee, started asking me about my career in Syracuse. I tried to downplay the disputes I had been involved in, but got caught up in the stories she was pumping out of me and became excited and animated. I worried that I might have come across as too combative, but at the end of the evening she sat beside me, put her arm around my shoulder, looked over at her husband, and said, "Jim, what do we have to do to get this guy here?" He smiled, and I felt an immense sense of relief. A few weeks later I received the job offer. He never told me what I would be expected to do, only that I would be an assistant professor.

We began preparing to move to Louisiana. I had trouble selling our house because of a foul odor that had developed and was especially strong in our living room. I discovered that it came from a dead rat under the floor of the front closet. After I got rid of the rat we finally found a buyer.

The night before we left Syracuse I had a dream. I saw shadows moving in the distance. When they saw me they came forward and stared, trying to puzzle out who I was or why I was there. I recognized my father, who had been dead for a dozen years. The expression on his face was one of calm resignation, as if the many small torments of life no longer had the power to distract and disturb him. I wanted to embrace him, but each time I tried he fluttered out of my hands. I asked him to remain still so that I could hold him in my arms, which I did infrequently in life because the time never seemed opportune, but he said, "Once the soul has left the body there is nothing that can be embraced."

I cried as I asked him why he had been such a stranger during my childhood, but he said only, "For a thousand years our family lived in one place, caring for vines that belonged to rich men. I went to America so that your fate would be different. It's a hard thing to survive in a new culture." He paused, and then asked, "Why are you here while you are still alive?"

"I don't know," I replied. "What is this place?"

"The souls of the dead must come here to atone for the sins they committed during life before they can enter heaven," he said.

He did not say what sin he had committed, only that it had occurred in his youth. He held some bullets in his hand which I thought was strange because he had hated guns. Then I remembered a story he once told me about finding bullets when he was a child and, in his ignorance, throwing them into a fire, the consequence of which he had never mentioned.

"How long must you remain?" I asked.

He replied that his time there depended on the prayers and good works offered on his behalf by the living. He brought tears to my eyes again when he asked me to do these things for him, because in life he had rarely asked me to do anything. He offered me no advice regarding what would merit the quickest release of his soul to heaven, but he told me that he was pleased with what I had accomplished in my life, which he said had already lessened his stay in that place.

"How did you die?" I asked.

"The evening of my death, after I ate my dinner and your mother had gone to visit her sister, I felt stabbing pains in my chest. I struggled from my chair to the front porch, where a neighbor saw me and called for an ambulance. But the men who came, seeing me stagger about, concluded that I was drunk and so made no haste, stopping once for a meal and once to visit a friend, and all the while I lay dying. When I arrived at the hospital, I was dead." After he said that, he departed.

Next, I saw the soul of Philip Handler. As he walked, he pushed a wheelbarrow that supported his penis, which was roughly the size of his leg and throbbed perceptibly with each heartbeat. When he recognized me and saw my pink cheeks he paused and pleaded for my help. "For the things I did on earth I must wander here, receiving no comfort from any soul until my sins against science have been disclosed to the living."

"What do you want from me?" I asked.

"Tell the world what I did while I controlled science."

"Even were I to try," I said, "no one would believe me."

"Please try. Perhaps some enterprising scholar, even yet unborn, will crack the nut of the Academy and reveal its inner workings, not all of which were crafted by me."

After saying this he turned and began walking back into the shadows from which he had come. But before he disappeared, he stopped, turned, and approached me a second time. After pausing, as if to gather his thoughts, he said, "Only after my death did I understand that science is the mother of good and evil. Because of me, false notions of science have taken deep root in human understanding, where they beset men's minds so that truth can hardly enter. Great troubles will ensue unless men are forewarned of the danger."

I promised Handler I would do what he asked, not out of pity, because seeing his fate did not lessen the revulsion I felt toward him. Rather it was because the corruption of science by opinions masquerading as fact that he had cultivated and propagated was the chief impediment to my success.

The next soul who appeared was Benjamin Cardozo, the venerable Chief Judge of New York whose portrait I had seen many times in its place of honor in the auditorium of my law school. Surrounding him were what at first appeared to be children, but which after some moments I saw were midgets, both men and women, fully mature and properly proportioned, but at half scale. They flocked around the judge and listened to him intently as they all walked together. Among them I recognized only Thomas Matias, the dishonest judge from the Public Service Commission in New York, but not immediately because he averted his eyes which made identification difficult. His sorry state would have been pitiable had I not known of the barrage of lies and distortions he had sent against me, so I offered him no pity. Instead I asked Cardozo why he was in that place.

"God sent me here," he answered, "to teach these wretches how to be judges, which they failed to learn in life when they occupied that sacred office."

Cardozo saw that I wanted to know more, and to benefit from his wisdom. He told his flock to sit on the ground and he sent Matias off to fetch water. While he was gone, Cardozo neither spoke nor moved, as if the water were needed to lubricate the gears of his mind. When Matias returned, Cardozo drank for several minutes, taking only small sips. Finally the water found the gears, and not in response to any question from me he said, "You are seeking knowledge, but your enemies will make your search hard because they hold a grudge against you for the troubles you caused. Even so, after many stressful encounters, you might arrive at the knowledge you seek if you contain your anger and strive to think clearly. A life's work stretches out before you on your path, with no likelihood of recognition or reward. After your death, your efforts will be dismissed or claimed by others, and all memories of your struggles will fade except among your sweet children, but then only for so long as they live."

I wished for a melding of his mind and mine so that my efforts would be unnecessary and my goal achieved in an instant, but I didn't make such a request because I knew that knowledge must come through struggle. He roused his flock, returned the half full cup of water to Matias, and prepared to depart. As he did, he said, "Do not continue your task except with a firm resolve to complete it."

Now came the souls of people who touched my life only lightly, but still left their mark. First I saw Denny DeMarco, who was reading Bullfinch's book on mythology. After him, in a succession that was so rapid it seemed as if they were traveling together, I saw my high school classmate Joe Spaeder, who died a few days before he would have graduated with me, and Bob Charles, who looked much as he did during the days of Sputnik, except that the cigarettes he had worn in his rolled-up sleeve were no longer there. He departed quickly, so I had no opportunity to ask how he had died at such a young age.

Next I saw Father Wallner, the vivid personification of the Jesuit philosophy that burned deep in my soul. He recognized me but couldn't remember my name so he called me "Agnostic," which was how I was known by the philosophy faculty.

"What did you do that displeased God?" I asked.

"I read books by Erasmus, Gibbon, Voltaire, Copernicus, Descartes, Kant, and Nietzsche that were forbidden by the Holy Office, whose commands I had sworn to obey."

"But the list of forbidden books was abolished after I saw you last and before you died," I said.

"Oath violators must be punished. I am an oath violator. Therefore I must be punished," he said.

"There is nothing wrong with that logic," I replied.

He nodded, tucked his hands in his sash, and then walked slowly into the fog.

Then I saw a stranger who abruptly asked me, "Is Mike Wallace still alive?"

I nodded, while trying to recall the context in which I had previously heard that voice. He responded with a lamentation that it had been his fate to labor in obscurity to make Wallace famous, and then to die while Wallace lived on past twice his age. Only in the middle of the stranger's lament did I place him as Richard Clark, who had organized my interview with Wallace but whom I had never actually met. Instantly I remembered our discussions, and his request that I prepare a report on the controversy over the hazards of powerlines. I had used forceful and plain language, less circumspect than if I had been on the witness stand. Knowing that the report would have been twisted by the lawyers for the power companies, I did not retain a copy so I had none to produce when they demanded it. They wailed in front of Matias, lamenting how injurious to their case was the absence of my report which would, they said, show that I was biased against power companies. Then they hit upon the ploy of seeking Matias's permission to ask Clark to provide the report. Matias, who was happy to help, authorized the lawyers to contact Clark.

When they finally reached him, in Alaska where he had gone fishing, one of them said, "In the interests of justice, I respectfully ask that you provide this court with a copy of the document given to you by Dr. Marino." Over the speakerphone Clark replied, "Go fuck yourself."

I had always delighted in telling the story, but I found that Clark did not remember it, or me. Then he departed, leaving me in a featureless, smoky-gray silence. I hoped that more souls from my past would appear, but none came. I began thinking about my homeward journey back to the land of the living, but I had no idea in which direction it lay nor knew any means to go there. I considered the possibility that I was marooned and would not be able to return to the land of the living and do the things I had promised my father. I thought of my wife and children whom I might never see until, in the natural course of things, they came to that place. I began to worry that I might have committed a sin that seemed trivial to me but that God viewed otherwise. I remembered the ammeter I stole in High School, the lie I told the judge in Virginia, the deception I practiced to gain my employment at Westinghouse, and my rejection of the intellectual patrimony of the great physicists, whose life work I had come to see as either trivial or destructive, not ennobling as I had imagined at the beginning of my journey. Perhaps hubris was my sin, and those great men had avoided me because they regarded my microscopic scientific accomplishments with contempt.

Now the gray smoke lifted as quickly as it had appeared and I saw the souls of Paul Dirac and Erwin Schrodinger. Dirac was wearing a long pointed hat that bent over to the side under its own weight, and tapered shoes whose tips curled backwards and had bells that jingled as he walked. Schrodinger's face was chalk white and he had a bright red spot on each cheek and a big red ball on the end of his nose. Dirac recognized me and asked, "Did you read Erwin's book?" I was so surprised to see him in such demeaning circumstances that I replied with a question. "Professor," I said in a respectful tone, "why are you here?"

He sighed deeply and said, "Minding only of the beauty of mathematics, I indulged myself and made no effort to improve the world." Schrodinger nodded as Dirac continued: "Had I joined my intellect with that of Erwin's, we might have accomplished much good. Instead I entertained myself with equations. As I did, the world around me burned and I never saw it."

I asked when they would be allowed to depart. Dirac answered, "When the Lucasian Professor of Mathematics at Cambridge again does something worthwhile in the world." Schrodinger said, "When my question about life is answered."

The soul of Trashcan Trischka suddenly appeared. He was accompanied by a girl so young that she had only recently learned to walk. He knelt beside her as if to protect her from danger. When he looked up and recognized me as one among the many graduate students he had tormented, he started explaining himself. "I saw my errors before I died, so my soul made swift flight to heaven. This beautiful child, who died on the day she learned to walk, cannot enter heaven because she has no use of reason. God commanded me to protect her until she attains that age."

Just as he finished speaking, the soul of saintly Albert Einstein appeared instantaneously, as if he had traveled on a beam of sunlight, except there was no sun in that place. He sighed an enormous sigh, as from one who has no hope, and we physicists stood in silence and stared at the greatest man of the century while the little girl stooped and grabbed the bell on the end of Dirac's shoe and shook it, delighting in the sound. Einstein placed his hands on my shoulders, but I felt no more than feathers. Then he lowered his head and cried so hard that his tears puddled on the ground. The little girl looked up and stared at the strange antics of the sobbing old man with the prune face and the white rumpled hair.

I said, "Professor Einstein, what did you do wrong? Did you make some monstrous error in your theories that misled a generation of physicists? Did you ignore your social responsibilities or sell your knowledge to a high bidder?"

"Visitor," he said, "my theories were perfect. I fought the military mentality. I lived frugally. Only after I died did I learn of my monstrous sin, which was of the same type but far greater than that of my brothers who stand here."

His crying forced him to stop talking, but when he had regained control he continued. "My view of the world demanded the highest possible precision in the description of relations, such as only the use of mathematical language can give. I therefore limited my subject-matter to the simplest events in the domain of experience and ignored all complex events that I could not reconstruct with total accuracy and logical perfection. Through perfect knowledge of the general laws that governed simple events, I believed that it would be possible to arrive at a description of everything by pure deduction. But in my drive to deny a dice-playing God I posited a world ruled by causal laws that left no place for novelty or spontaneous activity. I ignored the possibility of understanding from within, and never learned that the knowledge one has of his own acts of perception, thought, volition, feeling, and doing is entirely different from the theoretical knowledge encrypted in equations. I never understood that the world is stamped with the mark of radical uncertainty, so I was like poor Oedipus who could solve the riddle of the Sphinx but could not recognize who he was. Many men of science fell victim to my teaching that the highest form of knowledge was mathematical. Armed with that opinion, they created the horrors that I had opposed and opposed the social developments that I favored. I never saw that I nurtured what I hated and trampled what I loved."

I said, "It is a shame that your mind was infected with mathematics" and he answered, "Remember all this in your work, visitor. The mind of man is in free play and obeys no equation when it makes music or mischief. Study it, know it, but do not seek the law that strictly governs it as I sought the law that governs the planets and the stars. And put away in your heart this other thing that I tell you. When you come to the end of your journey, embrace mathematics; although it is a useless language for expressing the sublime, it is the only language of science."

With that, Einstein disappeared and the other physicists trudged off together, Trischka cradling the sleeping little girl in his arms.

Then Paul Gelling appeared. He was the only person I had seen smiling on November 22, during my first year in Syracuse. A snake sat on each of his shoulders and bit him on the cheek whenever he turned his face even the slightest amount, so all he could do was fix his gaze directly forward.

Next, in the distance, I saw twinkling lights that reminded me of faraway stars as they appear in a black night. The lights moved toward me and I recognized the face of Mr. Schmidt, although more wrinkled than the day I last saw him in Camden. Tiny light bulbs projected from his nose and ears, and grew from his head like hair, all of which flickered as he trudged along.

After him I was aware of Dr. Hart, who was driving a herd of goats. From time to time one of them would rear up on its haunches and charge, striking him with its horns and knocking him to the ground. He would sit there for a while as if dazed and disoriented, then slowly rise to his feet and continue his journey, only to be attacked again. I hurried away, both for fear of the belligerent goats and because I had no stomach to listen to even one more word from that perjurer.

I came upon a man chained at the waist to a post in a pond of whiskey. His arms and trunk were wrinkled and leathery, but not his gray face, which was that of a man not much older than I, except that it was so contorted in anguish it looked like that of a gargoyle on a medieval church. Every time he bent over to drink the whiskey he craved the level of the pond decreased so that it was just below his mouth. I approached him just after he had made such an attempt and asked, "What did you do in life that merited your treatment?"

He said, "My daughter was born with discolored skin and misshapen face, but with a sweet nature, as if God had compensated in one area for what he had denied in the other. But I saw only the ugliness, and turned to hard drink to ease my bitterness. As she passed from a baby to a child and from a child to a woman, I was there only sometimes in body but never in spirit, and she grew up hating men, whom she either deceives or destroys. The most pitiful are the war heroes who hobble on one leg and paint pictures while holding the brush in their teeth. My daughter thwarts the possibility of a cure for their lost limbs and delights in the gratitude that they, in their ignorance, express to her."

I understood who his daughter was, so I asked only when he would be freed from his torment, but he gave no answer.

I moved along and soon came across a huge but perfectly proportioned creature. Its face was concealed by long locks of hair and its vast, milk-white hands were extended upward, as in prayer. When it finished murmuring, it looked down and noticed me. I had never seen such a beautiful face. As I tried to decide whether to move closer or run away, it resumed murmuring, as if I weren't there. I felt both curiosity and compassion, so I approached the giant and during a pause in its outpourings I asked, "Who are you?"

"My name is Bobo," was the reply. I wanted more details but was afraid to speak. It motioned me forward, and when I was as close as I dared go it began telling me its story. "When I asked God to make mankind, I had imagined a contented species that possessed every pleasure and convenience. God made man and gave him many wonderful things, but electricity was not among them. I then gave electricity to mankind, with instructions that they should thank God for such a wondrous thing. I hoped that God would accept their offerings and that mankind would regard me with honor and devotion. But God took back electricity. It pained me to see mankind so deprived, and my anguish led me to give electricity to mankind once again. For this, God banished me and turned mankind into electrical beings, not something separate from electricity, as previously. Whoever enjoys electricity too often or too intensely is killed by it, but in some familiar way so that others knew how he died but not why. Now I am treated like a criminal by God and ignored by man. I cannot believe that I am the same creature whose thoughts were once filled with sublime visions of all that electricity could do for mankind. Where is the justice in this?"

At first I was touched by the expressions of his misery. But when I called to mind that he had disobeyed God twice, and when I thought about all the people who had died because of electricity, I became indignant. "Do not whine to me about your condition. You alone are responsible for it. You will get no sympathy from me," I said.

"It is not sympathy I seek," he replied. "It would do me no good here." He stood up as he said that and was carried away by waves and lost in darkness and distance. With that, my dream ended abruptly.

The next day Lin, I, and our children got into our car and set off for a new life in Louisiana. When we got to the outskirts of Syracuse we stopped, got out of the car, and waved goodbye.

Part III

◆

Detour: 1980-1994

*All that night I was carried along, and with the sun rising
I came to the see rock of Scylla, and dreaded Charybdis.
At this time Charybdis sucked down the sea's salt water,
but I reached high in the air above me, to where the tall fig tree
grew, and caught hold of it and clung like a bat; there was no
place where I could firmly brace my feet, or climb up it,
for the roots of it were far from me, and the branches hung out
far, big and long branches that overshadowed Charybdis.
Inexorably I hung on, waiting for her to vomit
the keel and mast back up again. I longed for them, and they came
late; at the time when a man leaves the law court, for dinner,
after judging the many disputes brought him by litigious young men;
that was the time it took the timbers to appear from Charybdis.
Then I let go my hold with hands and feet, and dropped off,
and came crashing down between and missing the two long timbers,
but I mounted these, and with both hands I paddled my way out.*

(12: 429-444)

18
Floating

We rented a house near Shreveport and began adapting to life in the South. Lin had a hard time. She found the summer heat oppressive and the attitude toward religion too aggressive. What bothered her most was feeling like a stranger. When I came home from work one day she was sitting on the edge of the bed crying. I asked her what was wrong and she said, "I went into the grocery store today, and nobody knew me."

Her outlook soon improved, especially after she returned to teaching, because she had a gentle manner and made friends easily. I put my effort into not making enemies by inadvertently playing into the southerners' stereotype of northerners as overly aggressive. I wanted to fit in so I would never again need to uproot my family.

Our daughter had only the vaguest sense we were in a different culture; when a little neighbor girl asked if we were Yankees, Lisa told her, "I think we're Phillies." Our sons acted as if we were on a holiday, which was the way they usually acted.

I began work as an assistant professor in the Department of Orthopaedic Surgery at the Louisiana State University Medical School. Jim Albright, my chairman, gave me a laboratory, an office, a technician, and a secretary. If I wanted something I only needed to tell our business manager. When he asked me where the money was supposed to come from, my stock answer was, "From the department." He would then say to Jim in an undisguisedly disapproving manner, "Marino wants to buy something else." Jim, however, always approved my requests. Finally, one day, after many repetitions of this scenario, Jim told him, "Whatever Andy wants, just buy it. Don't ask me." His generosity was like that of my father and Dr. Becker.

The Ph.D.'s on the faculty were expected to obtain grants from the National Institutes of Health. Those without grants did most of the teaching, and they rarely received tenure. For the first time in my career, at the age of forty-one, I faced the problem of getting grants.

The most successful obtainer of grants on the faculty was a famous physiologist named Dennis Thibodeaux. His project was to understand whether blood cells rolled, glided, or tumbled as they flowed in the capillaries, and with what concomitant biochemical changes. He had produced more than five hundred publications; I couldn't tell whether anything was ever settled, but there was no doubt that he was the personification of the successful scientist.

One morning on my way down in the elevator, Henry Vanderhide got on. He had several grants and had recently been appointed to the cardiovascular study section of the NIH. I had intended to ask if I could talk with him about how the NIH operated, but before I could raise the issue he told me that Thibodeaux was looking for me to ask me something about a legal matter. Then I asked my question and he invited me to come to his home that evening to talk about the NIH.

After I arrived, he showed me his collections of motorcycles and Japanese swords. Then I turned on my tape recorder and we settled down for our talk. I told him I knew there were two major parts of the National Institutes of Health – the twenty individual institutes and the fifty different study sections – but that I didn't understand how they were related to one another.

"Each institute," he said, "is devoted to finding a cure for a particular set of diseases. They get money from Congress annually and then decide what research projects they will support."

"Where do the study sections come in?"

"It's their job to pick the best grant proposals."

"So who's really in charge?"

"The Institutes decide what kind of research they want to fund, and the study sections decide which proposals for doing that kind of research are good science."

Remembering my experience with the radiology study section, and the pass-through that EPA arranged, I asked, "Can the Institutes spend their money on proposals that the study sections say are not good science?"

"Yes, if they're careful, and don't try to fund a proposal that got a really bad evaluation."

I asked him about the organization of the study sections and he told me they were headed by executive secretaries, who were not scientists but rather "political beasts." The secretaries select the members of the section, pick the chairman, and assign the proposals to each member, to be reviewed. I asked about the conflict-of-interest procedure to prevent friends from evaluating each other's proposals. He told me there wasn't much of an effort in that area, but that everyone was expected to behave ethically.

"Could a grant applicant lobby a committee member?" I asked.

"That routinely happens," he said. "It's considered part of the game."

"How does that work?" I asked.

"Well, if proposals from a laboratory normally go to a particular study section, it's very good policy for the applicant to make sure that some of the members of that study section are invited to his institution to give a seminar. That's something I never fail to do. They get to see you on your home turf and to put a face with your name. It's much more difficult to reject a proposal when you know the person."

"What actually happens at a meeting of a study section?" I asked.

"Typically, about a hundred proposals are reviewed. Each member receives copies a month before the meeting, and is assigned to review about 10% of them."

"Do the section members usually read every proposal?"

"They're supposed to," he said, "but almost always they read only their 10%. Even that is extremely time-consuming."

"It's hard to believe a committee could review a hundred proposals in a day."

"The meeting lasts two days," he said, "but there's a step that you're missing here. Some time ago it was decided at the highest political level at NIH that too much time was being spent discussing proposals that weren't going to get funded. So now each member is asked to make a decision before the meeting regarding whether a proposal he was assigned is in the top or bottom half. The proposals in the bottom half are automatically rejected."

"If half are rejected even before they're discussed, and the members usually read only 10%, the system is obviously unfair to many applicants."

"People just do the best they can," he said.

He went on to explain how the committee reviewed the proposals that survived the triage system. He said that the primary and secondary reviewers were asked for their scores. If they were close, the review was essentially done. If they were far apart, the two reviewers negotiated between themselves to arrive at a final score. In theory, the consensus was that of the whole study section because anybody could take part in the discussion. In practice, that rarely happened, so that the evaluation of each proposal was really done by only two people. They then write the critique that is sent anonymously to the poor applicant who never has a chance to correct any errors or misunderstandings.

Vanderhide asked me not to repeat what he had said.

I was amazed that NIH scientists were willing to behave that way at the command of the NIH authorities, because surely the scientists must have known that such obedience was wrong. I wondered how such an immoral system could have gotten started. I knew Vanderhide was not knowledgeable about that subject, so I returned to the line of questions that had brought me to his house in the first place.

"It's interesting that you're on a study section, but Thibodeaux isn't. Wouldn't that be crucial for him to continue his flow of funds?" I asked.

"Persons like him," said Vanderhide, "are tied in at a global level. Everybody knows them. When they get a call to serve on a study section, they say, 'I can't do it but take this guy,' and they recommend an associate professor from their department. That's what Thibodeaux did for me. Nobody has ever helped me so much, except my father."

I asked Vanderhide how Thibodeaux managed to stay on top for so long, and the answer was more or less what I expected, that Thibodeaux had mastered the process. He learned the rules, and he played by them. He hired people so he could incorporate their techniques into his grant proposals; that way, his experimental approaches remained state-of-the-art. He published prolifically. In addition to all that, he has a commanding personality. He is a presence. When he walks into a room everyone knows who he is and that he just walked in.

"Is there any limit to the number of NIH grants that someone may be awarded?" I asked.

"None that I know of," he said. "I have two, Thibodeaux has five."

I asked him whether he thought committee members had biases. He didn't like that word, and thought "agenda" sounded less sinister. He said a study-section member was likely to see all the proposals up for review in his field. If he wanted progress in his field, he needed to be an advocate for those proposals. Conversely, if he wanted to oppose particular approaches, he was in the ideal position to do so. After I heard that, I thought to myself that my word was probably better.

I asked Vanderhide what the characteristics were of good grant proposals. "Ask Thibodeaux," he said. "He has far more experience," and that's what I resolved to do.

Two days later Thibodeaux called and asked me to come down to his office to talk about an important matter. When I arrived, his secretary showed me in to a large room with a vaulted ceiling. I walked between two sofas that faced forward, toward Thibodeaux's desk which was at the far end of the room. Various mementos made it plain that he had enjoyed great success during his long career in science. I saw a photograph of him receiving the Outstanding Scientist award from Philip Handler at an annual meeting of the Federated Societies of Experimental Biology. A citation that expressed appreciation for his service as the president of the American Physiological Society was also in plain view, as was a granite obelisk with a brass plaque on its base that said, "To our teacher Dennis Thibodeaux, Professor of Molecular Physiology, in gratitude."

As if to put me at ease, he came out from behind his desk and motioned for me to sit on one of the sofas. Just then the phone rang. He returned to his desk and I heard him give advice concerning who would be a good choice for chairman of the cardiovascular study section. After that he came over and sat beside me and began telling me about his problem. Vanderhide had written a grant proposal which his wife submitted to the NIH, listing herself as the author. After she received the grant, a faculty member in her department whom Vanderhide had been mentoring complained to the dean that Vanderhide stole his ideas and put them in his wife's grant proposal. Thibodeaux was listed in the application as a co-investigator, and was worried that he might be sued. He wanted to know whether that were possible.

I realized I could put his mind at ease without any reasonable fear of being proven wrong if I said that I thought he could not be sued successfully, so I did so, and my advice had the desired effect. I used the opportunity to follow Vanderhide's suggestion that I ask Thibodeaux what a good grant proposal was, and he said that he would be happy to explain the matter to me.

He got up from the sofa, walked back and forth a few times, and then said, "First of all, the work proposed in the application must be important. By that I mean notable and prominent. The reviewer will be looking for an all-absorbing application, one that will put the others in the shade."

"I've heard others say the same thing," I said, "that the proposed work should not be insignificant, unimportant, or immaterial."

"Yes, exactly," he replied. "NIH will not fund work that is trivial or unworthy of consideration."

"I understand that 'importance' is an admirable quality in a grant application, but once an application is assigned to a reviewer, the investigator is at the mercy of the reviewer's judgment, and there's nothing that can be done to correct any misapprehensions. Still worse, I think, is the problem of defending against bias. When it comes to the radiation study section, that's quite a problem. They're like the Jesuits in old Paraguay."

"It's a common problem," he replied. "To deal with it you should cultivate the people who work at the institutes. They can tell you how the application should be shaped, and which study section will review it. They might tell you who the reviewer will be. Even if they don't, it's often possible to guess by matching the background of the members of the study section with the kind of work proposed in the application. Then, you invite the reviewer here to give a seminar, and pay him an honorarium. But after this is done for the first time, it should not be done again, or if it is, only rarely. Another thing you can do is ask a potential reviewer for help. People feel bound as much by the favors they do as by those they receive."

"Besides 'importance,'" I said, "what is the hallmark of a good grant proposal?"

"It should be novel," he said.

"But some people have their minds made up about the way things are," I said, "and it's almost impossible to induce them to change. I would have only the remotest chance of finding a mind on the radiation study section that was open to the possibility that EMFs can do anything other than cook tissue. It's hard to see how a novel proposal would be an advantage."

"Admittedly," he said, "too much novelty can disorient the reviewer and lead him to doubt the credibility of the proposal. The task is to discover what the reviewer sees as appropriate novelty. If you study your reviewer and get to know him personally, the task becomes quite manageable."

"Even if I succeeded," I said, "I'd find no satisfaction in pursuing his concept of novelty, rather than mine."

He extended his hands in a sort of shrug as he paced, and looked up at the ceiling; I was reminded of how a nun had once reacted after I had repeatedly failed to recite verbatim a formula in the Baltimore catechism. "There is a distinction," he said emphatically, "between the work that you propose, and the work that you actually perform. You propose what you think your reviewer will regard as novel, and then you perform what you regard as novel."

"Wouldn't that be fraudulent?" I asked.

"A scientist must be free to respond to new information and changed circumstances," he replied. "The NIH understands this, so it never insists that the work it funds actually be performed as it was proposed."

I agreed with that sentiment so I kept quiet, and he proceeded with the next thing he had to say. "An investigator applying to the NIH should be an expert regarding the proposed research."

"Does that mean he should know how to do the needed measurements and tests," I asked, "say, patch-clamping, or PCR, or electrophoresis, or nuclear run-on assays?"

"Not necessarily. Technicians can be hired to do that kind of work."

He paused for a moment, then continued with the characteristics of a good grant proposal. "You're far better off if you've done 90% of all the proposed measurements and tests before you submit the proposal. If you have data indicating that you will find the results you want, and all that's missing on every graph are only the error bars, then it's hard for a reviewer to argue that the experiments are flawed because you've already demonstrated that they're not. Then you can use the money to get the data for the next grant."

"Where does the money come from to start that process?" I asked.

"From friends," he replied, which I understood to mean people who, like him, already had NIH funds that could be rebudgeted as desired, so I asked no further questions along that line. Instead, I waited to see whether he would tick off another property of a good grant proposal, and he did. "It is risky to send in a grant proposal that has a clear connection with a recognized biological or medical problem. Ideally, the proposed work should be relevant to a previously unrecognized problem."

"That sounds counterintuitive," I said. "Why should an investigator avoid tackling what's already known to be a problem?"

"If he did, he might make an enemy of anyone on the study section who was approaching the problem in the orthodox way, and the others on the study section would be only lukewarm at best, because scientists have no faith in a new approach to an old problem until it has been proved by experience."

At this point he peered at me, as if to gauge how much I had understood. He was a keen observer, so he may have sensed that I felt I hadn't learned very much. I expected he would say he'd done all he could for me. Instead, after a few moments he took a deep breath and said, "Perhaps if you understood why the NIH is the perfect organization for promoting good science, you would be able to see what constitutes a good grant proposal."

"I like that idea," I said. "It might be easier for me if I first understood the big picture, because I'm sure that's reflected in the proposals that are successful."

I asked if I might respond to his points with questions or comments, and he readily agreed.

"Why does NIH necessarily produce good science?" I asked. "What is it that more or less automatically leads to that result?"

"First, the Institutes are controlled by a unique group of skilled professional managers who are better at what they do than other groups of managers."

"Why is that?" I asked.

"Consider the performance of other managers," he said. "Senators and representatives have done foolish things. Cardinals and bishops have made grievous mistakes. Company presidents and CEO's have created scandals. Generals and admirals have failed in their duties. But the managers of the NIH, who run a multibillion dollar industry that produces thousands of publications, are completely free of any hint of malfeasance or misfeasance. If I challenged you to find any criticism of the NIH in any respectable publication you could not do so. Their managers are wiser than any others."

"Certainly I hear less criticism of NIH than I do of the Pope," I said.

"Second, the members of the study sections, who make the judgments regarding which grant proposals are good and which are bad, are also tops at what they do. They are like well-bred dogs in the way they guard science. They are quick to see which proposals are good and to attack the bad ones."

"Like a dog which attacks strangers but is friendly toward those whom it knows?" I asked.

"Exactly," he replied.

"How does NIH get such dogs?"

"Future study-section members learn the skills and discipline needed to make good decisions by apprenticing in a laboratory supported by the NIH, where they receive a proper education."

"What sorts of things do they learn?" I asked.

"They learn the correct way to conceive the world. New phenomena are constantly being discovered, but the laws of nature that govern them are changeless. One is ready to sit on a study section only after he has been sufficiently incubated in this tradition."

"But new ways of looking at the world are constantly being suggested," I said. "I'm thinking, for example, of the views of Prigogene or Dubos."

"They pursued phenomenology rather than mechanisms, which are the timeless essence of nature. Their approach is unedifying."

"One day when I was in college," I said, "Father Wallner, who taught me logic and philosophy, drew a triangle on the blackboard, and also a profile of a man who was looking at it. Then he drew a miniature version of the triangle inside the man's head and said, 'Truth is when triangles are identical.' Is that the correct conception?"

"Absolutely," Thibodeaux remarked.

"May I explain the difficulty I have with that point of view?"

"Of course."

"Suppose there were two profiles, a Palestinian and an Israeli, and that each contained a miniature version of the triangle."

"Yes," he said.

"And suppose further that the triangle represented the observation of a boy throwing stones at an Israeli tank. In this situation we would say that both men see the same thing."

"Yes," he said, "by your definition."

"But the meaning of the observation, by which I mean the thing signified by the boy's act in throwing the stone, is quite different for the two observers. For one it is a warranted act of opposition against an oppressor, and for the other it is an act of rebellion against lawful authority."

"Certainly the two observers would interpret the act differently," he said.

"This example," I said, "shows that truth can be objective in one sense but subjective in another sense, and probably that the sense in which it is subjective is more important than the sense in which it is objective."

"The resolution of your difficulty," he replied, "rests in recognizing that science and human affairs are two different things. The meaning of what people do is always problematical and subject to debate, whereas in science, the data speaks for itself."

"So, you think that a regimentation of viewpoint is necessary?" I asked.

"Divergent viewpoints only confuse young scientists, so we don't do things that way. Everyone is taught that science is knowledge of how the world works, something intrinsically good."

"I think saying that science is something good in itself is a lie," I said.

"Perhaps," he replied, "but it is a noble lie because it fosters public support for science."

"Aren't you worried that telling lies is anti-scientific and therefore stunts the growth of young scientists?" I asked.

"Any such problem is overcome naturally, with the passage of time," he replied, "so knowledge continually grows."

"If NIH's point of view is the standard that everybody must follow, then how could we benefit from an innovator? Galileo, for example, advocated an opinion with which few initially agreed, but which finally won everybody's support. This implies that the criterion for good science should not always be the orthodox opinion."

"Such cases occur so infrequently throughout history that the phenomenon can be regarded as one of only theoretical importance," he said.

"What else goes into the proper education of a member of a study section?" I asked.

"Pursuit of scientific truth requires time," he said, "and therefore scientists must be provided with resources for their subsistence and their experiments. If it were too much, it would seem to outsiders as real money, necessitating some system of strict accountability. Too little, and no one would be attracted to a life in science."

"What, generally, is about right?" I asked.

"Between one and ten million," he replied.

"What are future study-section members taught to believe?"

"The Holy Trinity of science," he said.

I asked him what that was.

"That reductionism is the true method of science, that scientific knowledge is knowledge of mechanisms, and that the dose-effect relationship is nature's way of differentiating what is imaginary from what is real. The Trinity is the scientific ideal."

"According to that ideal," I said, "studies of the mechanism of action of an antibiotic would qualify as good science, but a clinical study to determine whether the antibiotic worked in patients would not qualify. That doesn't seem right."

"Applied science has its place, but it falls far short of the ideal. Drug companies should do clinical studies. It is appropriate work for that culture," he replied.

"It seems to me that any ideal is more or less just a personal opinion," I said. "Everyone at NIH may believe in the Trinity, and even that all scientists ought to believe in it, but still it remains only a belief. Prigogene's method or that of Dubos or many other prominent scientists whom I could name differ from the Trinity in many respects, yet all are admired by different groups of scientists. How can the question of what constitutes the scientific ideal be decided, except on the basis of personal choice?"

"We can consider something to be ideal if all of those who investigated it are in agreement," he replied.

"If there is such agreement, as with gravity for example, fine," I said. "Then I suppose that we could regard that as good science. But statements like 'EMFs are health risks' are basically different from statements like 'The planets are held in place by gravity.' I think that it's unrealistic to expect that there will ever be agreement among all those who have investigated EMFs. Perhaps, in cases like EMFs, whatever the real truth may be, the issue should be settled by a vote."

"What do you mean?" he asked.

"Whoever has the most votes wins the argument, and that opinion of 'good science' is put forward as the ideal. I don't see any other way, because there is no standard of good and bad science, except what the person using these words desires. So there is no question of proving or disproving whether NIH is ideal. The only question is whether you like the kind of organization NIH is and the ideal it espouses. If you do, it is good for you. If you do not, it is bad for you."

"Good science is good science," he said, as if repeating what he said would explain what he meant.

"But you can't talk about the idea of 'good science' without considering the purpose for which the science is performed. Pursuit of the most efficient processes to exterminate races of people considered to be inferior, for example, could never be good science no matter how much truth was contained in the data adduced in the effort. The problem of identifying good science using only reason or observation is insoluble. That is the ultimate reason we should resort to voting."

Despite the length of our conversation and the patience that Thibodeaux had shown me, I failed utterly to discover how that lover of molecules could hope to find happiness or satisfaction in a life devoted exclusively to the science permitted by NIH. What I thought was most worth knowing existed only at a higher organizational level than his philosophy allowed him to investigate. He reminded me of when I was a little boy, one hot summer evening out on my grandfather's farm. The place was alive with lightning bugs, and I was fascinated by their flickering light. But when I caught one and held it between my fingers, it no longer made any light. I thought its wings might be covering the part of its body where the light came from, so I tore them off, exposing its hind end which glowed, but didn't flicker. I blew on it and moved it up and down, but still it didn't flicker. Then I became fascinated by the glow. I took that part out of the bug and put it on my finger, like a ring, and ran to show my grandfather what I had made. He never said a word. He just looked at my finger, slapped me in the face, and walked away.

I knew I could not compete in Thibodeaux's world, where I would need grant after grant to support my research. I lacked his scientific pedigree and his knowledge of the practices of NIH. Even more, he drew energy from his belief that the NIH yielded worthwhile science, while I saw it as mostly useless knowledge, and I had no heart to serve that system.

19
Carbon Fibers

For several years my research mostly involved bone, tendons, and ligaments although, from time to time, I did some experiments with EMFs. Dr. Becker had loved bioelectrical research but in the end his passion for that work had brought him mostly pain. He had been forced to retire, and then his health declined. His fate weighed heavily on me. The hope for recognition that had fueled him and led him to suffer many outrages did not fuel me. My life was more than half over, and I wanted to enjoy what remained, without the controversy and travail brought by confrontations with powerful companies and government agencies. At least that's what I thought I wanted.

One day Angus Strover and Don Hourahane came to our department to talk about their research. Strover had grown up in Southern Rhodesia and moved to South Africa after his family had been forced off their farm. He was in his last year at Witwatersrand University in Johannesburg, where he was training to be an orthopaedic surgeon. Hourahane was an engineer who had left England for South Africa twenty years earlier to make his fortune. As a child he had been injured during the blitz and developed osteomyelitis, and had been cured by a new drug, penicillin. He was what my mother called a "talker." At one point during our conversation he grabbed his right wrist with his left hand and held his arms over his head.

"What are you doing?" I asked.

"I'm showing you that people from Africa have one arm shorter than the other, it's from swinging on trees."

Strover had done a surgical rotation in England where he met an orthopaedist named David Jenkins. According to Strover, Jenkins had been in a bar in Singapore and happened to sit on a stool next to a man who had just sold carbon fibers to a company that used them to reinforce tennis rackets. The salesman had ruptured a knee ligament years earlier; as a consequence of that injury his knee would give way unexpectedly. Jenkins explained that it was a common problem for which there was no good solution; reconstructing the ligament using pieces of tendons or ligaments taken from other places in the body often gave poor results, and there were no artificial knee ligaments. At this point the salesman gave Jenkins some samples of carbon fibers and said, "You ought to try this."

Jenkins had used the carbon fibers in experiments with sheep and then with patients, leading to what he concluded were successful results. Strover told me that ruptured ligaments atrophy, and disappear from the joint, but that the carbon fibers induced growth of a new ligament in place of the original, as if the fiber bundle was a riverbed along which new tissue flowed.

After he had returned home from England, Strover went to a local supplier to buy carbon fibers, intending to cut sections to length and implant them in patients. Hourahane, who had a small plastics company in Johannesburg, had gone to the same supplier the same day, also to buy carbon fibers, which he used to reinforce the products he made.

Hourahane said to me, "The clerk told me he had just sold the last of his stock to a doctor from Wits who was going to implant them in people, so I went there to try to find him."

"Why?" I asked.

"Because I wanted to stop him from hurting somebody," he said. "Carbon fibers sold for industrial purposes can't be implanted in the body because they have a special chemical coating that's poisonous."

Hourahane had found Strover in time, and they began working together on Strover's project. Hourahane obtained carbon fibers that had never been coated, and fabricated them into artificial knee ligaments which Strover used to treat patients who had an unstable knee joint. Hourahane had designed and built special instruments to facilitate the surgery because ordinary instruments would have damaged the carbon fibers, and he invented devices to attach the fibers to bone.

Strover's patients did well, and other orthopaedists in South Africa and Australia began using the carbon-fiber implant, instruments, and fixation devices, which Hourahane made and sold. Strover said that Hourahane's system allowed a cruciate to be repaired successfully with a greater degree of reliability than was previously possible.

Hourahane had patented his inventions and wanted to gain access to the huge American market, but he had not received much help at the other U.S. orthopaedic departments that they had visited. Some orthopaedists were opposed to the idea of using artificial materials to repair ligaments.

Others had committed to one of the companies that was already trying to bring an artificial ligament implant to market. Someone had suggested to Strover and Hourahane that they visit Jim Albright, who had recently published a book entitled *The Scientific Basis of Orthopaedic Surgery* and who had a national reputation as an innovative man and a lover of research.

During their visit Jim learned about the clinical results and became interested in the project. I initially had no interest, but my attitude changed when I saw Jim's reaction, and I offered to help in any way I could. I hoped to give back something to the man who had given me so much, and perhaps even to earn a profit.

Jim and I went to South Africa to see first-hand what had been accomplished. In Johannesburg we saw Strover implant carbon fibers in the knee of a man who had been injured in an automobile accident. It was deft surgery, done in a fraction of the time normally needed when such injuries were repaired using tendons taken from some other area of the patient's body. At Baragwanath Hospital in Soweto a surgeon told me that his carbon-fiber patients were able to walk the day following surgery. In Pretoria I met Hourahane's biggest customer, an Afrikaner orthopaedic surgeon who told me that before he had started using carbon fibers he had never been able to successfully repair anterior cruciate injuries. At a hospital for gold miners in the Transvaal I visited an orthopaedist who had been especially meticulous when implanting the carbon fibers, and his results were the best of all – although, of course, not perfect.

Hourahane proposed that we go into the business of making and selling the carbon-fiber implant system in the United States, and Jim and I agreed. I incorporated our new company in Louisiana and became its president, with the major duty of doing what was necessary to obtain a license to sell the implant. Hourahane agreed to wire me the money I would need.

I went to the Food and Drug Administration in Washington, D.C., to learn exactly what was required for the license. When I entered the building I saw a huge room filled with desks piled high with stacks of paper; people swirled around like smoke in a wind tunnel. Someone directed me to a conference room and told me to wait there. The room was drab, almost depressing. Most of the formica on the edge of the table was missing, and there was a half-inch step between the halves. Ripples in the rug ran at right angles to a path that had been worn by the traffic in countless previous meetings. I sat in the least comfortable chair, no two of which were alike, directly opposite a picture of President Bush. After a while a dark-skinned man with a repellent body odor entered the room, along with three other people.

"I am Nirmal Mishra," he said. "I am in charge of applications for artificial ligaments, these people are on my team. How can I help you?"

"I would like to sell a carbon-fiber knee implant," I said. "What do I have to do to get a license?"

"Give us evidence that your device will be safe and effective," he replied.

"What kind of evidence?"

"You must prove that it is strong. You need to do animal studies. Then you must prove that it works in people and doesn't have any side effects."

"I'm unclear about exactly what I need to provide. For example, I can put the carbon fibers in a machine and measure how much force it takes to break them. Is that what you mean about proving strength?"

"Yes, that would help, but you should do other studies as well."

"What studies?"

"Studies that show how strong it will be."

I began to feel uneasy. "How strong does it have to be?" I asked.

"Strong enough so that it will be safe and effective," he replied in an impatient tone.

"Suppose it took a hundred pounds to break the carbon fibers, would that be strong enough?" I asked.

"The situation can't be oversimplified. The device must be strong enough that it won't ever break, however strong that is, well, that's your answer, that's the number you want."

He seemed annoyed, so I lied and said, "I think I understand what needs to be done."

I began to tell him about what I thought was important data from over three hundred patients in South Africa who had received carbon fibers and had generally done well. But as I was speaking he started shaking his head from side to side.

"We can't rely on the opinions of surgeons, they often don't admit bad results. You will need to do a clinical study in the United States, according to our specifications," he said.

He gave me a booklet entitled *Guidance Document for the Preparation of Investigational Device Exemptions and Pre-Market Approval Applications for Inter-*

Articular Prosthetic Knee Ligament Devices and said, "Follow this." Then he rose from his seat, and the people on his team immediately shot up as if they were connected to him by springs. He walked toward the door, and when he reached it he turned and said, "You need to continue this dialogue with us so that we can tell you what we expect;" I quickly agreed to do so.

As soon as I returned home I planned a study in which the anterior cruciate ligament in the knee joint of goats would be cut out and replaced with carbon fibers. When I called Mishra's expert in charge of animal studies to obtain his approval I told him, "We will assess the quality and strength of the new tissue that grows."

He was a friendly young man who had just graduated from college. "That sounds reasonable," he replied.

"How long should I let the goats live before I kill them and test the ligaments?" I inquired. He asked for my opinion, so I told him that I thought three months was long enough, and he agreed.

"I expect to have some problems because no one else has ever done the kind of surgery we are planning, but in most cases we anticipate that the implant will hold up well," I said. "Realistically, that's all that could be expected."

"Yes," he said.

When the experiment was over I reported the results, but this time the young man was less realistic. He flipped through the pages of data and said, "Two goats didn't do well."

"Yes, but that's more or less what you would expect."

"Nevertheless, the fact that some goats can get bad results indicates that the device can be risky."

I thought about telling him there are always risks, but I said nothing for fear that I might give offense.

The strange mechanical properties of carbon fibers intrigued me. They were stronger than steel when pulled at each end, but broke easily if bent at an acute angle. Their apparent ability to grow a new ligament seemed marvelous, and I took it as my goal to understand this process. I hoped that moving the subject of artificial ligaments from the realm of speculation and anecdote into the realm of science would relieve Mishra's anxiety about granting me a license, so I did many experiments on rabbits, mice, and rats. During that time I also did experiments that Mishra had insisted upon, though I saw no merit in them.

One day the secret of how the carbon fibers worked became apparent to me. I discovered that the tissue which grew around the fibers didn't actually stick to them, but rather that each fiber could slip out of its tube of tissue as if the fiber had been coated with grease. The reason the carbon fibers strengthened an injured ligament was that the new tissue that grew around them joined the ligament's original tissue and insertion points on the femur and the tibia, thereby reinforcing the injured ligament. Each implant contained forty thousand carbon fibers, so there were forty thousand tubes of tissue that grew to strengthen the injury site, many times more than grew in response to the artificial materials being developed by competitors who were using other materials. The strength of the carbon fibers was not what mattered, but rather the strength of the new tissue. Moreover, most materials could spontaneously trigger an attack by the immune system but carbon never did, so carbon fibers were twice blessed.

My good feelings about the knowledge I had gained were tempered by discouragement about the time and money wasted in performing all the experiments Mishra had required. Had he and his team left me alone, I could have gained my understanding more quickly and inexpensively; their suggestions and opinions – really commands – were rarely useful. He called those on his team "scientists" – they weren't, but that didn't stop them from giving me rote advice concerning how experiments should be performed, like medieval monks braying out memorized psalms they don't understand. Mishra himself often preached to me about "risk." One time he sweated for an hour and stunk up the place even more than usual, but succeeded only in showing that nobody could objectively define "risk," although of course he thought he had done it. Nevertheless I always did as I was told by Mishra. I hated the way my desire for the license made me act. It was as if I were limp, like my carbon fibers.

After many delays, Mishra finally approved my plan for a clinical study to determine whether carbon fibers were safe and effective for repairing the anterior and posterior cruciate ligaments in the knees of humans. The plan had to conform to his *Guidance Document*, so there was much in the plan that had little to do with science but rather was intended to generate interlocking patterns of data that would make it difficult to fabricate data favorable to my interests which, it seemed, the FDA assumed any company trying to obtain approval for a device would do.

The clinical data I collected on each of the patients who took part in my study was supposed to come together to support Mishra's judgment regarding the issuance of a license, but I didn't understand quite how, which put me in the position of a man searching for something but having no way of knowing whether what he found was what he was looking for.

"How will you know from the data I collect whether the implant is safe and effective?"

"I'll know when I see the data," he replied.

During the next year I collected thousands of pieces of information about each study patient, some of whom received carbon fibers and others of whom received conventional ligament reconstruction with tendons harvested by the surgeon from elsewhere in the body. I had to hire bookkeepers to keep track of the data, and I struggled constantly with the orthopaedic surgeons who performed the periodic follow-up examinations on each patient, which I needed in order to evaluate whether the carbon-fiber treatment was successful. They resented the length and prolixity of the follow-up form Mishra had required, and often omitted data they thought meaningless but Mishra considered important. His term for each blank line was "protocol violation," and he construed every such instance as evidence against the idea that the implant was safe and effective.

As the study went on I saw things that could have been done differently and likely would have helped the patients, but Mishra denied me permission to make any changes in the experimental procedure he had approved at the inception of my study. He too learned of things that could have helped the patients in my study, knowledge he obtained from studies being done by my competitors, but he refused to share it with me because, he said, it was against the policy of the FDA to require competitors to share scientific data. "Why should they pay for information for you to use in your business?"

While I was performing my study, which was taking place at my institution in Shreveport, Brooke Army Base in San Antonio, and the University of Iowa in Ames, Hourahane continued to sell carbon fibers to orthopaedic surgeons in South Africa, Australia, Canada, Israel, and Europe, which he was free to do because the system in those countries placed primary responsibility for the choice of medical treatment on the doctor and his patient, not the government. Many hundreds of patients received Hourahane's carbon fibers with generally good results, at least in the opinions of the surgeons and the patients.

As the implant became progressively more popular, Hourahane began receiving phone calls from veterinary surgeons who expressed an interest in using carbon fibers. The application he judged to be most promising involved the treatment of lameness in racehorses. Although the topic had been far from my mind, I learned that when one of the large tendons in the horse's leg stretched too much, the result could be swelling that deformed the leg into a backward-pointing bow. Then the horse couldn't walk, much less run.

I asked how he planned to repair the horses' injuries.

"I remembered that a young man in Louisiana discovered that no bonding took place between carbon fibers and tissue, and that the tissue that formed along the carbon fibers was aligned by them. I'll make a pointed cannula that can enter the tendon through a quarter-inch incision and then pass up to the top of the tendon. The vet will be able to pass the carbon fibers through the injury site by means of the cannula."

South African veterinarians implanted carbon fibers in more than fifty racehorses and show jumpers, and many of the horses returned to competition. I then performed two studies in the United States in which veterinary surgeons implanted carbon fibers in bowed tendons of thoroughbred racehorses, and did standard therapy in another group of injured horses – more than 100 horses in all. The results showed that use of carbon fibers was an effective treatment; I published a report about the experiment in the *Journal of Equine Veterinary Science*.

Meanwhile, the other companies that were trying to develop a ligament implant avoided all basic science research, and did only the experiments Mishra told them to do – or at least what they thought he had told them. One by one, the companies received an education at the public meeting the law required prior to a licensing decision. That's when each company learned its path had been only a big circle.

The first presentation was made by a company whose implant was made of polyethylene. After the company had presented results involving mechanical tests of its implant, a panel of experts Mishra had appointed to the FDA's review panel told the company's president, "We think the device is not strong enough."

"Compared with what?" he asked.

"Compared with a normal ligament," he was told.

"But that's not what we are trying to do. The patients we operated on had no ligament left. We were only trying to improve the patient's condition, not return him to mint condition. We don't know enough yet about how to make a device that's as strong as the original ligament. We have to take this a step at a time."

"I think you should make a better device and then do more studies," the panel chairman replied.

At the same meeting, other companies requested approval for their devices. One company described a tongue depressor, which was said to be an improvement over present models on the market because it was thinner and shorter, making it more suitable for depressing the tongues of children. Another company described an adhesive bandage that was said to result in less pain upon removal from the patient. Both devices were approved by the panel.

The last presenter that day described an implantable total-knee prosthesis, and illustrated its use with graphic slides that showed it being placed in a patient in France, where it had been developed. The surgeon cut off the knobby ends of the femur and tibia, and replaced them with devices made of titanium and polyethylene, which were attached to the bone with a special glue. The presenter then said, "We request approval of our implant because it is substantially equivalent to an implant that was legally sold in the United States prior to May, 1976," and the panel quickly approved the application. At that time I did not understand why manufacturers of such lowly devices as tongue depressors and band-aids were striving to improve their product, yet the manufacturer of a device whose use requires major surgery would go to great lengths to emphasize that his product was no different than one that had been marketed for a long time.

Six months later, a company that had developed an artificial ligament made from the tendons of pigs presented its invention to the panel. After a spokesman had finished explaining the company's animal and human studies, the panel chairman told the company president that patients might eventually reject the implant because it was foreign tissue.

"We treated the implant to reduce that possibility," the spokesman told the panel, "and we have not had any instances of bad reactions in patients."

"Have you looked inside the knees of these patients to see what is going on?" a panel member asked.

"Of course not," the president said. "We couldn't justify operating on a patient who had no problem simply to see what the ligament looks like."

The panel told him that using animal tissue was a bad idea because it might activate the immune system.

"Why didn't you tell us that when we started?" the company president asked. "Why did you wait until we had worked for three years and spent $5 million?"

He might just as well have saved his breath because the panel ignored him. Why Mishra had assembled such a panel of fools had not yet become apparent to me.

Other companies at that meeting who were seeking approval for their devices fared much better. In short order, the panel approved a new design for a hospital bed that had larger wheels, making it easier for the nurses to move, and a new formulation of plaster of paris that hardened into casts 20% faster. The sponsoring companies assured the panel that there were no risks associated with their products as long as they were used according to label instructions. The panel then approved a total hip consisting of a cobalt chrome stem designed to be hammered down into the bone's canal after it had been exposed by cutting off the top part of the bone, and an articulating ceramic component that was attached to the pelvis. Again, a company spokesman told the panel that the hip was "substantially equivalent to a device that had been marketed prior to the enactment of the device law." By then I'd learned the reason for that formula. Scientific evidence of safety and efficacy was not needed because any implant on the market before the 1976 device law had been deemed to be safe, like BHT. Consequently, for the sake of consistency, a new implant that was exactly like one of the old implants also had to be deemed safe.

Several months later the device panel held another meeting. The first presenter was a company seeking a license to sell a Teflon ligament. The panel chairman told the spokesmen for the company that small particles could flake off the device and cause an inflamed and painful knee.

"That's why some of the patients in your study did poorly," he asserted.

"Some patients in every surgical series do poorly," the company representative replied. "That alone is no reason to deny us a license." But that's what the panel did.

The last presenter in the morning session asked the panel to approve a total knee implant. Hoping to gain an advantage in the marketplace, the company had advertised in trade journals that an innovative porous coating on the surface of the implant would encourage bony ingrowth, leading to a longer-lasting implant. A controversy ensued when some panel members expressed the opinion that the coating was technologically impossible in 1976, and therefore the FDA should not deem the implant to be safe. In the end, the company convinced the panel that the claim had been made by its marketing department, and shouldn't be taken seriously. The company got its approval without needing to run the FDA gauntlet I was running and that any manufacturer of a new implant is forced to run. None had ever survived, but I hoped I would be the first.

While sitting in the audience I met people from various companies that hoped to someday obtain a license for their products. A young woman from a start-up company in Phoenix told me about a device she had invented that was powered by a watch battery and produced a tiny EMF that made fractures heal twice as fast as normal. She had gone deeply into the biology of bone healing and learned that the EMF did not alter the amount of proliferation of the stem cells of bone caused by the injury, but rather increased the rate at which they differentiated into the kind of cells that actually build bone. She showed me a report entitled "A new theory of the biology of bone healing." Across the title page Mishra had written "Not recognized," which he underlined twice. The opinion of others at that meeting who also heard her story was that it was madness to present novel scientific results to the FDA.

Of all the companies trying to get a license to sell ligaments, Hexcel Medical was the most aggressive. Its product was also made from carbon fibers, and the company kept trying to pin down Mishra regarding exactly what evidence would convince him they were safe and effective. Somebody from the company visited or called Mishra every day but, according to what I was told, he never allowed himself to be trapped into giving an answer.

Thinking to strengthen its position in its dealing with Mishra, Hexcel hired David Jenkins as one of its experts. He had learned about the toxic coating on carbon fibers that were sold for ordinary purposes such as reinforcing tennis rackets, and he attempted to remove it with a solvent. Unfortunately, he had not rinsed the carbon fibers well enough to remove all of the solvent which, it turned out, was more toxic than the coating. Some of Jenkins's patients developed chronic inflammation due to the solvent residue, and their fate came to light in a paper published by a surgeon named Dandy who had examined Jenkins' patients. To prove his point, Dandy himself implanted carbon fibers that contained solvent residue; unsurprisingly, his patients developed chronic inflammation.

At the FDA meeting where Hexcel presented its scientific evidence, the panel chairman asked where the data was that showed chronic inflammation in the patients in the study. But Hexcel, like me, had obtained pure carbon fibers, thereby obviating the error committed by Jenkins and then aped by Dandy in the so-called interests of science. Nevertheless the panel accused the company of hiding data that showed bad results, and Hexcel was denied a license.

I avoided the mistakes in judgment and the errors in science I perceived in the way Hexcel had gone about its effort to obtain a license. I had discovered the biological basis of the action of carbon fibers implanted in the body, and I had proved that they could be used successfully in animals. Hexcel Medical had done neither. My system for implanting carbon fibers included specialized surgical tools for grasping the carbon fibers and threading them through the joint, and specialized devices for attaching the implant at each end to bone. Hexcel lacked these instruments; consequently the surgeon often broke some of the carbon fibers when he grasped them with ordinary surgical forceps or routed them over the sharp edge of an unchamfered bone hole. Hexcel had no means of attaching the carbon fibers to bone, so the surgeons had attached them to soft tissue using sutures, which often resulted in a fatally weak repair. Worst of all was the design of the clinical study. Carbon fibers were used to repair an instability anywhere in the body, from the foot to the shoulder, and without any control group. My study, in contrast, specified that only ligaments in the knee would be repaired using carbon fibers, and I had included a control group consisting of patients who received standard surgical treatment. Thus the ability of my clinical study to reveal whether carbon fibers were effective was focused on a single problem, not diffused across many different problems. It was the first controlled study in the history of orthopaedic implants.

When I reached the point where I had a year of follow-up data on the last of the 145 patients in my study, I evaluated the data and I saw that I had proved my point. I went to Washington to show my results to Mishra. We met in the same drab conference room, but the contrast between my surroundings and how I felt inside couldn't have been more dramatic. I thought I had reached my goal, and that my rewards would soon be coming. When he entered we exchanged greetings, and then I told him of my intentions.

"I am ready to make my formal application, I have all the requisite evidence."

"I think longer follow-up is needed to insure that carbon fibers are safe and effective," he replied.

When I heard those words I started to see flashes of red and blue light. I could see that he was continuing to speak, but I couldn't hear anything. Then I couldn't see him, or feel anything, as if gravity had ceased and I were floating in space. After a while, how long I didn't know, I could feel the uncomfortableness of the chair, and I recovered sufficiently to blurt out, "But you said one year's follow-up would be enough. I have it in writing."

"Well, the state of the art changes. We now know that implants which look good after one year may not be good after two years. The health of the public is at stake here. We must be certain."

"You said one year's follow-up would be enough," I said again. "I have it in writing."

"I make the rules, Dr. Marino," he said, "and I can change the rules."

Obtaining the follow-up data had always been difficult and expensive, but now it became a nightmare. Some of the patients had been athletes at the University of Iowa; after graduation they moved all over the country, making it difficult to locate them and arrange for a physical examination by an orthopaedic surgeon where they lived. The situation was even worse with the servicemen who had been stationed at Brooke Army Base in Texas when they entered the study. The patients at my institution frequently failed to return for the yearly examination because they felt well and saw no need to take the trouble to do so. Some patients at all three locations sensed our desperation to obtain the follow-up information, and demanded to be paid large sums of money for allowing themselves to be examined.

Hourahane finally reached the point where he could no longer supply money to maintain the study. So I told Mishra that I would make a formal application for a license, as I had the right to do under the law.

"Why don't you put together all your data and present it to me informally, and we'll take it from there," he replied.

The meeting with Mishra and his team took place in the same room as our other meetings.

"Well, what do you have?" he asked, and I began at the beginning.

"My first animal experiments were on mice. We wanted to show that carbon fibers were well tolerated, so we implanted them in muscle and fat, and near nerves, and we found that the fibers were well tolerated by all the tissues."

"How long were they implanted?" he asked.

"Three months," I replied.

"Suppose you left them in longer. What would have happened?"

"Probably nothing, but I don't know because that's not what I did."

"So you don't know if there would have been problems."

"I had no reason to implant them for longer than 3 months."

"Safety is a paramount concern. It's your responsibility to answer all the questions."

"What questions?" I asked.

"My questions."

I continued. "In rabbits, we removed the Achilles tendon and replaced it with carbon fibers; nylon was the control. We discovered that the amount of new tissue that grew depended on the surface area of the material. Carbon fibers have several hundred times the surface area of nylon, so they produce several hundred times more tissue. The additional tissue added strength to the injury site. That is our rationale for carbon fibers."

He rocked back in his chair and looked up at the ceiling. Through his usual frozen smile he said, "It was an interesting experiment, but what does it prove as far as safety is concerned?"

Before I could reply he said, "What else do you have?"

"In goats we took out the ligament and replaced it with carbon fibers."

"Was the result as strong as the original ligament?" he asked.

"Of course not," I said. "The purpose of the study was …." He interrupted, "Well, that's what people will expect."

"Well, we'll just tell them the truth," I said, but he waved me off saying, "They won't listen. What else?"

"We thought that carbon fibers would work in injured tendons, and we proved it in racehorses."

"That kind of a study has to be evaluated by veterinarians," he said.

I told him it was, and that it was published in their journal. He nodded, pursed his lips, and asked a question he surely knew I could not answer, "Were there any carbon particles in the lymph nodes?"

"I don't know. We didn't look. The horses didn't belong to me. They were very expensive animals, and their owners wouldn't let me operate on them just to remove tissue. I did examine the lymph nodes in other animals."

"But not in the horses, right?"

"Right," I said.

"What did the human studies show?" he asked.

"By any reasonable interpretation of the data, the patients who received the carbon fibers did better than the control surgery," I replied.

"What about side-effects?" he asked.

"The number of infections and complications was the same in the carbon-fiber and control groups."

"What about the strength of the carbon fibers. If you pull on them, how much force does it take to break them?" he asked.

"About 100 pounds," I replied.

"The average strength of the cruciate ligaments is 500 pounds, so the fibers are much weaker than normal tissue," he said.

"That's comparing apples and oranges. The original ligament is gone, so the strength of what remains is zero. We shouldn't make the perfect the enemy of the good."

"How much strength do you get eventually?"

"I can't possibly answer such a question unless I remove the new ligament from the patient and test it."

"How about the animal studies?" he asked, and I gave the only answer that was possible. "Animal studies can't be done to answer that kind of a question because people and animals are different."

"What scientific basis is there for the hypothesis that the ingrowth of tissues bears a load or has any strength to it? What's the basis for that hypothesis?"

I plowed the same ground again, telling him, "We showed that new connective tissue grew in animals. We showed that it added mechanical strength and that it led to improved function, as evaluated on the basis of the ability of racehorses to return to racing. Then we did the clinical study. The results couldn't have been as good as they were unless there was new tissue."

"You have not given us any scientific data to show that in fact there's any tensile strength to that stuff that grows in. Sure, fibrous tissue grew, but what's the strength of it?" he asked, even though he knew exactly what my answer would be.

"I can't answer all that," I said.

"Well, my question is pertinent," he said, "because you're saying that arbon fibers in there and suddenly tissue grows in, and I'm saying is the proof?"

He paused for a moment and then said, "It's too bad, because you've got independent ideas of your own, I almost envy you. Nevertheless, carbon fibers cannot be accepted by the Food and Drug Administration because such an implant might lead to harm. The review process must be very conservative. Everything must be understood in that light."

"What light?" I asked.

"Any device licensed by the Food and Drug Administration must be free of all risk. Risk leads to harm, and that leads to disrespect for the Food and Drug Administration."

That was such a shocking thing to say I could not believe my ears.

"It is impossible to make a device that has no risk," I said, and after I caught my breath I continued, "Besides, that's not what the law says. It does not say 'possible risks,' it says 'probable risks.' And even if there are probable risks, the law says that I'm still entitled to a license if the probable risks are balanced out by the probable benefits."

"And those are exactly the judgments I shall make," he said, "so that we have happy people." When he said that he motioned for his team to leave the room, so only he, I, and the secretary I had brought to take notes remained.

"Do you know what it takes to make people happy?" I asked.

"Eight hours' work not doing anything very arduous, then beer and television. Then they're happy. They really don't need anything more. What could they ask for?"

"They might ask a doctor to fix whatever ails them," I replied.

"Technically, I suppose, those kinds of problems could be fixed. But would the people be any happier? I don't think so. That experiment was tried. We gave them artificial hips and knees, and what was the result? They began to complain about other problems, and to demand that doctors do even more. My files are stuffed with applications for devices that their proponents claim will do wonderful things. Thousands of such applications. There's not a part of the body that someone hasn't proposed replacing with something made of plastic, metal, or some new space-age material."

"Why don't you license the devices?" I said. "Then we could see if they perform as advertised."

"For the sake of the people themselves," he replied. "It would be sheer cruelty to fix one problem because there will always be another, without end. We have to think about the stability of society."

"What do implants have to do with the stability of society? What do you mean?"

"Devices create instability by fostering the craving for more and more scientific advances. To do what? To defeat aging and death, which is ultimately impossible. Implants like yours are a menace to the stability of society, so the science that breeds them must be kept chained and muzzled."

As I sat there, too stunned to say anything, he launched a long soliloquy after which he abruptly left the room. I had heard this speech somewhere else. I couldn't remember where, but in my mind I could fill in the argument as I heard it from Mishra.

"People," Mishra said, "had a strange idea about scientific progress. They imagined that it could go on indefinitely, regardless of everything else, as if scientific knowledge were the highest good and truth the supreme value. But ideas began to change and the emphasis shifted to universal comfort, which can keep the wheels turning steadily; truth can't do that. What the masses really wanted was comfort, not scientific knowledge. In spite of that, unrestricted medical research still continued. Big business went right on talking about scientific knowledge as if that was what they sought, but in reality it was only profit which the corporations achieved when they acquired ownership and control of the science of medicine. People finally had enough. They were ready to stop enriching the doctors and the corporations, so the FDA began controlling things. It hasn't been very good for scientific knowledge, of course, but it's been very good for society. What's the point of fixing one pain or ache when you realize that such knowledge only leads to an infinite regress. But one can't have something for nothing. The stability of society has got to be paid for. You will help pay, Dr. Marino, because you happen to be too interested in knowledge."

I complained to President Bush about the FDA. I never received a response, but on the day of my panel meeting the chairman began by saying, "I would like to note for the record my strong exception to the charges you made in your letter to President Bush regarding the integrity of the FDA." The panel members all nodded in agreement, like puppets. Then they took turns chastising me.

The first said, "I am concerned about the very limited amount of animal data, and I am not convinced that Dr. Marino has provided all of the clinical data."

The second said, "I have not seen any probable benefits. I have seen some possible benefits, but not any demonstration of any real benefits."

The third said, "I believe we are dealing with very flawed data, and that there may be long-term problems due to unforeseen difficulties."

The fourth said, "I believe it makes no sense to put foreign materials into the body when perfectly acceptable surgical procedures are already available."

The fifth member of the panel, the chairman, said, "I find this submission does not meet the minimum scientific merit necessary for an experimental study."

That was how my grand effort to sell a medical device ended.

When I had started studying science I thought that it was the truest, most certain thing in the world. I had turned my attention to Dr. Becker's project because it seemed more valuable and important than anything I knew about. As wonderful as it would be to put a man on the moon, how much more wonderful would it be to understand how electrical forces made life. I was certain that there was a canonical method for gaining an understanding of the intricacies of bioelectricity, and that my task was simply to learn that method, and then implement it, like a batter learning how to swing a bat. After I had confronted the issue of health risks from EMFs, I saw that there were intractable differences of opinion and that it would be impossible to resolve the issue with the certainty or reliability that I naively had thought characterized everything that could be called "science." Nevertheless, I had still believed there was such a thing as a canonical method, one that scientists generally agreed upon, even if they disagreed about the meaning of the results they found. But in the wake of my defeat by the FDA, I understood that there was something in science that was deeper than method – there was desire, the reason why anyone makes any effort in the first place. Everything begins with desire. It influences both the choice of a method and the meaning attached to the data spawned by the method.

The heart and soul of the culture at the FDA, its very essence, was fear – fear of the consequences of failure. Now, I could fail; I could be wrong. And it would come as no great surprise to anyone. People would say, "Well, he tried, but it didn't work out. He thought he knew how the body would react to carbon fibers, but he was wrong. He just didn't know enough about biology." On the other hand, if the FDA approved an implant and then some people were injured because it failed to perform as expected, everybody would be shocked. They would wonder how such a thing could happen, and then you would hear a litany of complaints: "Incompetent bureaucrats;" "They must have been paid off by the company;" "If they really knew anything about science, they wouldn't be working for the government." Everyone at the FDA knew that this was how the public would react because that was how it had reacted in the past. How reasonable, how inevitable, that a culture should develop at the FDA to deny approval to *any* implant, whether of carbon fibers, or Teflon, or polyethylene. An implant that was not approved could not be sold, and an implant that was not sold and placed in the body could not fail and thereby injure a patient. The fault lay not with Mishra and his team. My enemy was the culture that they had imbibed. If someone ignorantly thinks that the science of knee-ligament implants ought to be as predictable as the science of levers or computers, then it is quite understandable when such a person reacts with bewilderment and disgust upon learning that a carbon-fiber implant ruptured and left a patient worse off than before the surgery. The true problem lies in the imagination of such a person, because that is where myths are found. The high opinion that people had of the reliability of knowledge in the life sciences was what had spawned the fear that had shaped the culture at the FDA, necessitating not only my defeat, but that of any proponent of a knee-ligament implant.

 Knowledge for Herman Schwan was information that could be sucked out of an equation which, itself, was the law that all things followed. Thibodeaux knew nothing of equations or laws, nor did he care about them. For him the so-called method of science was something that was dominated by assumptions, that is, principles he found desirable – the most grievous one being that anything *real* in biology must reappear in machine-like fashion, must be replicable on command. Mishra differed from Thibodeaux chiefly in that he had different desires. For Mishra, the stability of society was the paramount concern, whereas for Thibodeaux knowledge of mechanisms was everything. After having spent my whole life working in science, and approaching fifty years of age, a coherent picture of what science really was had finally emerged. It was as much about what we wanted the world to be as it was about the world in itself, maybe more. There had always seemed to be another mountain that I had to climb in order to be in a position to look down into the valley of understanding. Now I felt I was there, and I could see that they were going to have to rewrite the textbooks.

After I recovered from my experience with the FDA I realized that my years of dogged effort had unanticipated salutary consequences. First, they had been appreciated by those in my department, particularly Jim Albright, earning for me a repository of good will and affection. The experience I gained doing biomedical research was another benefit. The many studies I had designed and conducted, some better than others, had taught me the trade of doing research on animals and human beings in the only way that skill could be acquired. Those experiments led to many publications, sufficient in number for me to earn promotion to full professor with tenure, which occurred over the lone dissenting voice of Thibodeaux, who thought me undeserving because I had brought in no money from the National Institutes of Health, which by his lights was the only true mark of success in science, and the measure of one's worth as a scientist.

20

Power

───◆───

A lawyer called and told me about one of his clients, a man named Albert Goadby who was involved in a dispute with the Philadelphia Electric Power Company over the construction of a high-voltage powerline. The case had started in the normal way. The company decided it needed to move energy between two points, so it drew a laser line on the map, certified the route to the Pennsylvania Utilities Commission, and began construction. In the meantime, Goadby was building a house on a three-acre woodland he owned, ignorant of the powerline that was headed his way. One day an agent for the company appeared and told Goadby a powerline would pass over his house, so he had to move it. Goadby took down the house, filled in the foundation, and began rebuilding several hundred feet away from where the towers for the powerline would be located. When he was well past the point he had reached earlier, the agent returned and said that the company had changed the route, and that Goadby had to move his house again. Then the company condemned a 0.7-acre right-of-way through the middle of his property; ironically, the land they took did not include the original location of the house. Soon after that Goadby read about me and EMFs in an article in the *Readers Digest*, and he contacted Joseph Romano, the lawyer who had called me.

Romano couldn't have been out of law school for more than a year or two. I didn't know what had prompted him to take the case, youthful exuberance perhaps. Whatever the reason, it wasn't a wise business decision because Goadby had no money, and the company was politically powerful. What Goadby wanted most was to save his house. His next most important desire was to be able to build a third house somewhere on his land where he would be safe from EMFs. But the powerline would be so huge that its EMFs would blanket the entire three acres. So no conceivable result of the litigation could satisfy either of Goadby's desires. The best he could realistically expect was to be paid full value for his property, all of which Philadelphia Electric had effectively taken; but the company had offered to pay only for the 0.7 acres upon which the powerline would sit.

When Romano had complained to the commission that the powerline would be unsafe because it would subject Goadby to EMFs wherever he was on his land, the commission said that he should have raised the issue when the company had first presented its plan to build the powerline. But Goadby hadn't read the *Reader's Digest* article at that point, and so the only people who knew about EMFs were the company and the commission, and neither had raised the issue. Goadby lost that round.

Romano had also made no headway in state court. Judge after judge had ruled that the issue of EMF health risks should be decided by the commission, which was the expert in EMFs as far as the courts were concerned. Romano then made the desperate choice to go to federal court and argue that Goadby's constitutional rights had been violated – it was like *Mr. Smith Goes to Washington*.

I met Goadby the day before the trial. He was a tall, taciturn man, with only a ninth-grade education. For him, reading the *Reader's Digest* article was like throwing gasoline on the simmering fire of resentment he felt toward the power company. "I don't want to be no god-damned guinea pig for them," he told me.

After he had been divorced and laid off from his job, he retreated from life and lived on his land, which had a creek and many tall oak, maple, and hickory trees. The house was an echo of the one he had built and then torn down. He used boards from old barns to build the house, and had carefully located them so he could place the square-headed nails in their original nail-holes. The house meant everything to him. But from out of nowhere, Philadelphia Electric appeared. It destroyed what he had built, instigated another cycle of destruction, and salted Goadby's land with EMFs so that there was nowhere for him to live.

Romano's contention in the federal lawsuit was disarmingly straightforward. The EMF from the powerline would actually flood all the land, thereby depriving Goadby of its safe use. So Philadelphia Electric was actually taking more than 0.7 acres, it was taking the entire three acres. Goadby's constitutional rights to equal protection and due process had been denied in the state courts because he had never been given the opportunity to be heard on the health risk issues. Unless the judge recognized the validity of Goadby's constitutional claim, the issue of risks due to EMFs would be irrelevant, as it was when Goadby's case was in the state courts.

The trial was held before Judge Clarence Newcomer in downtown Philadelphia, not far from Independence Hall. Goadby was the first witness. He spoke plainly, with a demeanor that suggested the angry desperation of an animal caught in a trap. His grammatical lapses and poor syntax only added to his credibility. He came across as someone who probably wasn't smart enough to make up the stories he was telling about how Philadelphia Electric had deceived him.

On cross-examination Philadelphia Electric's lawyer handled Goadby with kid gloves. Romano was asking the court for an injunction to stop the powerline so that there would be time for the EMF issue to be heard. But the only uncompleted section of the line was the portion over Goadby's land. No court had ever stopped Philadelphia Electric, so the lawyer simply had no reason to grill Goadby.

When I testified the first thing I told the judge was that EMFs from the powerline would cover Goadby's land. Then I told him about some scientific studies that had been done using EMFs of comparable strengths. I had many published studies to pick from, but I chose only seven; two were mine, and I arranged with Romano to put one first and one last in the presentation sequence.

My study about how EMFs affected fracture healing in rats was first because we wanted to make it clear that what I had really been trying to do was use electrical energy to treat diseases, and that I hadn't started out to go after power companies. The Wever study was next because it showed that extremely weak EMFs could affect human body rhythms. A questionnaire study was third; it showed an association between EMFs and cancer in children, which I knew would get the judge's attention. I picked two animal studies to show the diversity of the effects that could be produced by EMFs; one study involved effects on the heart, and the other on the brain. I also chose a Russian study because it showed that EMFs affected growth in rodents, which was the same result I got in the final study that I described. Romano would ask me a question like, "Have you read the study by Dr. So-and-So?" or "Have you performed a study involving EMFs and such-and-such a biological endpoint?" "Yes," I would reply, and he would ask, "Would you please tell the court the results of that study and how they relate to the situation on the Goadby property?" Then I would do the rest. When I answered I looked at the judge as much as possible to emphasize that I knew what I was talking about.

That night I reflected on my testimony. Philadelphia Electric had provided the electricity for the lights that burned all night when I was in college learning about science. Now, there I was, a key part of a court case in which the company was accused of really bad faith.

In court the next day we didn't have to wait long before we got an indication of what Judge Newcomer's decision would be. Off the record, in response to some procedural questions, he told Philadelphia Electric's lawyer, "I'm going to see to it that what you did to Goadby you will never again do to a citizen of this state." Romano's eyes got as big as saucers and all the blood drained from the face of the lead counsel for the company.

We went on the record, and Judge Newcomer said: "This Court finds that Dr. Marino's opinions and hypotheses are persuasive and credible. At most, defendant's cross-examination of Dr. Marino showed that there is disagreement in the academic community about the conclusions that can be drawn from studies made of the effect of low-level radiation. Dr. Marino's opinion was that this exposure, which would be of a constant duration, posed a health hazard to plaintiff."

In a slow, measured cadence delivered at a level that seemed as if he were shouting, because the courtroom was completely silent, the judge said, "There is a distinct possibility that the easement taken by Philadelphia Electric may in fact and in effect be considerably larger than 150 feet wide. By virtue of the apparent encroachment of possibly harmful electromagnetic waves, the taking here may be twice or three times as broad in physical scope as revealed by Philadelphia Electric."

The courtroom was still and stone-silent as the judge peered at the row of lawyers sitting at the counsel table for the defendant power company and said, "Plaintiff is entitled to due process of law in the act of that taking. A necessary element of due process is that plaintiff be given an opportunity fully and fairly to make out his claims, in this instance against a public utility. Of course, if the taker conceals or fails to disclose the true extent of the taking all subsequent hearings findings are a sham. This court concludes that those dimensions may have been a misstatement, and that both Philadelphia Electric and the Public Utilities Commission knew or should have known that Albert Goadby was entitled to try to prove that the scope of the right-of-way sought was greater than stated by Philadelphia Electric. Because he was not informed of the nature of the taking, he was not alerted to make such arguments and the hearings on his claims were meaningless."

Romano's most formidable hurdle now loomed. The court had to inquire into the nature of the harm asserted by the plaintiff and balance it against the likelihood of significant harm to the defendant should the injunction be granted. The entire powerline had been built, except for the portion that was supposed to cross Goadby's land. The question therefore was whether the losses suffered by the company and the inconvenience suffered by the energy consumers in Pennsylvania that would be incurred if the injunction were granted was disproportionately greater than the loss Goadby would suffer if the injunction were not granted.

Judge Newcomer said, "This court is mindful of the equities present here. Plaintiff is not a wealthy man. His land is his only valuable asset. Defendants propose to raze his trees on approximately 20 percent of his land and to suspend overhead high voltage transmission lines. They will do so, having failed to reveal to plaintiff the possible nature of the injury he is to endure. The Court must also consider the public. It is true that electricity rates are affected to a minuscule degree and temporarily there will be a greater risk of power shortages, but this Court cannot conclude that the public is well served by a ruling that would contribute to the diminution of its rights. We are considering here the fundamental rights of citizens to be secure from arbitrary government action. Surely the interests of the public are better served by the protection of its basic liberties than by a callous sacrifice of the rights of one for the marginal convenience of many.

"Having made these findings and these observations, the Court will hereby order that Philadelphia Electric Company is enjoined from entering upon the land of Albert Goadby until such time as he has been heard in the appropriate administrative hearing of the Public Utility Commission to state his claims regarding the encroachment of an electromagnetic field upon his land by the proposed powerline or his claims have been heard in this Court."

Then, in accordance with normal legal procedure, Judge Newcomer asked for recommendations from the lawyers regarding how high a bond he ought to set. A lawyer for Philadelphia Electric jumped to his feet and demanded a bond of $5 million. Romano told Judge Newcomer that Goadby's only assets were his land, a truck, and the unemployment check that he would receive in about a week. Romano suggested that $200 would be an appropriate bond.

The lawyer for Philadelphia Electric replied swiftly, "I would find a bond of $200 totally unacceptable. Your Honor has to consider the cost to each rate payer, which includes most people in this courtroom, and the possibility that, should it ultimately be found that Mr. Goadby's action was brought frivolously, he could be subject to an action by the rate payers to refund their losses."

"I have heard enough evidence to be convinced it is not frivolous," Judge Newcomer replied.

The lawyer continued, "That may be so, but Your Honor has to consider the ramifications of your decision, and the fact that it will ultimately be brought to another forum. I think that is inevitable. I think Your Honor probably knows that."

"I wouldn't be surprised," Judge Newcomer said.

The lawyer continued his argument for a large bond, the effect of which would have been to end the case immediately, because Goadby had no assets. The lawyer told the judge, "Mr. Goadby is a property owner. What is wrong with him posting his property as security?"

The judge said, "I have considered this aspect of the matter before I came into the courtroom. If Mr. Goadby were to pledge his home, and if he should fail to prevail in this matter, he would lose his home. That to me is such a chilling effect on his right to really have an opportunity to have due process that I can't believe that the interests of justice would be served by that kind of an order. I am not for a moment going to delude myself with the idea that whatever amount I set on this bond, in the event that plaintiff fails to prevail, that it is going to compensate Philadelphia Electric for its losses. It is just not possible for a man in his position to do that. I am going to do what I think the interests of justice requires under these circumstances, bearing in mind that we have a very important constitutional right involved here of a citizen. I am going to set the bond in the amount of $500."

When Newcomer announced his decision, Goadby broke down and cried. The lawyer for Philadelphia Electric called the ruling unprecedented but made no other comment as he slipped away. For the first time in the United States, a power company had been enjoined from completing a powerline because of potential health effects. Romano, in contrast, hop-stepped down the steps of the federal courthouse, unable to contain his joy, and said to a gaggle of waiting reporters, "For once the little guy has defeated the omnipotent Philadelphia Electric. They were so sure of themselves. They never thought we had a chance. But we convinced the judge that we were right."

To another reporter he said, "They can never again not tell a homeowner about the scientific information on health effects. We destroyed the mystique that Philadelphia Electric is God and can never lose. The decision by the judge shows that there is equal justice under the law, and that the little man can be the giant."

Goadby said only, "I was proven right. I did it not for myself but for the public. Now I'm tired."

Two weeks after Judge Newcomer made his decision, it was overruled by Judge John Gibbons of the United States Third Circuit Court of Appeals. He accepted the company's argument that whatever risk there might be to Goadby was far outweighed by costs to the company, which it claimed were $350,000 per week. Romano appealed to the U.S. Supreme Court and Justice William Brennan effectively overturned Judge Gibbons's decision, but then Brennan quickly reversed himself and allowed the powerline to be built over Goadby's land. At that point Goadby, who was on the edge of a nervous breakdown, just threw up his hands and quit. Neither he nor Romano had any heart left to continue the fight.

For me too it was a bad time. Goadby and a few thousand other Pennsylvanians who lived along the right-of-way had lost a piece of their freedom. I couldn't decide who was better off, Goadby for knowing, or them for not knowing. The state and the company had dominion over the people, and they exercised it to take the right to be free from disease from some people so as to further the convenience and wealth of others. It was true, as Philadelphia Electric's lawyers said, that many people would benefit from construction of the powerline. From their perspective, freedom could be reduced to arithmetic, and causing cancer was an acceptable outcome. Incommensurate values had been balanced off against one another. I felt that something dirty was going on, and that I had blundered into it beginning the day I put rats in an EMF and decided that something bad had happened to them. It was as if the Utilities Commission and Philadelphia Electric and the courts all got together and planned to take away people's freedom. I knew there was no conspiracy, but the effect was the same because all these institutions had self-organized around the same principle. Nothing I said in court was untrue, it was simply irrelevant. It just didn't matter. But it seemed to me that it *should* matter. It was all too confusing.

21
Experts

When British Columbia Hydro, which was constructing a powerline on Vancouver Island, notified John Marton of its intention to take some of his land, he raised concerns about possible health risks, especially cancer. The company gave him a copy of a letter by a questionnaire expert named David Savitz in which he had said there was no proof that powerline EMFs caused cancer. Marton asked BC Hydro whether he or the company had the burden of proving safety, but received no reply.

In the meantime Sophie Antigone, another property owner, had become suspicious when the company would not guarantee that the powerline would not pose a health risk to her family. She had begun searching for information on EMF health effects, and came across various scientific articles that appeared to support her concerns about EMFs, including some articles that I had written. Antigone, who was a young mother with no education, complained to a local government board that the powerline might cause cancer, and she organized a citizens committee to work toward rerouting it away from populated areas. The committee filed a complaint with the Canadian provincial ombudsman.

At a meeting with Marton and Antigone, BC Hydro reiterated its position that there was no scientific evidence to indicate a conclusive relationship between EMFs and an increase in disease. Nevertheless, the company offered to purchase their properties at fair market value on the condition that they not pursue their complaint. Marton immediately accepted the offer. Antigone insisted that it be extended to everyone whose property was adjacent to the right-of-way, but the company refused.

In defiance of the scientific advice she had received from the company, Sophie made presentations about EMF health concerns to an elementary school parents' group, a teachers' association, and to the school board, and all three groups requested that BC Hydro reroute the line because of possible health effects. But the company, citing the increased costs, turned down the requests and said it saw no substantial evidence that the powerline EMFs would be a health risk. The company then hired the Environmental

Information Corporation, a consulting firm that specialized in reviewing research on EMFs, to give lectures to the residents about powerline EMFs. The EIC experts said that living near powerlines was no different than using microwave ovens, and that many blue-ribbon committees had concluded after thorough investigations there was no reason to believe that exposure to EMFs posed a risk to human health.

Despite these steps, public pressure on BC Hydro continued to rise, and the ombudsman turned the matter over to the British Columbia Utilities Commission. The company attempted to defuse the situation by extending the buy-out offer to all the residents along the powerline, as Sophie Antigone had requested earlier. The offer extension, however, had the opposite effect because many people who lived beside existing powerlines interpreted it to mean that BC Hydro recognized EMFs as health risks. The commission was then flooded with questions and complaints, and it responded with press releases attributing the furor to misinformation and lack of information. But that didn't stanch the public's concern, so the commission decided to hold a public hearing with John McIntyre, the commission chairman, as the judge.

British Columbia Hydro announced who their lawyers and experts would be, and McIntyre appointed an attorney named Karl Gustavson to represent the commission. The local residents, however, had no money to hire a lawyer or an expert. British Columbia Hydro had initially promised to pay for the services of an expert chosen by Sophie's committee, but the company withdrew its offer when it learned that I had been picked. The committee then organized bottle drives, telephone contacts of local businesses, and a flea market to raise the money to hire me. Donors were assured of a seat at a public lecture that I had been scheduled to give. The committee could not afford to hire both me and a lawyer, but the provincial government donated the services of a lawyer who, I soon learned, was worth exactly what the committee paid him.

I arrived on the island and was taken to a dinner in my honor where we had salmon that had been cooked ten different ways. Afterwards, at a school auditorium, hundreds of island residents gave me standing ovations, both before and especially after I had finished my lecture.

That night I decided to have a drink at my hotel before I went to my room. As I sat on a couch in the lobby a stranger approached and sat opposite me, and after a few moments he asked, "Did you notice anything unusual about this island?"

"What do you mean?" I asked.

"I've been here for three days but I haven't seen any houses with aluminum siding."

"I hadn't noticed that," I said, "but I've only been here for a few hours."

"I used to install aluminum siding," he said.

"I know nothing about that," I said. "Is it a good business?"

"It was," he said, "but people switched to vinyl, so I decided to try something new."

"What was that?" I asked.

"I went back to school and got my Ph.D.," he replied." I taught for a while but now I work for a company called EIC. I've been there almost two months, specializing in a new subject called electromagnetic fields. They're all over the place, especially near high-voltage powerlines."

"Is there money to be made in electromagnetic fields?" I asked.

"Yes," he said, "for people who can talk well about them."

"Are electromagnetic fields good or bad?"

"I'll be giving a speech about that tomorrow," he said.

Just then a man came running up to us. He was extremely thin from side to side but not from front to back, like a herring. "Dr. Marino," he said, "I'm Karl Gustavson. I represent the commission. It is grossly improper for opposing experts to talk to one another. You should immediately stop talking with Dr. Erdgas." Upon hearing that, Erdgas departed.

The next morning, at the hearing, Sophie Antigone told McIntyre that BC Hydro had not been forthright in explaining the potential health hazards of the powerline. "We are not radical people, and we are not trying to hurt anyone. But we feel as if we are part of a massive experiment, and we choose not to be experimented on."

When Erdgas took the witness stand, BC Hydro's lawyer performed a *voir dire* to persuade McIntyre that Erdgas was a true expert. He testified that he worked for EIC, but didn't mention that the company was owned by former employees of the Electric Power Research Institute or that EPRI was EIC's biggest client. He said he had earned a Ph.D. from the University of Oklahoma and, after completing fellowships in neurobiology and pharmacology at the National Institutes of Health and at Cornell, he had become an Assistant Professor at Johns Hopkins, which he left to work for EIC. He didn't reveal that fellowships were no big deal because any

Ph.D. could get one, or that he had been asked to leave Hopkins because his research had been below par. He said he had conducted research on the electrical properties of cells, but he didn't make it clear that the work had nothing whatever to do with powerlines, or EMFs. At that point the lawyer offered Erdgas as an expert on the subject of the biological effects of electromagnetic fields. McIntyre said, "He certainly seems qualified." The lawyer then asked Erdgas to give his opinion about the suggestion that powerline EMFs were hazardous to health. In response, Erdgas gave one of the best pro-industry speeches about EMFs I had ever heard.

"At the outset, I think it would be helpful to indicate the methods of analysis that I used to evaluate the scientific evidence. First, I identified the available scientific research that was relevant to the question of whether powerline EMFs can produce biological effects of any kind. The gold standard consists of data from laboratory and animal studies, because people won't participate in studies if they are expected to donate tissues for analysis. Of course it's not possible for me to discuss every single study, but I have considered all the EMF studies in forming my opinion.

"Second, I analyzed the reports to determine whether the experimental methods used were correct, the data was reliable, and the conclusions drawn from the data were sound. The criteria I used to make these judgments were those used routinely by experts. These include insuring that the results of a given experiment were internally consistent, quantifiable, replicable, and that rigorous statistical analysis was used to test whether the effect was real or due to chance.

"Third, I considered collectively the body of the individual studies that were analyzed in order to assess their consistency, reliability, and coherence.

"Adhering to this three-step analysis is particularly important because many of the EMF studies have been inconclusive and contradictory. It's not difficult to discern why. The conduct of experiments in this arena requires use of sophisticated engineering concepts, fundamental principles of physics, and thorough knowledge of biology. If one of these factors is missing, the results obtained will be artifactual.

"Although a broad range of responses to EMFs by animals, isolated tissues, and cells have been reported, few of the reported responses have been replicated. Of those that have been replicated, none are serious enough to be considered hazardous. Some of the animals sensed the presence of EMFs because of stimulation of body fur, much in the same way that we sense static electricity when combing our hair in dry winter weather. Graves showed that such reactions come and go in an instant, like the Cheshire cat which disappeared, leaving only a smile.

"Reduction of melatonin from the pineal gland has been reported sporadically in rodents and monkeys. However, the melatonin levels returned to normal after a few days.

"Graham performed thorough and well-controlled studies of possible effects of powerline EMFs on human performance and physiology. He evaluated physiological, sensory, neural, motor, perceptual, and cognitive function including respiration, heart rate, visual acuity, focused attention, short-term memory, time perception, information processing, and decision-making. No significant effects were observed on sleep, appetite, sexual activity, cognitive and physical functions, anger, fatigue, confusion, and depression, blood pressure, or temperature. There was a small change in heart rate, only about 3 beats per minute, which can be caused by a person taking a deep breath. There were a few minor changes in brain waves, but they were so slight that they fell within the normal range. Thus, powerline EMFs do not appear to affect animal or human health, or mental perception or performance.

"Animal experiments have been performed to examine the possible effects of EMFs on reproduction and fetal development. As a whole, these reports provide no conclusive evidence that EMF exposure constitutes a reproductive health hazard. Marino was the first investigator to conduct experiments in this area. He reported increased fetal mortality in animals exposed to powerline EMFs. However, Phillips, Graham, Seto, Sykov, and Rommerein all were unable to reproduce his results. Either Marino is wrong, or all of the other investigators were wrong, and in my opinion the former conclusion is more reasonable.

"A number of responses have been documented in cells and tissues. They include altered calcium levels in the brains of chicks, decreased RNA levels in the salivary glands of flies, enhanced DNA synthesis and enzyme production in various types of mammalian cells. These findings, however, have been extremely difficult to apply to the resolution of uncertainties about human risks associated with EMFs. The strength of the EMF used was well above that produced by powerlines. Moreover, there were no dose-response relationships. That is, the cell, tissue, or animal effect does not increase correspondingly with increased exposure level. The absence of a dose-response relation is contrary to the known mode of action of toxic agents.

"It is clear from all this evidence that there is no scientific basis for claiming a link between exposure to EMFs and health risks to humans."

When Erdgas finished his story McIntyre said, "I don't profess to be able to understand this at all. I never got past high school biology, so I couldn't possibly understand this science. But I want to say that I think your testimony was thorough and complete. Thank you very much."

For the longest time, the poor lawyer who represented the residents remained slouched in his chair with his head pointed down and one hand cupped over his eyes. Finally, he arose and walked toward the witness stand and began his cross-examination of Erdgas, who looked more confident and self-assured than when he had first taken the stand, if that were possible.

"Dr. Marino's research was published in well-respected and prestigious journals. Doesn't that indicate that the results were correct?" he asked.

"Not necessarily. Publication is evidence of merit, not perfection. The purpose of publishing articles is to allow other scientists to see and evaluate them."

"Then is it your opinion that the powerline will be safe?"

"There is no indication of potential health risks of concern."

"Doesn't it matter how long someone is exposed to the EMF?"

"I don't like to make sweeping generalizations, but a health risk is usually related to the amount of exposure only if something is proven to be a potential hazard to begin with, which is not the case with EMFs."

"Is it fair to say the causes of many diseases are unknown?"

"We have different amounts of information on different diseases, but there are very few where we know all the factors."

"How can you say that the powerline is safe if the causes or potential causes of various health effects are unknown?" the lawyer asked.

"If we don't know what causes leukemia, that doesn't mean that this powerline causes it, unless we have evidence," Erdgas responded, and then smiled broadly. After his testimony ended I told the hapless lawyer, "Congratulations, you had him eating out of the palm of your hand."

When I took the witness stand, that sorry excuse for a lawyer who represented the committee asked me to give my opinion regarding whether the EMFs from the proposed powerline would be hazardous to health. I said, "If you look at the many hundreds of studies, and you discount the ones that lead nowhere, that is the industry-funded studies, you're led to the conclusion that powerline EMFs are biological stressors. It is well established that chronic exposure to stressors promotes disease because it taxes the body's adaptive capacity. So, I expect that people who live beside the powerline will become sick more often, earlier, or both, than would otherwise have been the case. Exposing people to these EMFs without their consent amounts to involuntary human experimentation. If a proposal were made in a medical school to expose people to exactly the same EMFs without first obtaining informed consent, the proposal would never be approved. Maybe it would be approved in Nazi Germany, but not in America."

At that point McIntyre, who had been more or less somnolent during the day's proceedings, spoke up and asked me, "Could an honest and objective scientist conclude, based on the evidence to date, that EMFs from powerlines do not pose a health risk?"

"That would be against the weight of the evidence. To reach that conclusion you must ignore part of the data," I replied.

"Dr. Erdgas told us that it has not been proven that EMFs cause cancer or other diseases. Isn't that a true statement?" he asked.

"A better question would be to ask what Dr. Erdgas means by that statement. The laboratory definition is that a cause is a factor linked to an effect that doesn't occur when that factor is absent *but all other conditions remain the same,* or nearly so, which is a situation that can be created in the laboratory. Outside the laboratory there can never be such a cause, however, because the conditions in the world cannot be controlled. For cancer, therefore, we can only speak of predisposing factors, not laboratory-type causes. If Dr. Erdgas meant to apply the laboratory definition to the world outside the laboratory, his statement is true, but meaningless because such proof is impossible. If he intended the proper meaning of cause, his statement is false."

"Well I suppose we have a situation where there are conflicting opinions," McIntyre said.

"It's not the scientific situation that is the cause of the conflict," I said, "it's money. Every study that Erdgas relied on for support was paid for and run by the power companies. It's been my experience that if a power company has anything to do with a study, if it designs it, pays for it, analyzes it, or pays to have it analyzed, then the study is tainted and worthless."

"My paycheck comes from the provincial government," McIntyre said, "and BC Hydro is a Crown corporation, but that doesn't mean I can't be fair to its opponents. I do not understand your reasoning."

"It's not a matter of reasoning," I said, "but one of power. He who has the gold has made the rules."

When McIntyre had finished, the lawyer he had appointed to represent the commission, Karl Gustavson, began his cross-examination by announcing to me, "During the course of your evidence you've made a number of what can only be characterized, even in the most conservative terms, as highly controversial statements. Some would characterize them as inflammatory and potentially slanderous or libelous. You've made allusions to Nazi Germany, branded a host of scientists as little more than dishonest prostitutes, and made allegations regarding international –"

"Now wait a minute," I said, "wait a minute –"

"Allow me to finish my question and I'll let you –"

"This is ridiculous," I said. "I didn't make any slanderous statements, and I'm prepared to back up my testimony if you want to go into it. If you do, then ask me questions. Don't give a speech. I didn't come here to listen to your opinion."

"I want to ask you a serious question," he said.

"That's better," I said.

"You've made allegations concerning a conspiracy to suppress information. You said that certain studies had been kept secret and couldn't be accessed."

"What are you referring to?"

"Did you not say that power companies had worked together to suppress certain scientific data?"

"No, you're misrepresenting my testimony. It's possible that the decision to suppress scientific data was made independently by each company."

"Did you not say that certain studies had been kept secret?"

"Certainly. Anyone who knows the first thing about the powerline EMF controversy knows that. And that's not half the hanky-panky the power companies have done."

"Well, what I'm concerned about is that many of the statements that you've made have been cast in extremely strong terms. Have you given any consideration to the effect of that kind of statement on your position as a research scientist, and to the weight that people will attach to the results of your research, given the strength of your convictions and the way you phrase them in this kind of forum?"

"I told the truth," I said. "If you can't handle the truth, that's your problem. Although there are exceptions, the general rule is that the studies performed by industry are rigged to support whatever conclusion best serves the industry."

"Rigged studies implies that people are deliberately seeking to disguise or to hide the truth, or to falsify results. Scientists wouldn't do that," he said.

"You just don't know the territory, and you don't know what you're talking about," I said. "Richard Phillips received many multi-million-dollar contracts, which supported his lifestyle and thirty employees that answer to him. In return he provided power-company witnesses like Dr. Erdgas the ammunition to argue that powerlines were safe." With that, McIntyre brought my testimony to an abrupt halt.

As far as I could tell, no minds were changed as a result of the hearing, and the affair ended at more or less the same point it had been before I became involved. Some of the residents along the right-of-way accepted BC Hydro's buy-out offer and went to live elsewhere. The company quickly resold the land, subject to the provision that neither the new owners nor their heirs or assigns would object to powerline EMFs.

Just as the powerline conflict on Vancouver Island was ending, a similar problem was beginning in Palm Beach County, Florida, where some businessmen had built a gated subdivision of expensive houses. Their spokesman was a world-class diver who had won many Olympic medals, and his reputation had helped to sell many of the houses. A high-voltage powerline ran along one edge of the subdivision into a switchyard; there, power was delivered underground to the homes. The homeowners had young children, and the businessmen offered to donate land to the school board of Palm Beach County provided it built an elementary school on the donated property, which contained the powerline and switchyard. Land in the county was expensive, so the board accepted. The board built Sandpiper Elementary School beside the switchyard, and used the open space under the powerline as the playground.

Sherry Robinson was the mother of one child who attended Sandpiper. The Robinsons once had a son, but he had died from a brain tumor. At that time they had lived near a high-voltage powerline and an Air Force radar installation. Sherry had not immediately connected her son's tumor with EMFs, but after a while she had begun to wonder about it. She bought an EMF meter, and began to take measurements at Sandpiper. As she moved from place to place around the school, the meter needle danced wildly. When she placed it on a desktop, the needle wandered randomly during the course of a day. On the playground, underneath the wires, the needle became pinned against the stop on the right side of the meter.

Sherry told other parents about the measurements. Many of the fathers were dismissive because they knew that Sherry was no expert in measuring EMFs, much less in interpreting their significance. Some of the mothers, however, had read about a possible connection between EMFs and cancer, and they particularly worried about the implications of various questionnaire studies that linked childhood cancer and powerlines. When some of the apprehensive parents approached the school board, the superintendent told them that engineers at Florida Atlantic University had measured EMF values at Sandpiper that were comparable to those found in the average house. The superintendent also said that twenty-four schools in Palm Beach County were located next to powerlines, and that there had not been any problem with unusually high numbers of cancer among the children. When Sherry pointed out some alarming news reports, including my interview with Mike Wallace on *60 Minutes,* that had occurred more than a decade earlier, he told her that his children had attended the schools in Palm Beach County and that they had all survived. "Kids are used to EMFs," he said. "They've grown up with TV's, microwave ovens, electric blankets, and clock radios." When the power company told the parents that it would not consider undergrounding the powerline without conclusive evidence of harm, some of them petitioned the school board to allow their children to attend a school that was not located next to a powerline, but the board denied their requests.

The parents continued to complain, so the superintendent hired a local physician to determine whether the EMFs were harmless or harmful. He concluded, "I certainly believe that there are some effects on humans from exposure to EMFs. The question of whether or not this exposure causes disease is really the problem. To date, there are *no* scientifically reproducible data that EMFs cause cancer or any other human disease." The physician's opinion did not satisfy the parents, so the superintendent decided to hold a workshop to educate them about EMFs and powerlines. The list of scientists who might be hired to run the workshop included me, but the superintendent chose Phillip Keine, an expert in questionnaire studies.

At the workshop Keine told the people that the school was as safe as an individual's home, and that he would not be fearful in sending his children there. The parents, however, continued to worry about what the consequences might be if the children sat in an EMF all day and played in the even stronger EMF under the wires. When the school board refused for a second time to allow children to transfer to another school, the parents sued the school board. The power company volunteered to defend the school board and hired a famous lawyer named Patty Ryan. I agreed to testify on their behalf *pro bono,* and to encourage other volunteers to appear in court on behalf of the parents.

I did not ask David Savitz because I knew that most of his money for conducting questionnaire studies about EMFs came from the Electric Power Research Institute which, in my experience, had never permitted anyone in its herd of EMF contractors to testify in court. The questionnaire expert I did ask to testify was Nancy Wertheimer, who had written the famous report about powerline EMFs and childhood cancer. But she declined, as did Dr. Becker. They both had testified in an EMF case in Wisconsin where a federal judge had treated them contemptuously, almost calling them incompetent and ignorant. With their eyes fixed on the stars, those well-meaning but naïve people had stumbled into every kind of trap imaginable that had been set for them by the lawyers who opposed their testimony. I asked Harris Busch, the pharmacologist who had testified with great success against a power company in Houston that had built a powerline next to a school, and Allan Frey, who had become well known for his interest in the health consequences of EMFs. I asked Stephen Smith who was famous for his theory of cyclotron resonance which some thought could explain at a deep level how EMFs affected cells, and Ross Adey, who was the acclaimed inventor of the "window" theory by which EMFs were said to affect cells. All of those whom I contacted were unable, for one reason or another, to travel to Florida to testify on behalf of the parents. I went alone.

As I walked to the courthouse with a group of the parents on the first day of the trial, we saw a hawk attack a pigeon and knock it down. "That's a good omen," one of the mothers said. "It means we will win our case." Some of the other mothers smiled in hopeful agreement, but one said, "Maybe it means we will lose." One of the fathers, a computer engineer, said, "It doesn't mean anything."

Soon after the trial started, Patty Ryan called his expert, the same Phillip Keine whom the school board had chosen to run its EMF workshop. He was an old man with dark eyes, hair the color of lamp black, and a deeply furrowed brow; his demeanor seemed that of a melancholy man.

"What position do you hold?" the lawyer asked.

"I'm professor and head of the Department of Epidemiology at the University of Alabama," he replied.

"What is epidemiology?" Ryan asked, and Keine replied, "Epidemiology is a branch of science that tries to uncover the causes of diseases by actually studying people as opposed to studying animals."

"How do you go about that work?"

"By conducting surveys and analyzing questionnaires."

"Are there laws or equations for doing this, as in engineering?"

"No. It's all based on common sense and experience."

"What education do you have in epidemiology?"

"I have a doctorate from Harvard."

"Are you also a physician?"

"Yes. I earned an M.D. degree from the University of Vermont."

"What is your specialty?"

"I am board certified in preventive medicine by the American Board of Preventive Medicine."

"What sorts of studies have you made in relationship to electromagnetic fields?" Ryan asked.

"I have spent a considerable amount of time reviewing the literature and trying to come up with an overall picture of what is going on here," Keine replied.

"Were your efforts in relation to environmental or occupational hazards due to EMFs?"

"Well, I wouldn't call them hazards. Hazard, to me, implies something that it is known to be dangerous."

"Do you have opinions concerning powerline EMFs and cancer?"

"Yes. In my opinion there is no evidence that powerlines cause cancer. Even if there were, we could never know about it. And if we could, we could never prove it."

"Please explain the basis for your opinion," Ryan asked, and then riveted his attention on Keine, as if he expected Keine to disclose the secret of the universe.

"If powerlines caused cancer, there would be evidence of it. Some authors said that they thought there was a slight increase in cancer cases, but others reported the opposite result so, overall, the work amounts to nothing. Furthermore, production of electric power is ten times greater than it was in 1950, but there has been only a fivefold increase in cancer, which also shows that it is unrelated to EMFs. But even supposing that powerline EMFs caused cancer, we cannot do an experiment to prove it because no one can observe EMFs actually causing the cancers. If that were possible, whoever said that he saw it might be lying, so there is no reliable way to communicate the information. Therefore any effects of powerlines on people are unknowable, and even if they were known, they couldn't be proved."

"Is the Wertheimer study one of those surveys you referred to?"

"Yes."

"What was her conclusion?"

"She said that the likelihood of cancer was higher the closer the children lived to powerlines."

"What about the Tomenius study?"

"He claimed that there was some association between cancer risk and powerlines. However, the closer the home was to the powerline, the lower the risk. So this work went directly against the earlier one."

"Does that result make any sense to you, based on what we know about the laws of physics?"

"No. EMFs decrease in intensity as you move away from a powerline. So, under the laws of physics, the greatest risk for cancer should be among people who live closest to the powerline."

"What about the McDowell study?"

"It was a good study. He showed that the mortality rate for people who lived near powerlines was identical to that of the general population."

"What about the Savitz study?"

"He found an extremely modest relationship between living near powerlines and cancer risk, but no relationship between actual EMF measurements and cancer risk."

"I don't understand that. How could there be a relationship between living near powerlines and cancer, but not a relationship involving measurements?"

"The implication is that the EMFs could not have caused the cancer."

At that point Patty Ryan held up his hand as if to say, "Stop," adopted what was surely a feigned expression of surprise because everything he did in a courtroom was always well scripted, and said, "Isn't that illogical?"

whereupon Keine answered as they had planned.

"Not at all. Powerlines attract lightning. Maybe lightning caused the cancer. Who knows? There are a million possibilities. Even if there were a link between powerlines and cancer, which there isn't, the EMFs couldn't be responsible."

"Why not?"

"Because there are no powerline EMFs in the homes near powerlines, thanks to vectorization. That explains Savitz's results."

"I don't think I understand that, doctor, could you explain your answer?"

"Well, EMFs don't just come from powerlines, they come from all electrical appliances. Now, EMFs are vectors, which means they point someplace as, for example, the earth's magnetic field always points to the north. If you take a vector that's pointing in one direction and add a vector of similar strength but pointing in the opposite direction, the two vectors cancel out by means of a process called vectorization. I think that EMFs from electrical appliances canceled out the EMFs from the powerlines, and consequently the people who lived near the powerlines weren't exposed to EMFs. That's why Savitz couldn't find any relation between EMFs and cancer."

"Building on that idea, doctor, we all know that in some places the power companies have run more than one powerline in a right-of-way. Is the EMF greater if there are more powerlines in a right-of-way?"

"Multiple powerlines produce EMFs that tend to cancel out by vectorization. Typically, therefore, more powerlines means less EMFs."

"If the power company built enough powerlines, could the EMFs disappear altogether?"

"It's certainly possible."

"How many studies are there relating cancer to EMFs?"

"Probably more than fifty. The important thing to recognize about these studies is that we're dealing with human beings, and all their diversity, and that epidemiologists are merely observers because they don't really have any control over the exposure circumstances. As a result, epidemiological studies are notoriously unreliable. The really crucial question, I believe, is whether or not, considering the totality of the studies, there is any consistent link between EMFs and cancer. In fact, there isn't."

"Are you saying that maybe twenty-five studies show some correlation and twenty-five studies do not show it?"

"Yes, it's about an even split. But those that showed a link are not very convincing."

"What about those that did not show a link. Are they convincing?"

"Yes. They were good studies. They settled the matter, so there is no need to keep digging."

"What studies were not good?"

"Well, we've just talked about the Wertheimer, Tomenius, and Savitz studies. They are the major ones."

"What's an example of a good study?"

"McDowell."

"What are the major problems with the bad studies, those that appear to be positive?"

"For one thing, they don't have measurements of the EMF itself, so they just assume that because people lived near powerlines they received high EMF exposure. Another problem involves whether the amount of disease that occurred was really greater than what was expected due to chance. Epidemiology has some pretty serious problems because of its non-experimental nature. It's all just based on filling out questionnaires. Human nature being what it is, people who do these studies sometimes come up with interpretations that are at the edge of their data, or even beyond it."

"Doctor, what is beyond it?"

"Fantasy," Keine replied.

"Doctor, what about laboratory studies of EMFs on amoebas or cells or animals, or clinical studies of bone healing? Do you think these effects are real?"

"Some probably are. The important question is whether they are harmful."

"Do you know of any circumstances where harm was proven?"

"No," Keine replied confidently.

By the time the trial ended for the day, the spirits of the parents were at a low ebb. Keine's performance had impressed the judge, as anyone could plainly tell from the expression on his face, like that of a sports fan looking at his favorite athlete. What made the judge's reaction even more ominous, and intensified its meaning, was a newspaper article he had given to the lawyers. There had been a case in Georgia in which epidemiologists had reviewed twenty surveys and concluded that the birth defects of a little girl named Katie had not been caused by the contraceptive jelly her mother had used. Some pharmacologists and geneticists, however, had reached the opposite conclusion based on their analysis of laboratory studies. The judge had accepted their evidence and ruled in Katie's favor. The article that the hero-struck judge had given the lawyers was not the decision of the Georgia court, but rather an editorial in the *New York Times* which had excoriated the judge's decision as an "intellectual embarrassment" because it was against "the best scientific evidence." Only a fool, the editorial suggested, would take the word of laboratory scientists over that of an epidemiologist in a matter relating to human health.

The daunting task of assaulting Keine's testimony fell to the lawyer who represented the parents, a youngster named John Smith who was still imbued with the idea that there was a social purpose for practicing law beyond that of simply making a living. In that respect he reminded me of Bob Simpson, whom I remembered fondly from our work together when we had fought to dispute the rosy picture of EMFs that had been painted by the power companies in New York. But unlike Simpson, Smith had not prepared for the assault, and it was impossible for me to remedy that problem in the few hours we had together. So I quickly wrote a cross-examination for him to carry out that required no effort or knowledge on his part.

Smith began by asking, "Dr. Keine, do EMFs cause cancer?"

"There is no convincing evidence," Keine replied.

"Do herbicides cause cancer?"

"The data does not support that link."

"Does asbestos cause cancer?"

"Only in people who smoke."

"Does saccharine cause cancer?"

"Perhaps in laboratory rats but not in people."

"Do insecticides cause cancer?"

"There is no scientific basis to believe in that."

The rest of the cross-examination revealed that Keine had testified on behalf of many different companies, saying that their products didn't cause cancer notwithstanding the evidence that suggested otherwise, and that he had been paid $400 an hour in each of the cases. By the time he left the witness stand it was apparent from the judge's body language and comments that his respect for Keine had evaporated, which pleased the parents. In the beginning the mothers had pressed the case against the school board while the fathers humored what they perceived to be their wives' emotional but irrational fears of EMFs. The minds of some fathers had softened following Keine's direct testimony because his story had seemed to them to be an instance of an opinion determining the evidence and not one where the evidence determined the opinion, which is the way they had thought science always worked. But even for those men, it was more like a whiff of suspicion rather than a definite taste of something bad. Keine's performance during his cross-examination, however, led many of the fathers to voice the same emotions as their wives.

Although the parents had been pleased, I was ashamed of myself. Keine was a man trapped by his perceptions. No expert really wants to think that his science is incapable of adding meaning to the world, but that was exactly the situation in which Keine found himself. He had followed his perception of the world right off the edge, into nihilism. But even though his thinking was extreme, it was pure, and for that reason perhaps worthy of respect or at least of being confronted directly, not by means of eristic. Keine's other testimonies could have been grossly biased, as we suggested in the cross-examination, but his testimony regarding the EMF surveys could still have been true. Thus the judge was deceived twice, and a direct confrontation that could lead to the best truth was avoided, as it had been since the EMF issue had first arisen.

When I took the witness stand I tried to refute Keine as best I could, considering I was not speaking to him. I told the court that the proper way to make judgments regarding the hazards to humans was to evaluate the results of honest studies on animals and, in the limited area where they are permissible and not abhorrent, studies on human beings. I said, "Professor Keine shuns the correct methodology and instead resorts exclusively to analysis of surveys which he concludes lead to an aporia. Since his arguments apply to every factor in the environment suspected of causing disease, he showed only that the kind of activity he practices can never lead to certain knowledge, which was only a tiny point because nothing can. I accept his conclusion, but it is not relevant to this case. The issue here is not whether EMF survey studies are conclusive, but rather whether they add any weight against the school board's claims that the powerlines will be safe. Clearly, they do."

I thought my presentation was decidedly sub-par. For one thing, Smith had not gone over my testimony sufficiently and was therefore not in a position to maximize the impact of what I had to say. There were gaps in his presentation, and he asked many unnecessary questions, resulting in wasted time and momentum. We lost arguments that we should have won as, for example, when the judge would not allow me to answer questions regarding the process of vectorization. Smith himself did not understand what it was and therefore could not defend me against a side-bar argument by Patty Ryan that I was not qualified to testify about the physics of EMFs.

Another example involved the key concept of our case, that the average field in the school was much higher than that in other schools. I had prepared charts and illustrations that depicted this fact, with the idea that they would leave a lasting impression on the judge. Unfortunately, Smith did not lay a proper foundation in court for these exhibits, so the judge refused to accept them into evidence; that decision cut short my testimony because Smith had no props to use.

Smith's shortcomings was only one of the problems. The obsessiveness of the parents wore me out, psychologically. They always wanted to talk about the case, and since I worked on their behalf for free it cost them nothing to hound me. I was fatigued and tired, and concerned that I might not make my flight home. Smith asked me to stay for the remainder of the case to help him, but I returned to my real life.

While I was going about my business in Louisiana, Ryan presented an engineering expert who testified that he had measured EMFs at different locations inside Sandpiper, and also inside other schools in Palm Beach County that were not near powerlines. He told the court that when he had averaged the results he found that the children at Sandpiper were not exposed to more EMFs than children at other schools in the district, a fact that he explained on the basis of the vectorization gimmick. He never said how he chose the locations within each of the schools for his measurements which, of course, was the dirty trick he used to make the averages come out the way he wanted.

Soon thereafter the judge issued a decision in which he ordered the children to be kept away from the playground directly under the powerline, but allowed the school buildings themselves to be used without restriction because, he concluded, there were no unusual risks due to EMFs thanks to vectorization. He said that if the power companies built another powerline, perhaps vectorization would allow the children to use the playground again.

Soon after the case in Florida had ended, I was contacted by a lawyer from California who had brought a suit against the Pacific Telephone Company on behalf of his clients, Meyer and Muriel Silverman. They lived in a large house on scenic property in Riverside County, and the company had built an antenna on adjacent land and begun sending out a microwave beam, part of which passed directly through their house. The company told them that living in the beam was nothing to worry about, and that they would be just as healthy as they had ever been. But the Silvermans didn't trust the company so they sued under a theory of inverse condemnation, claiming that the company had taken away their right to be safe in their own home and therefore that they should be paid for their loss.

When the Silvermans' lawyer asked me for my opinion, I told him that there were numerous Russian and U.S. reports of biological effects due to many types of EMFs, including microwaves, that fields could affect the body's chemistry, and that I interpreted these facts to mean that an exposure to EMFs was a potential hazard, even though nobody could say exactly what would happen to the Silvermans or when.

He hired me to testify in court, and so notified the phone company. Shortly thereafter I was called by Don Justesen, who frequently emerged when there was an issue involving EMFs. When I had been fighting Philip Handler, Justesen had sent me a letter inviting me to Washington to meet with a man on Handler's staff so that "we can solve your teapot tempest." It was Justesen who, when he was president of the Bioelectromagnetics Society, had encouraged me to deliver what he called my "philippic" against Richard Phillips, thinking, I supposed then and still supposed, that I would be defeated in our battle of speeches. It was Justesen who came to Syracuse in the summer of 1980 and testified under oath that yet another tower planned for Sentinel Heights, where Dr. Becker had already found a high rate of cancer, would be "perfectly safe." He told the court in Syracuse that some people think otherwise because "human beings are a suggestible lot, witness the success of voodoo." It seemed that Justesen was shadowing me, as he had also shadowed Milton Zaret.

"I understand that you will be testifying in California in a case involving a cell telephone tower," Justesen said.

"Yes," I said, "unless the case is settled. How does that concern you?"

"I think that kind of testimony would be bad for you and bad for the country," he said. "These are still dangerous times."

"I don't see that my testimony on behalf of two people who are worried about being forced to live in a microwave beam has any larger significance than whether they should be compensated and allowed to move somewhere else."

"It's the principle, Andy," he said. "We are talking about a minuscule microwave beam, and it's not good for the country to foster a fear that, somehow, that would be a hazard."

"Wouldn't you be worried about consequences of living in the beam, if it were you?" I asked.

"Let me tell you what I did," he said. "I exposed myself for thirty minutes to an EMF that was strong enough to raise my body temperature by more than half a degree, and an hour later my body temperature had returned to normal and I suffered no ill effects. If I can withstand an EMF of that strength and suffer no ill effects then I think we can say there won't be any damaging effects due to microwave beams at the nanowatt level, which is what the tower will produce."

"So your opinion is that the Silvermans would get used to living in the beam."

"Certainly, like diamond miners in South Africa. If you or I had to do that work we wouldn't survive the first day. But they adapt. Even if there were effects produced by the minuscule beam, prolonged stimulation leads to acclimation, so the Silvermans would adapt."

"Well," I replied, "take the Russian reports, for example. They studied workers who were exposed on the job over a long period, but didn't find acclimation. On the contrary, they found that the workers suffered from chronic fatigue and tiredness, what was called a neurasthenic syndrome."

"Russia is a different story," he said. "In a socialistic environment, the state is required to take care of you from the cradle to the grave. So the state sets up safety levels to prevent disease and minimize health impacts. I have to agree that they are very safe, but in many instances, that number is not really enforceable. It's a goal to which the bureaucrats aspire, more or less pie in the sky. There is no real science there, only fear about what could be."

"This is the way I look at the situation," I replied. "The telephone company is run by businessmen, and they surely have access to more facts than I do about what sorts of medical problems might be caused as a result of living in a microwave beam. It would have been foolish for them not to have inquired into the matter, which has become an increasingly popular topic everywhere. Surely they made discreet arrangements with private outfits to research the topic. If I had been a lawyer for Pacific Telephone, I would have recommended such an inquiry, and I would have counseled the necessity of constructing effective firewalls to shield adverse results from prying eyes so that not even the possibility of concern could be detected by those outside the company. It would have been my professional responsibility to give such advice. So, I'm confident that the phone company actually received that kind of advice from its lawyers, which are the best money can buy. Now, had the evidence obtained supported the company's position that living in a microwave beam was safe, the company would surely have produced the scientific studies attesting to that supposed fact. But they have produced no such studies, which suggests to me that they have evidence locked away somewhere that leads to the conclusion that EMFs from the antenna aren't safe."

"Assuming, for the sake of argument," Justesen said, "that evidence exists which could be interpreted as indicating a hazard, you have to agree that EMFs aren't responsible for some kind of a rampant epidemic. At best, there are some isolated cases. Look at what the cost would be to try to prevent them."

"What cost?" I asked.

"I can tell that you have never been on a Navy warship," he said.

"True," I replied.

"If you go on an aircraft carrier and look around you will see a thousand antennas, more than half of them operating at any given time. They are the means by which all of the weapons and communications systems are able to function. Without the antennas, the ship would be useless. The sailors on the ship are constantly immersed in their EMFs, and at levels vastly greater than the minuscule levels on the Silvermans' property. What kind of a message do you think it sends to the sailors when you say there is a hazard on the Silvermans' property?"

"I suppose it tells the sailors that they too might have a similar problem, only worse," I said.

"Do you think such a message is a good thing, particularly considering the absence of clear scientific evidence of a problem?" he asked, more or less rhetorically. I told him that I just didn't see things the way he did, and I would not quit the case. Shortly thereafter the phone company announced that Justesen had been hired to testify on its behalf.

At trial, he began by telling the jury that he was a tenured professor of psychiatry in the school of medicine of the University of Kansas, and also the head of the Committee on Man and Radiation which, he said, was sponsored by engineers to educate and enlighten the general public about EMFs. When he started talking about EMF studies, he said there were two categories of problems, excessive exposure and misinformation. Excessive exposure came from EMF-producing equipment such as heaters and sealers in the plastics, leather, and lumber industries. All that these workers needed to know was that, when they felt heat, they should move their hands away from the EMF beam.

"Misinformation about EMFs," he said, "is a far bigger problem. If faith can move mountains, false beliefs can mount movements in which the shared illusion of danger creates psychological and even somatic disabilities in sensitive individuals. The false belief that low levels of EMFs have destructive affinity for biological systems is widespread, and has been fostered and perpetuated by accounts in the popular media. The fact is, however, that there is no objective evidence of a clear and present danger for the general population from chronic exposure to EMFs at current environmental levels."

I did not expect that the Silvermans' lawyer would make even a dent in Justesen's testimony during cross-examination because he was a hard nut to crack. It was also plain that the lawyer himself was not looking forward to the confrontation, and viewed it more as a task than an opportunity. I therefore proposed to him that I be withdrawn as the designated expert for the Silvermans and then appointed as counsel for the purpose of conducting a cross-examination of Justesen. I knew that the court would accept anyone with a Ph.D. or an M.D. as an expert, and that consequently it would not be difficult to hire someone to give the testimony I would have given. But the ability to deconstruct Justesen's testimony was a rare expertise, but one that I possessed. He leapt at my suggestion and asked permission to approach the bench, where he and the lawyer for the company held an extended side-bar conference – the upshot of which was that the judge admitted me to the bar in California for the purpose of conducting the cross-examination. At that point the lawyer for the company asked for an adjournment until the new expert was named and the company could prepare to cross-examine him, and the judge granted the request. The next day the phone company settled with the Silvermans.

• • •

How natural it was that experts should appear to be authoritative witnesses. Erdgas, Keine, and Justesen had academic degrees, so when they said in the courtroom that there is no reason to suppose that people exposed to EMFs will be more likely to get sick, they knew their words would be respected without any inquiry into why they spoke them or how they knew them to be true. Their message was inherently negative – that the research implicating EMFs as hazardous was meaningless – and people found this view comforting. All I could do was try to describe the meaning of that research, and in the process I probably appeared vague and alarmist.

They could pick and choose their evidence to support their message to the court; all they had to do was ask themselves what would help or hurt their case, and choose accordingly. By avoiding anything that might complicate their task, it seemed as if they had reached their conclusions inexorably. The choir I belonged to was quite different. My approach to scientific explanations was nothing like theirs. My answers to questions about the significance of experiments or about the conclusions to which they summed were never certain but always open to debate because the body of work was so vast and interconnected, and because, after all, it was only science and therefore inherently imperfect. Nevertheless, I believed in my heart that my views were far better and more defensible than theirs, even when the circumstances damaged my cause, as when I was confronted in court with the research results of Richard Phillips. When I laughed at that, I must have seemed frivolous. When I tried to explain why his work deserved only scorn, I must have seemed mean-spirited or biased.

These experts could deliver persuasive cases because they were permitted to give speeches and were never interrupted and asked to defend the particulars. Whenever I had the opportunity to directly ask them how they could maintain that it is right for people to be exposed to electromagnetic fields, even children and the elderly, I never got a satisfactory answer. But juries and others who sit in judgment are predisposed to believe whatever is said by anyone who sounds scientific and is not plainly biased, and lawyers do not take the trouble to teach the jury that experts deal in opinions, not exclusively in facts as is generally but wrongly supposed.

Nobody knows how this fault developed, but it was probably something like this. One day, word spread throughout the countryside that a woman had given birth to a sheep. The uproar greatly troubled the king, so he summoned his ministers to discuss how best to calm the people. Someone said: "My liege, the people distrust the opinions of the clergy. Let us choose scholars and doctors to investigate the matter," and so he did. It so happened that a woman had indeed given birth to a sheep, and that upon seeing the monstrosity and believing it to be the work of the devil, her family slew both. So, when the scholars and doctors held their meeting in the great hall and asked all who had evidence concerning the matter to come forward, the woman did not appear and the sheep was not produced. They then announced that no man's knowledge of the supposed event was more than hearsay and that no book in the great library recounted an observation that a woman had given birth to a sheep, and concluded that no such event had occurred. In the following years, the king called on the scholars and doctors more and more to resolve issues that concerned the people, who accepted all manner of opinions that they offered because it was presumed that they knew the world in a special way that was beyond the province of ordinary men. And so the people fell into the habit of never expecting nor demanding an answer in plain language from the learned men to the plain question, "How do you know?"

The danger in making scientists the objects of hero-worship arises because there are two kinds of scientists, neither of which deserves worship and only one of which deserves respect. The true scientist is free to pursue truth as best he can, but the counterfeit version has relinquished the free spirit he must have had when he was young because everybody starts off wanting to understand the truth. As he got older, he fell into grievous error. Some, like Keine, came to believe in nothing and hence to deny everything except the perfectly obvious. Others decided that truth, whatever it was, didn't matter as much as wealth, like Erdgas, or as the respect of authorities, like Justesen.

Keine, Erdgas, and Justesen had M.D.'s and Ph.D.'s, so what was lacking in their character that made them counterfeit versions of true scientists? It was that they lacked justice. This is where they failed. They ignored the reality that opinions about the presence or absence of a health risk due to EMFs always involve an element of justice. They were unscientific because they opined as if there was a right and wrong that was solely within EMFs themselves, and therefore that justice was irrelevant.

Experts are part of the modern world, so it will always be necessary to expend effort to distinguish between the two kinds. It is a deep problem because experts are chosen by lawyers, lawyers are chosen by clients who are desirous of winning, and sooner or later, everybody is a client.

22
Masters of Energy

◆

The Public Service Commission in New York had shocked the power industry when it ordered EMF limits at the edge of the right-of-way for new powerlines. No government agency had ever previously told a power company how it ought to design its overhead powerlines. The decision in New York was the start of a national movement. During the next two decades almost everyone heard something about how powerlines were a health hazard. They no longer seemed just part of the natural environment, like trees or mountains, but more like machines that might be dangerous.

The potential dangers of powerlines and the character of the men who built them seeped into the culture. In the movie *Ohms*, Ralph Waite and Dixie Carter were conservative Midwest farmers who rallied their neighbors against an arrogant power company that, with the help of the governor, was pushing a powerline through their farmland. In *The Distinguished Gentleman*, Congressman Eddie Murphy secured passage of a law intended to fund honest scientific research into the link between cancer and powerline EMFs; the premise was that the government had to run the program because the power companies couldn't be trusted. There were exposé-type books about EMFs, a series of articles in *The New Yorker*, stories in *Time* and *Newsweek*, and countless newspaper articles, all of which added to the cloud of suspicion that hung over the power companies. In addition, what looked like the start of a groundswell of litigation had started. Not just people like Goadby were bringing suit, but more powerful plaintiffs like a school district in Houston that had won a multimillion dollar judgment against a power company because of health risks from a powerline near a school.

Against this backdrop, power-company executives and experts organized an international meeting in Toronto to discuss the problem of the public's fears of EMFs. They invited me to give a speech. I agreed, thinking that the meeting would be a good opportunity to argue my case.

When I arrived I saw well-dressed, middle-aged men drive up in limousines and disappear down a long hall. I followed them, but guards stopped me and asked for identification, which I provided. They couldn't find my name on the approved list of attendees, so I was forced to wait several hours until they received confirmation that I had been invited.

Before I could enter the auditorium I passed through a metal detector. I knew it emitted strong EMFs, but my brief exposure was the price for the opportunity to speak directly to the management of the power companies. I wondered how the woman who operated the equipment, and therefore experienced the EMF more or less constantly, would have weighed the risk had she known about it.

By the time I got into the meeting the first morning session was over, and people in the audience were greeting one another and talking. Their nametags revealed that they were from Central Maine Power Company, Southern California Edison Company, Philadelphia Electric Company, Virginia Power, Ohio Edison Company, Omaha Public Power, San Diego Gas & Electric, Boston Edison Company, Allegheny Power Service Corporation, Pennsylvania Power & Light Company, Los Angeles Department of Water and Power, Jersey Central Power & Light Company, Interstate Power Company, United Power Association, American Electric Power Service Corporation, Texas Electric Service Company, Pacific Gas & Electric, Bonneville Power Administration, Central Power & Light Company, Wisconsin Public Service Corporation, Arizona Electric Power Cooperative, Platt River Power Authority, Texas Utilities Generating Company, Illinois Power Company, New York Power Authority, New England Electric, Niagara Mohawk Power Corporation, Duke Power Company, Tennessee Valley Authority, Carolina Power & Light Company, Consolidated Edison Company of New York, Alabama Power Company, Georgia Power Company, Detroit Edison Company, and Tampa Electric Company. There were also representatives from power companies in Canada, England, France, Italy, Sweden, Taiwan, Germany, and South Africa.

I walked around the circumference of the auditorium looking for the desk where I could obtain my nametag, which was no simple matter because the crowd was so large. Some of the men whom I had seen drive up in limousines were sitting onstage, talking to one another. After a few minutes they began to mingle in the crowd. One of the last people to leave the stage was Judy Klein, whose nametag indicated she was the Director of Public Relations for the Florida Power & Light Company. I asked her who the men were, and she said they were the top management of five power companies; following the scheduled speeches, they had been invited to form a panel and comment on them, and when the morning session ended they had begun a discussion among themselves on the topic of building high-voltage powerlines. She was excited about what she had heard, and was heading back to her room to transcribe her notes while the discussion was still fresh in her mind. I told her that I was Andy, from Louisiana, and that I too was interested in the topic. I asked if she would tell me what the various speakers said. She agreed, but warned that she had recorded only the passages she thought particularly memorable.

The first speaker had been Carl Rodriguez, who was a director of her company. His point was that powerlines were not only practical, but also inspirational. "The problem," he said, "is that we have a lot more people in Florida than ever before, and they consume energy at a significant rate. We must supply them with abundant energy, or there will be inflation as millions of people compete for an insufficient amount of energy, which would slow progress, or stop it altogether. We need to concentrate on generating more electricity and building more infrastructure. It's a huge enterprise. We need improved technology so that we can move energy efficiently over large distances and pull power off the grid whenever and wherever it's needed. We need bigger towers and new kinds of conducting cable, perhaps wires made from carbon fibers. I'm told that carbon wires would be six times stronger than steel and would carry a thousand times more current than copper. Just as we were inspired by the generation of engineers who designed and built the magnificent powerlines we have today, so too we need to inspire a new generation of engineers, a new Sputnik generation that will be of tremendous benefit to this country and to the world. This bold new enterprise will be good business, good politics, and most important, good for the soul. Our industry has a proud record and it must, and shall, continue to intelligently fulfill the promise of the slogan that first attracted us to this business, 'Live Better Electrically.'"

According to Klein, the next speaker was Hank Novak, a honcho of some kind at the New Jersey Power & Light Company. He emphasized that there was a right way and a wrong way to build powerlines. "In urban areas, it is important that powerlines be at least forty feet above the ground so as not to hamper fire engines whose ladders project high into the air. In the countryside, metal fences and large metal buildings must be electrically grounded so that no one inadvertently receives an electrical shock when touching the structure. The shocks are not harmful. They are like the sensation experienced when you walk over a rug and then touch a metal object. Nevertheless, shocks experienced along a right-of-way due to improperly grounded equipment can be annoying, and in rare cases, they can be dangerous. Theoretically, it's possible that a farmer fueling a tractor near a powerline could generate a spark thereby igniting the fuel. Another potential problem that ought to be attended to by a forward-looking power company that wants to do things correctly involves people who have pacemakers. There has never been a reported instance of a problem but, in principle, it is possible that powerline EMFs could disrupt the functioning of implanted pacemakers. The public should be warned that anyone with a pacemaker should avoid going near powerlines."

Klein went on to describe the comments of Norman Cook, who was in control at the Southern California Edison Company. "The previous speakers have ignored something important about building powerlines. They failed to consider how people perceive the risks of EMFs, and how the government responds to that perception. When people make decisions about risk, what counts most is the number of people killed or injured. But there are other important considerations such as whether the benefits and risks are imposed on the same or different people, and whether the risk is voluntary or involuntary. There is nothing irrational about such views. They reflect concerns about things like freedom, justice, and democracy that people hold to be important. By paying careful attention to public perception of risk, we can head off controversies before they occur, or at least before they gather momentum. People judge the probability of something happening by the ease with which they can imagine it happening or think of examples of it happening, neither of which people can readily do with regard to EMFs.

"Clearly, we want to try to avoid awakening any perception of risks. This is best done by means of a reliable flow of appropriate information. Audiences perceive the power companies' vested interest in concluding that powerlines are biologically benign. Therefore presentations by highly reputable blue-ribbon committees that corroborate this conclusion help to dispel the public's propensity for disbelief. The use of trustworthy institutions such as the National Academy of Sciences is a good way to be seen delivering information. Another cost-effective step to relieve public concern over new powerlines is a well-designed communication program. We should present a human face so people will feel less threatened. The message itself should be tailored to suit the audience, and it should be true. For farmers we should discuss research on milk production, crop yields, and livestock health in the powerline environment. For environmentalists we should emphasize the minimal nature of the impact powerlines have on the environment. Urban audiences have various safety concerns, and we should address them. Even though the objective is to educate the public, what we say will probably be less persuasive than the fact that we say it. You can't really educate people in the few minutes that they will talk with you, so the message content is less important than the simple fact that the power company has a knowledgeable staff that projects a caring attitude about public concerns. In this way, the power company is associated with the virtues of education and public service. These positive attributes provide a welcome contrast to the 'faceless juggernaut' portrayed by opposition groups.

"Governmental regulation of risk occurs in fits and starts, and is often based on political considerations rather than on attributes of the risks themselves. Government agencies can typically handle only a few tasks at a time, and usually only when a wave of concern sweeps through the media and through political circles. Then there is a lot of action after which, almost independent of whether the solutions generated are reasonable or effective, things quiet down and the agencies go on to worry about other issues. Care must be taken to assure that the political consequences of the public's perceptions lead to rational criteria for constructing powerlines, if there must be any governmental criteria at all."

The fourth speaker had been Robert Harvey, who was vice-president for research at the New York Power Authority. He told the group about the origins of the EMF health controversy, and how best to handle the arguments of the two scientists who had caused all of the problems with EMFs. "In 1974, when my company and the Rochester Gas and Electric Company tried to build high-voltage powerlines, a new issue developed: the claimed certainty of the effects of EMFs from powerlines. The specter of stress was raised by a biophysicist named Marino and a physician named Becker, who said it would cause a lot of disease and illnesses in people. At a lay level, what stress essentially means is that somebody's on your back, you're trying to do everything you can, the work's piled up, the phone's ringing, bills are overdue, and all those sorts of things. It's a general overall reaction. That's what 'stress' really implies. This is testable, so we hired experts from the University of Rochester and the University of Pennsylvania to do just that. What they found is that, after an initial reaction, the animal says 'that's it.' When they tried to run this response again, they didn't get it. The animals responded once, at a very moderate level, and then they returned to baseline very quickly. It's like the Cheshire cat in *Alice in Wonderland* which would appear without warning and then vanish. Is this indicative of stress? The answer is no. There's no evidence of it. Marino and Becker can throw out vague terms if they want to, vague notions of some kind of illnesses based on emotionality, but that's not science. Science says that on a scale of zero to ten, the likelihood of a health problem due to powerline EMFs is barely a one. Now, I don't know where science will be next year, or two years or three years down the road. Someone may find scientific evidence to back up the charges against us. Right now, however, no one has proved there is a health risk. The scientific evidence today does not point to any particular disease state that is caused by EMFs. No one can point to any real problem."

Next, Rice Harris, the senior vice-president and general counsel of the Houston Power & Light Company had told the group about how the legal environment hampers power companies in their delivery of what is undeniably the most important factor in the progress of society. He was far from being discouraged by the aberration in the law that victimized his company because the law has self-correcting mechanisms that are available to those with the desire and resources to invoke them. Consequently, things will come out right.

He prefaced his speech with a story that showed what it's like to be in the eye of the storm. "I felt like the frog who sat across from the fortune-teller who looked into a crystal ball and said, 'You're going to meet a very beautiful woman. She is going to be very interested in you, fascinated by you, and she's going to get very close to you.' The frog said, 'Where am I, in a singles club?' and the fortune-teller said, 'No, in a biology class.'" Then he described his case, which he said was impossible to understand in rational terms because it was based mostly on emotion. "We were in the process of acquiring seventy miles of right-of-way for a 345,000-volt powerline. Less than a mile of the right-of-way lay across a tract of about a hundred acres owned by the Klein Independent School District. The School District had an elementary school and an intermediate school on that tract, and was in the process of building a high school. We sought to acquire a one-hundred-foot strip of right-of-way adjacent to the schools. After unsuccessful negotiations with the school district, we took possession of the land and built the powerline. The school district then sued us, claiming we had 'grossly abused its discretion' in condemning their property. The theory of their case was that the schools were rendered unusable for school purposes because of the health effects of EMFs emanating from the powerline. Their witnesses told the court that it was scientifically possible that the children would be more susceptible to cancer as a result of the EMFs. Our witnesses said it was not possible. The verdict of the jury was 5–1 in favor of the school district, which was awarded $25,104,000 in damages. This was a first in Texas! We filed a mandamus proceedings in the Court of Appeals, and also appealed the judgment of the trial court. In our appeal we are raising sixty-five points of error involving the evidence and the way the judge handled the trial, as well as errors concerning the jurisdiction of the court. It's only a matter of time before the verdict is totally set aside. The ultimate result will be that the company will pay the school district only for the one-hundred-foot easement across its property, a matter of no more than a few thousand dollars."

According to Klein, Harris concluded his remarks by emphasizing the difference between science and law. "The purpose of science is to discover the truth about nature. The purpose of the legal system is to resolve disputes, so there must be a winner and a loser."

The break between the morning sessions ended just as Klein finished recounting what she had heard. When the platform presentations resumed the first speaker was Asher Sheppard, who had started his own EMF consulting company. His speech was like a description of reality by a bad poet. When he came down from the stage I said to him, with a smile, what I had told him the last time we had spoken: "You seem to shift back and forth so as to always support your client's position."

"That's harsh," he said returning my smile.

"When you worked for Chauncey Starr and Robert Flugum you didn't exactly do definitive experiments, and your speeches about the idea of EMF risk weren't exactly crystal clear, but when you wrote the decision for Judge Matias there was no mistaking what you intended to say."

He replied with uncharacteristic earnestness. "You're a tenured professor. I struggle to make a living and support a wife and three sons." Then he walked away.

I finally got my nametag, but I then had difficulty striking up a conversation so I took it off. By that time, however, it was too late.

Richard Phillips spoke in the afternoon, immediately before my talk. I was surprised to learn that he had severed all connections with the Electric Power Research Institute and had gone to work for the U.S. Environmental Protection Agency. But that surprise wasn't half of what I felt when I heard his speech.

"What is the likelihood that electric power facilities pose a hazard to human health, both long-term environmental and/or occupational exposures? I've been in this field for many years. I've been walking that narrow line, as Dr. Marino well knows. Based upon the data we have right now, I think the probability is high. The reason I say that is we do have interactions and effects in the nervous system, at the cellular level, in circadian rhythm, in all probability on reproduction and development, and possibly on cancer."

As we passed, him descending from the stage and me ascending, I made the sign of the cross in the air, like a priest giving absolution. He smiled.

I could see the audience was braced for my talk, but I surprised them. I would speak, I said, as if I were the head of a power company and had been asked by other top managers for my views about building powerlines, and then I began.

"There are those who claim that living near powerlines helps to make people sick. The notion is something like, 'If you live near a powerline you're not going to be as healthy as you otherwise would have been.'

"Sick in what way? In every way. There are many articles in the literature that link field exposure with many kinds of cancers. But not just with cancer; also with suicide, blood disorders, infertility problems, and other problems.

"How is it all supposed to happen? The only credible theory is based on stress. The idea is that when you are exposed to fields, it taxes the ability of your immune system to deal with the myriad other factors in the environment, and so your chance of getting sick is increased.

"Are the claims true? It depends on who you ask. 'No,' according to our blue-ribbon committees, but the opposition is driving this controversy and it answers with a resounding 'Yes.' They may be right.

"What are the legal problems? There are many lawyers in the United States, they could sue us under any of a litany of legal theories, and they work for contingent fees. Consider the analogy with medical malpractice litigation. It did not exist prior to 1960, and since then it has fundamentally altered the practice of medicine. Tobacco litigation and asbestos litigation are further examples. All the major factors that caused or contributed to the rise of litigation in these areas are present in the powerline area today, and things could explode at any moment.

"How have we responded? In the New York hearings, we said that there are no health risks because there are no EMF effects. Even though our own consultants have disavowed that position, our lawyers nevertheless continue to make that tired old argument. And when we do concede that there are effects, we say, 'An effect is not necessarily a hazard.' Yes it is, by presumption, until proven otherwise. Our position is inconsistent. It looks like bad faith, and it's not good for business.

"The EMF issue is not a scientific problem. A scientific question is something like, 'Does a cell make more protein when an EMF is applied?' That's not like the question, 'Are people made sick by being exposed to a powerline?' which is basically a political or social question. We should establish a national organization of generalists, not scientists, and charge it to elaborate a reasonable, self-consistent position. For many reasons, scientists are incapable of this task. For one thing, their thinking is too limited and too mechanical. More importantly, it's not their call because the question is fundamentally social, moral, and philosophical. The organization should listen to the people who disagree with us, not hide from them, oppose them, or seek to destroy them. It should ask, 'Why do you think what you think?' If someone makes no sense, if he's inconsistent or incoherent, then he's dismissible. But if his argument is rational, if his evidence is good, if it hangs together, then it's got to be factored into our position.

"The organization should stake out a coherent position for us, one with borders that responds to the evidence as it comes in. It's not in our long-term interest to build a false case. We should be truthful. We don't want to give away the family jewels, but we also don't want to deceive one another."

When I finished, the session moderator asked the panel of bosses for their comments. Carl Rodriguez spoke first.

"You suggest," he said, "that we shift our goals away from a narrow focus on expanding the industry and efficiently supplying power in the interests of our shareholders, and toward the goals of a broader constituency. You emphasize less the supplying of power, and more the noncommercial consequences you think are socially beneficial. Well, we manufacture and deliver electric power, the single most important factor in the progress of society. That is how we enrich society, even though our contribution is sometimes underappreciated."

Then, inspired by his own words, he looked at the audience and said, "We are like a band of brothers who stand together and fight for something worthwhile, something noble. Our companies feed society, so that it can grow. The very word 'company' comes from the Latin 'panis,' which means bread. We provide electricity, which is the bread of our society. Life is no more imaginable without it, than it is imaginable without bread."

Hank Novak tried to speak next but he started to hiccup so strongly he couldn't make himself understood and had to leave the stage, so the analysis of my talk by the power company authorities was continued by Norman Cook.

"It's only common sense that we should develop a variety of good-neighbor policies, but it would be irresponsible to go further and transform ourselves into the kind of creature that we are not. Our responsibility is to conduct business, not to do the work or emulate the behavior of biologists or physicians or government officials. We make our contribution through invention and innovation as we search for efficiency and economy in the task of providing electrical power. We are not a government, a charity, or a scientific institution that concerns itself with whether EMFs cause disease."

Robert Harvey was next. He remained seated as he spoke because, I supposed, his big rear end wouldn't slip between the chair arms if he rose.

"I'd like to be the first to say that Dr. Marino has presented a very interesting case. It has absolutely nothing to do with science, but it's very interesting. The problem is the vagueness and uncertainty of his claims. There is a great deal of danger here for the industry, and we would be foolish not to take it seriously. But scientific commissions have gotten together and dealt with the science, and made their findings available to state agencies and regulators. Their decisions were made on the basis of scientific evidence, not on the basis of emotion.

"One other point, that generalists not scientists should settle the issue. This is a palpable absurdity. EMF powerline research is an intrinsically arcane area of knowledge, and nothing good will result by subordinating it to laymen, who are incapable of comprehending it. Analysis by trained experts is required. If you put your faith in political scientists, economists, historians, philosophers, and others who cannot interpret data objectively, you're asking for more trouble than you can handle."

The final comment was made by Rice Harris; it was cryptic in its detail, but I got the drift.

"The emotional approach described by Dr. Marino," he said, "has been tried in court, and has even occasionally prevailed. Attorneys who specialize in this area, Patty Ryan for example, don't approach the topic that way."

In the evening there was a dinner fine enough for kings. Candles in crystal candlesticks on each table perfectly matched the cream damask tablecloths. Silver bowls held arrangements of daisies and roses, and the food was served on beautiful red and gold china. The menu consisted of lemon-basil seared striped bass served with sauvignon blanc, pepper smoked beef filet served with cabernet, and assorted sweets served with muscat.

I sat with executives from the Bonneville Power Administration, the Tennessee Valley Authority, and the Central Electricity Generating Board in England. They discussed among themselves how best to prove that powerline EMFs don't cause cancer. The man who ran the Bonneville Power Authority said there should be a questionnaire study. "When people see that there is no cancer epidemic they will understand that powerlines are nothing to worry about." Someone pointed out that it would be necessary to require that the people stay in their homes, to avoid the problem of confounding causes. The man in authority at TVA suggested that could be done by building prisons along powerlines, but then he realized that the metal bars would interfere with the EMFs so he withdrew the idea and suggested that people might be paid to remain in their homes. The head of the Central Electricity Generating Board objected to the cost, and said the proposal was impractical. He described a study his company had undertaken in which subjects lived for twenty hours a day in a specially designed room where they were exposed to simulated powerline EMFs. They were allowed one hour a day for exercise and showers, but were required to urinate in the exposure room so that their urine could be examined scientifically for the presence of stress hormones. Despite this strict requirement, there were many instances in which the subjects urinated in the shower, thereby depriving the scientists of important data.

The after-dinner speech, given by an official from a Canadian power company, was entitled "Corporate Responsibility Regarding Risks of EMFs." He said that when people read about the occurrence of a cluster of cancer cases near a powerline, it's only natural that they would think that there was a cause-and-effect relationship. Somehow, that sounds to them like a more reasonable explanation than does coincidence. But people are not stupid, he said. They can be made to see that the law of chance functions in exactly that way. As anyone who has ever gambled knows, events of a similar kind can and do occur in clusters, even though they are entirely unrelated to each other, and do not arise from a common cause.

The wine continued to flow after the speech ended, but I had had enough of power companies for one day so I headed for the door. Before I got there I was intercepted by three men. Their nametags indicated that they represented an electrical construction company in New York, but they told me that they were attorneys involved in an important EMF case, and one of them, Jack McGurk, asked me to join them for a nightcap.

The youngest of the group told me that he had not heard my talk because he had taken his wife to the zoo, and he asked me to summarize what I had said. I told him, "I said that powerline fields made people sick, probably because they cause stress. People can't rely on blue-ribbon committees or industry research because they are rigged. The chief problem faced by the industry is the potential legal impact. The industry should honestly reappraise its position and take into consideration the health of people exposed to EMFs." The lawyers were particularly interested in my ideas about the potential legal impacts of EMFs, and we agreed to talk later about the subject.

As I started to leave for a second time I bumped into Judy Klein. She had been on her way to the head table to join Chauncey Starr, the founder of the Electric Power Research Institute, Carl Rodriguez, some men I remembered from the Department of Energy who now held sway at various power companies, and several young women, none of whom I had noticed at the sessions of the meeting. The behavior at the table was somewhat rowdy; we saw one man wearing a bra on the top of his head. Klein changed her mind about joining them, and we struck up a conversation. She complimented me on my talk, and said my perspective was disarming, and helped to ease the hard feelings some of the big shots felt toward me. She didn't say that I had won anybody over, but she thought I had done the best job possible. Then she asked what I thought of Rodriguez and the others.

"They believe their companies are good because they're large, profitable, efficient suppliers of an essential product. They don't give a moment's thought to the dark side of their activities."

I told her about an African wildlife film I had once seen. There was a scene in which an antelope was running for its life, being pursued by a lion. Suddenly, the antelope started to give birth, and a lion grabbed the baby antelope even before it hit the ground.

"You see the power companies that way?"

"Yes," I said, "raw nature. A manifestation of some basic motivation."

"They appear that way to me too, sometimes," she said, "like the lions, only more intelligent."

"I would say more cruel," I said.

"They try to excel in a competitive world," she replied with a sense of resignation.

I agreed with that assessment and said, "They never seem satisfied. It's always more and more, bigger and bigger. They have a test facility in Massachusetts where they are testing a one-million-volt powerline, and I'm told that they have plans on the drawing board for a five-million-volt powerline whose EMF will be ten miles wide. I don't think it's only greed or arrogance that motivates them. The very nature of what it's like to be creative is involved. They are always ready to move on to the next project. It's a matter of self-expression. From desire to satisfaction, in an endless circuit."

"To me, that makes them very admirable. They can do anything they want. Isn't that something to be admired and respected?" she asked.

"No," I answered.

"Why not?"

"Because they injure human beings and don't care two cents about the harm they produce. Their engineering accomplishments don't wipe away their injustice. Men who are admirable don't hurt others, or themselves."

"How can you say that they are hurting themselves?"

"They have families. When they perpetuate the story that EMF pollution is harmless, they increase the health risks for their own families."

"Don't you envy their clout?"

"The emotion I feel toward them is anger."

She asked why, but before I could answer, Rodriguez came over and tried to induce Klein to return to the head table with him, but she wanted to hear my answer. So I gave it as Rodriguez stood and listened. "Because the men who head power companies are unjust toward the unsuspecting people who live beside high-voltage powerlines. It's the worst sin that a man can commit against another."

"Me too?" Rodriguez chimed in. "Am I unjust?"

"You don't look to me to be any improvement over the other company presidents," I shot back.

Klein tried to defuse the situation. "Suppose, under pain of death, you were required to either live next to a powerline or deceive someone else to do so. Which would you choose?"

"I think it's worse to be unjust to someone in that way than to experience the injustice," I said. "I would move my family into the house next to the powerline and take the necessary steps to shield out the EMFs, which I know how to do, but most people do not."

Rodriguez laughed, and Klein also thought that I hadn't given a very intelligent answer. She said, "I don't know many people who would agree with you."

"Can we at least agree on this point?" I said to her. "A man who intentionally injures an innocent person commits an unjust act."

"Yes," she replied, but Rodriguez only rolled his eyes.

At this moment Chauncey Starr, looking far older than his 74 years, came barging into our conversation; he was carrying a glass of wine, and accompanied by Adolph Gelling, his second in command and apparent heir to the leadership of the Electric Power Research Institute. Starr tried to get Rodriguez and Klein to rejoin him at the head table, but when he recognized me he screwed up his face, and the added wrinkles made him look even uglier. I took advantage of the moment by repeating what I had said to Klein a few moments earlier.

In response, Gelling descanted on the will to power of the power companies. "When you call the advantage of the power companies an 'injustice,'" he said, "all you actually do is drag up a common fallacy. It's only natural for the strong to have the advantage over the weak. Talking about justice in a college philosophy course is a good thing, but when someone gets older and still talks that way, he looks ridiculous. Was it adherence to justice that led Carnegie or Edison to be successful? Perhaps they dabbled in that notion as young men and as old men, but certainly not during their productive years. What you call justice is nothing but a rule made by the weak, the majority, to prevent the strong from dominating them."

Gelling paused for a moment, then continued, and Starr continued to nod approvingly. "Anyone in his right mind spends his life going after what he wants. Those who have courage and intelligence are able to get what they want. This is impossible for most people. To conceal their impotence, they claim that wanting greater profits is shameful. It is their own cowardice that leads them to praise justice, which is no more than an impediment to those with natural gifts and talents that enable them to accomplish great things. The truth, Dr. Marino, which you claim to seek, is that happiness is the power and ability to get what one wants. Everything else is an aberration, the unnatural slogans of the weak. You should thank God for the power industry. It is a beacon that shows the way to the future."

• • •

I had never been able to discern what actually motivated the men who ran the power companies, the Masters of electrical energy. Everyone knew it was cheaper to build powerlines in the air rather than in the earth, but more than economics had to be involved in the way powerlines were designed because the companies were regulated monopolies, so they always earned a profit, irrespective of their costs. My field of view expanded in the light of the meeting, and I began to think about those complex men in terms of their own values and outlook.

For the Masters, powerlines were far more than just highways for electrical energy. Powerlines had symbolic meaning – industrial development, progress, modernity itself – and they had no desire whatever to bury that symbol. On the contrary, they wanted to build bigger powerlines at higher voltages, without limit, thereby continually reaffirming the technological greatness of the companies and their role as the engine of progress.

The Masters had never answered to anyone concerning how they built powerlines, so I supposed it was natural for them to regard my claims as an interference with their prerogatives. But they also resented the nature of my message. I did not present clear, persuasive, unambiguous, replicated evidence that EMFs were health hazards. From their point of view, this fact meant that EMFs were safe, or at least ought to be regarded as safe until science definitively proved otherwise. What Dr. Becker and I had said about the implications of the animal studies, the meaning of the questionnaire studies, and the mechanistic role of stress simply made no sense to them within their discourse.

Then I got even closer to the truth. The Masters were conducting a vast unplanned experiment on unknowing subjects. I struggled to understand how they could embrace that course of action, which seemed to me to be evil. What explained their decision to design powerlines with complete indifference to the health of those who would live in the transported energy? Only hard men could do that to innocent people. I felt there had to be something more than misplaced pride, a sense of resentment, or an idiosyncratic view of the meaning of scientific data. Eventually, I came to understand that the peculiar morality of the Masters was responsible for their behavior, their belief they had the natural right to do whatever they wanted. That perspective enabled them to ignore objective morality or morality rooted in humanism in favor of a morality that was entirely self-referential. They viewed getting what they wanted as "good" and not getting what they wanted as "evil." They were driven toward technical excellence, not as a goal to be achieved but as a push to accomplish something they perceived as noble. That push, that will, was what motivated them. It was the will to make the companies as profitable as they could possibly be.

After I understood why the Masters behaved as they did, I felt strangely at peace, even though I thought their morality was scandalously egoistic. My prospects for converting them to my view of EMFs looked hopeless, but still I felt fine. Whether or not their will ultimately prevailed wasn't my problem. I was doing exactly what I thought I ought to do. I was still in control of my own life.

But while the Masters had been going about asserting themselves, they had begun to hear ominous rumblings. The public was becoming alarmed, regulators were beginning to be concerned, and plaintiffs' lawyers had started to show interest. Something had to be done. I was not so naïve as to think that they would change their moral perspective based on my speech at the meeting, but I did think that their behavior could improve for reasons that were consistent with their code of conduct. Perhaps they might begin to protect the public health not because it was just, but because it was something good for the companies, something that perpetuated their hegemony and allowed them to remain as the alpha baboons of EMFs.

There was a troop of baboons that lived in a zoo. They were governed by an alpha male who did whatever he wanted. When he wanted the comfortable spot in the tree he would take it, and any baboon that happened to be there was forced to scamper away. When the colony was fed the alpha baboon ate his fill before any other baboon could eat. If he wanted to mount a female, he mounted. If he wanted to throw things, he did. All the other baboons resented the alpha, but they were powerless to oppose him.

Then one day word began to circulate in the colony that a new zookeeper would be coming who despised the alpha baboon because of the power he held over the colony. Over the next few months the alpha's anxiety level progressively increased because of rumors that the new zookeeper planned to have him destroyed. He remained in a position of absolute power, but at the same time began to exhibit self-restraint, like the beta males and the females, hoping that the other baboons would speak well of him when the new zookeeper came.

The gist of my advice to the Masters had been that they should modify their behavior, like the alpha baboon. Then the companies would appear reasonable to the people, the betas and the females, who would then say some good things about the companies as the truth about EMFs emerged. The companies might remain dominant because they appeared moderate, even though in their heart they were the same as they had always been.

Everything depended on the new zookeeper. If he did not appear, the alpha baboon would stop acting as if he were just another baboon and would resume his previous behavior.

Why should I even hope that a new zookeeper *could* appear? The attitude of the Masters toward science was more or less the same as that of everyone else. There was a shared belief that scientists were invested with powers over the natural world like those exercised by priests over souls. This belief was what made it possible for the Masters to do as they pleased. Amazingly, people bought the daydream that scientists had the ability to identify true and certain knowledge. Science had billed itself as "enlightened," free from the dark superstitions of the past. What it actually did, however, was develop its own myths. Not only was there the myth of the wisdom of scientists, according to which they could see through the mist into the truth of things, but also the myth that there always was a truth about nature that could be discerned if only we looked in the correct fashion, like little green men. The myths were new with Newton, Maxwell, and Einstein. No ancient scientist ever had such access to truth. Only the prophets had it, by virtue of their contact with God. Through a process of historical development, the gaze of the scientist replaced God as the source of "truth." Technology spawned by science reinforced the myths, and they became internalized. Finally it was not even necessary that individual scientists demonstrate the penetrating power of their gaze – it was automatically ceded as part of one's status as "scientist." What had become hidden, in this magisterial process, was any sense of the complexity of nature, so the ordinary person did not understand that the best science could ever hope to do was understand living beings imperfectly and uncertainly, and to achieve knowledge about them that was only slightly better than a guess.

The structure that had evolved was ripe for exploitation by anyone with the will and the financial resources. Robert Flugum, Governor Carey, Samuel Koslov, and Paul Tyler had such a will, and so did the Masters of energy. The Masters staged innumerable dramas – committees, commissions, meetings, seminars – that were no more than puppet shows where experts, whose jaws were worked by the Masters, offered reassuring risk assessments. In this way the Masters induced people to discipline themselves so that they believed the way that the experts desired them to believe. The will of the Masters was thus given effect not by means of a gun or a whip, but by defining it as "knowledge" and embedding it in the discourse that people accepted as rational and used to shape their behavior. It wasn't the companies that corrupted science, it was science that corrupted the companies. The Masters were not responsible for the scientific discourse they were presented, that glass globe which surrounded them and limited their thinking about the power of science.

It was as though we were in a historical cul de sac. The better, more honest truth about powerline EMFs that was available was not consciously ignored by the Masters, they simply could not see it. Even if a new zookeeper appeared it probably wouldn't matter because the Masters, the brainwashed brainwashers, wouldn't see him.

23

Razzle-Dazzle

My old adversary in New York, the Power Authority, decided to buy cheap Canadian hydro power and build a 345,000-volt powerline through Oneida, Herkimer, Otsego, Delaware, Sullivan, Orange, Ulster, and Duchess Counties so that it could sell the electricity in New York City, where the rates were the highest in the U.S. As usual, the Masters at the Power Authority took whatever land they wanted and built the powerline as they saw fit. Their press releases, which said there was no persuasive evidence that powerline EMFs caused cancer or any other disease, were incorporated into articles in many small-town newspapers. Most landowners along the route settled for a nominal amount as compensation for the land that the Power Authority had taken, but some turned to local lawyers for help. They soon learned that dealing with the Masters was arduous, and one by one the cases disappeared, leaving about fifty landowners who were all represented by Jack McGurk, the man I had met the night I gave my talk in Toronto. He offered me ten times what I had been paid to speak in Toronto if I would appear in court in Goshen, New York, where the case would be tried, and explain why I thought the powerline would be a health hazard.

At the same time the Electricity Commission of New South Wales in Australia decided to build a 500,000-volt powerline between two of its substations. The Masters drew a straight line between the substations and then announced their intention to acquire a hundred-foot-wide strip of land ninety miles long. Landowners in a shire west of Sydney objected on health grounds to the siting of the powerline, and asked that it be built in a corridor that already contained a powerline rather than on privately owned residential and agricultural land. The Masters denied the landowners' request, pointing to the conclusion of the World Health Organization that there was no possibility of danger to health. Peter Peterson, a solicitor at the Sydney law firm hired by the landowners, contacted Ross Adey, a transplanted Australian who had become a famous EMF researcher at the Brain Research Institute of the University of California at Los Angeles and later at the VA Hospital at Loma Linda. Adey told his former countrymen that he sympathized with them but declined to become involved. Peterson then hired me to testify about my perspective on the EMF health issue.

While these events were taking place, the Masters of the Central Power & Light Company in New Jersey decided to build a powerline to connect the Freneau substation in Matawan with the Taylor Lane substation in Middletown. They anticipated opposition based on the EMF health issue, so they published a booklet that said, "The suggestion by some that a powerline of this size is of any health concern has never been substantiated. The scientific community finds little or no controversy regarding the conclusion that there is no established relationship between powerlines and public health. If you would like more information about the conclusions drawn from these studies we will provide you with a summary of the assessment of Dr. Phillip Cole, a Harvard Medical School-trained doctor." The Middletown officials wanted the powerline built underground but the Masters refused, so the parties decided to settle the issue in court. Melvin Greene, the attorney for Middletown, asked me to testify.

Besides the offers from McGurk, Peterson, and Greene, I accepted offers to testify from lawyers in Texas, Pennsylvania, and Connecticut, all of whom represented clients who found themselves on the opposite side of a dispute with a power company regarding the health risks of EMFs. It had become apparent to me that the issue could never be decided on a strictly scientific basis, like determining how strong a building needed to be to resist a wind surge of a stated magnitude, because far more than the results of measurements and calculations was involved in formulating an answer to the question about EMFs. I hoped I would meet a Bob Simpson in one of my cases, and that together we would produce a legal precedent concerning what it meant to say that powerlines were health hazards.

I had a fairly good grasp of the spot I was in. If you are the proponent of the idea that something is a health risk, you are like the wife who goes to court to get a restraining order before she is beaten. Should the court issue the order to obviate the possible harm, or deny the request because whether she will ever receive a beating is conjectural and speculative? Should she be required to have at least one knot or black eye before she can get a restraining order? Suppose her husband then says, "I didn't do it. It must have been somebody else?" There are no easy answers, but there needs to be a rule.

McGurk called and told me that Patty Ryan would represent the Masters at the Power Authority. I soon learned that Ryan would represent the Masters in all the other cases in which I had committed to testify. He was a flamboyant toxic-tort lawyer who had a national reputation for defending chemical companies. His success was based on skillful courtroom histrionics and on hiring the most authoritative experts. I told McGurk, "Ryan will hire world-renowned scientists who will tell the court that there is no persuasive evidence that EMFs caused cancer or other diseases, and therefore that the powerline will be safe. What will be your theory?"

"Their experts have their opinions," he said, "and my experts disagree. I'm not trying to get the court to declare the powerline a health hazard. I just want the judge to recognize that scientists disagree on the issue. Then I think he'll have no choice but to order the power company to pay damages to my clients."

"Wouldn't it be better to depose the witnesses and inquire deeply into the basis of their opinion?" I said.

"That would be expensive," he said, "and too hard to do."

As I prepared for a skirmish in Goshen, Ryan prepared for global nuclear war, although the details of what he was doing didn't become public until later. By letter, phone, and direct contact, either personally or by agent, Ryan approached the cream of the American scientific establishment, professors at Harvard and Yale, scientists at the National Cancer Institute, and other eminent experts, and asked, "What do you think about the claims of health hazards from powerline EMFs?" Some who responded were strident in their belief that the evidence of risk was not persuasive, none more so than a group of physicists who taught at Harvard, which was a hotbed of discontent over the national prominence to which the issue of EMF-induced cancer had risen. Richard Wilson, the chairman, and Nicolaas Bloembergen, a Nobel laureate, both said that the notion of adverse health effects from powerline EMFs was inconsistent with the known facts of physics and biology. Professor Robert V. Pound, who had been awarded the National Medal of Science, and Sheldon Lee Glashow, another Nobel laureate, went even further and concluded from what they had read that EMFs cannot cause cancer. Walter Gilbert, still another Nobel laureate, said that the absence of a well-understood mechanism by which powerline EMFs could affect biological systems precluded the existence of any such interactions, and that he thought this fact should put to rest concerns raised by some apparently positive questionnaire studies.

Robert K. Adair, a physics professor at Yale, had much the same opinion. Further research on EMF health effects should be prohibited, he told Ryan and anybody else who would listen, because it diverts resources from real environmental problems. Adair was a baseball fan and an expert on the physics of the curve ball, which he used as a metaphor for the social significance of EMF research. Rosalyn Yalow, a Nobel laureate who taught at the Albert Einstein School of Medicine, and Glenn T. Seaborg, who had discovered plutonium and abbreviated it *Pu* as a joke, were two other enthusiastic endorsers of Ryan's position, as was Allan Cormack, a professor at Tufts University and a Nobel laureate.

The experts voiced their unchallenged opinions, in newspapers, magazines, on TV, and on radio, that everything I said about EMFs was completely wrong. Out of the sea of uncertainties, these experts selected those presuppositions, studies, and methods of analysis that led to what they believed was the right conclusion, which was really the object of their desires. Each of the experts was like a child who goes into a candy store and walks past the jellybeans and lollipops and points to the Hershey bars. "I want that." But they were all old men, or in the case of Rosalyn Yalow an old woman, and none would agree to undergo the rigors of participating in trials and being cross-examined. It wasn't long, however, before Ryan had signed up prestigious experts who did agree to testify in court. His team included Margaret A. Tucker from the National Cancer Institute, Richard Bockman, Professor of Medicine at Cornell University Medical College, Edward Gelmann, Professor of Medicine, Anatomy, and Cell Biology at Georgetown University School of Medicine, Aaron Stuart from the National Cancer Institute, and Roswell Boutwell, Professor of Oncology at the University of Wisconsin. Ryan also signed up Herbert Terrace, a professor at Columbia University, Jan Stolwijk, the head of the Department of Public Health at Yale, Ken Zaner, a professor at Harvard, Lucius F. Sinks, from the National Cancer Institute, Edmond A. Egan, Professor of Pediatrics and Physiology at the State University of New York in Buffalo, Darwin R. Labarthe, a professor at the Baylor College of Medicine, and Michael Repacholi, who was the head of the section on EMFs at the World Health Organization. None had any experience conducting EMF research.

The first of the six trials was the one in Goshen. It took place in Judge McCabe's courtroom.

•••

McCabe's bench is like a high pulpit in the front of the room. The witness box is at his left, as are the jury seats which remain unoccupied because it is to be a non-jury trial. The clerk's desk is at the right. There are two long tables, one on each side of the center aisle, with chairs for the lawyers and others on their teams. A high rail, with center passageway, shuts off the plaintiffs and others in the audience. None of the Masters is present.

Judge McCabe is fussing with papers on his bench; he is in his sixties, and needs a shave. The clerk patiently holds the Bible waiting for court to begin. The bailiff is standing, gavel in hand; the stenographer is sitting demurely behind her machine. Ryan is in motion, always working. Slicked-back black hair that glints in the morning sun streaking through the high-arched windows, an Armani suit, a white-on-white shirt with a line of fine lace, classy but not effete. He has a happy camaraderie with everyone in the courtroom. A knowing look to the bailiff, a shake of the head toward the clerk, or a smile for Judge McCabe, as if to say he knows that McCabe's not taken in by all this stuff about EMFs. I'm sitting at the counsel table, beside McGurk.

Court begins. Ryan calls his first expert. I see the clerk hold up a Bible, but all I hear is, "blahblahblahblahblahblah... truth...truth...truth... help you God."

"What is your name?" Ryan asks.

"Aaron A. Stuart," is the reply.

"Where do you work?"

"At the National Cancer Institute, in Bethesda, Maryland."

Ryan extends his arms, looks directly at the judge, and slowly repeats, "the National Cancer Institute."

Q: What do you do?
A: I study the molecular causes of cancer.
Q: What is your present position at the National Cancer Institute?
A: I'm chief of the laboratory of cellular and molecular biology.
Q: Dr. Stuart, where did you get your undergraduate training?
A: At the University of California at Berkeley.
Q: And in what subject did you receive your degree?
A: In chemistry, and I also got a Masters in biochemistry.
Q: After that did you attend medical school?
A: I attended medical school at the University of California in San Francisco.

Q: Did you have any training after medical school?

A: I did a fellowship at the University of Cambridge in England, and then an internship in medicine at the University of California.

Ryan listens carefully to each answer and registers the proper emotion, respect.

Q: After you graduated from medical school and you completed your fellowships, et cetera, did you then start to work?

A: I went to work at the National Cancer Institute and eventually I became chief of my own laboratory.

Q: Have you received any recognition for your work in your fields of expertise?

A: I have received many awards, including the Meritorious Service Medal from the Public Health Service, the Rhoads Memorial Award from the American Association for Cancer Research, and more recently I received the Paul Erhlich prize. It's named after a Nobel prize winner. I, myself, have been nominated twice for a Nobel Prize.

Q: But you haven't won one?

A: *(smiling)* No, but I'm only 43.

Ryan smiles, and nods. McGurk is looking out the window.

Q: What does your laboratory, the one that you are head of at the National Cancer Institute, do, just in simple terms?

A: We try to understand the molecular basis by which normal cells become cancer cells.

Q: Have you personally conducted molecular and cellular genetic research on the causes of cancer?

A: Yes, I have.

Q: Could you tell me roughly how many times your research has been published?

A: Roughly 340 papers.

Q: I assume you probably don't get to be head of a laboratory unless you have done that, do you?

THE COURT: Depends. Sometimes politics, you know.

Ryan and Judge McCabe exchange smiles. McGurk sits at the counsel table, bored.

Q: Is it common in your field of expertise for researchers to investigate new or novel, different, unusual, bizarre hypotheses about cellular and molecular causes of cancer?

Ryan is almost apologetic for asking this question but, after all, it is what the trial is all about.

A: We certainly look at a variety of potential causes of cancer, yes.
Ryan, even more apologetically.

Q: Do you review suggestions about what causes cancer, regardless of how crazy they might seem?

A: Yes. One of the mechanisms is to peer review research before it's published.

Q: And you don't get paid any extra for this?

A: Not much, for that.

MR. RYAN: *(with admiration)* That's what I thought.

THE COURT: That's a comparative term, by the way, "much."

MR. RYAN: These guys do a lot of free work, not like us lawyers.

Judge McCabe laughs out loud. Ryan waits until the judge has fully composed himself before continuing.

Q: And at the time you were retained to testify in this case, did you have any professional opinion in your area of expertise regarding the effects of EMFs on cancer?

A: None.

Q: Where did you get the material that you examined to prepare your testimony?

A: I first did a computer search of this area, and your office sent me various references.

Ryan motions to two of his assistants and they come forward with a ball and stick model of the DNA molecule. The balls, painted blue, yellow, green, or red, are connected together with sticks to form a complex twisted structure. Ryan requests the model be marked as Defendant's Exhibit Y.

Q: Before I get into the meat of this, I'd like to ask you a few basic questions about molecular genetics, just so we can understand this better. I don't want to make this too long or in too much detail, but maybe you could give us the benefit of a few principles that we need. Could you describe for us in simple terms what a cell is?

A: Sure. Of course. I love to talk about that. Cells make up our bodies. The cell has a nucleus, which is made of DNA. DNA is the genetic material.

When the cell divides the DNA doubles, and half goes to each of the two daughter cells. It's very important stuff.

THE COURT: Like little computers.

THE WITNESS: Actually like a little computer program.

Ryan, pointing to Exhibit Y.

Q: Tell the judge what this is.

A: It's a model of DNA, but enlarged. If you really had the entire DNA in our cells on this scale, it would stretch a hundred thousand miles. Watson and Crick got a Nobel Prize for discovering it.

The four kinds of colored balls form a ladder. The sequence of the colors determines what a gene will do.

THE COURT: The color of your eyes, or when your hair will fall out?

THE WITNESS: Exactly. Or you get farsighted.

THE COURT: You mean like Mr. Ryan.

Ryan and Judge McCabe exchange smiles. McGurk sits at the counsel table, reading a newspaper that is concealed in a law book.

Q: Let's come right back now and let me ask you how all of this relates to the research that you do at the National Cancer Institute?

A: Changes in that DNA molecule are responsible for making a normal cell become cancerous.

Q: If something, EMFs for example, can't change the DNA molecule, does that mean it can't cause cancer?

A: Yes, absolutely.

MR. McGURK: *(stirring slowly)* Objection, your honor, I think that's just speculation.

MR. RYAN: *(outraged)* He's the expert and that's his opinion, so it must be a fact.

THE COURT: Overruled.

Q: Now, just because a scientist is evaluating a pet idea about causes of cancer, does that usually tell us that where there is smoke, there is fire, something is going on here?

A: Not necessarily at all.

Q: Can you give us an example of a situation like that?

A: There was a report of a potential link between drinking coffee and pancreatic cancer. Then, better studies essentially knocked out that idea.

THE COURT: *(leans forward)* Doesn't that happen almost every day in the year? Some new substance that will cause cancer is announced by someone somewhere or another? I remember back far enough one time where they said mother's milk gave cancer.

THE WITNESS: Exactly! A lot of reports can come out, and what you see is they go away, and you never hear about them again.

Q: This pancreatic cancer thing kicked around for awhile?

A: Probably, yes.

Q: *(with a wave of the hand)* Did the scientific community finally come to some conclusion that there really wasn't anything going on there?

A: Yes.

Q: Now, roughly, how long was it before that was decided, more or less – I guess it would be better to say that view came to be accepted – is that a fair way to say it?

A: Yes. The initial observations or report and the eventual knocking down of that report really occurred over a span of, say, three or four years.

Q: Well, now, when it was clear that it was accepted that nothing was going on there, can you give us some idea how much research had been done on that subject?

A: Only a few studies that were required to simply knock out that particular sort of red flag that was waved for a little while.

Q: You mentioned, I think, that some cancers are caused by environmental agents. Can you give us, just quickly, two examples?

A: Certainly. X-radiation and smoking. We try to tell people smoking is bad for you.

THE COURT: You just said "we" try to tell people.

THE WITNESS: I work at the National Cancer Institute, so I'm always trying to help people's health. I always tell people that smoking causes cancer.

THE COURT: Remember that, Mr. McGurk.

Ryan smiles broadly and McGurk clears his throat.

MR. McGURK: Yes, Judge.

Q: Let me ask you this: Can you quickly give me kind of a one-sentence summary that tells what all this comes down to?

THE COURT: Before you do that, could I ask something? If cells divide in a certain way, could changes inside the cell be a cause of cancer in future cells that would be then made from that?

THE WITNESS: That's where cancer comes from, your honor. It's got nothing to do with powerlines. If the change happens in a particularly critical place in a gene then the cell goes bad, and we get cancer.

THE COURT: In other words, it can just start by itself. It's what we laymen would say, well, it's just an act of God?

THE WITNESS: Right. Long before we had pollution we had cancer. It was well known in the times of the ancient Greeks.

THE COURT: Like the poor have always been with us.

Ryan, ever ready to seize an opportunity, departs from his script.

Q: What the judge is referring to, is what you call spontaneous cancer, I guess?

McGURK: Objection. He's leading the witness.

RYAN: *(blandly)* Strike it out.

Q: Is that the area that, what the judge is saying, what's that called?

A: That would be spontaneous cancer.

THE COURT: How do x-rays or smoking cause cancer?

THE WITNESS: Your honor, think of x-rays as bullets that hit the DNA and blast it to smithereens. Think of smoke as some chemicals that get stuck inside DNA and block it from working.

THE COURT: Then if you don't do that to DNA, you don't get cancer?

THE WITNESS: Exactly. And I feel certain that powerline EMFs can't do that, your honor.

Q: Let's move on. What animal studies have been conducted with EMF? Have there been some?

A: Actually, there have been quite a few.

Q: For example, I think there are some studies by Marino and Becker dealing with mice, et cetera.

A: There was this initial report by Marino and co-workers that argued that mice exposed to electromagnetic fields showed some evidence of stunting of growth. That raised a question of whether there might be an effect of EMFs on these animals.

Q: Were there later studies to replicate these studies?

A: There have been a good number of very carefully performed studies that actually were better conducted and designed.

Ryan raises his left hand, with his index finger pointing up, the signal to Stuart to answer, "No."

Q: Were they able to replicate the results that were reported by Marino?

A: No.

Ryan, in mock surprise:

Q: No?

A: No.

Ryan looks directly at me, reproachfully.

Q: Let me ask you this: Rather than to go through all of these studies that were not able to replicate Marino's findings, just take them all together, roll all of the studies together, including Marino's and tell us, taken together, what do they show?

A: There is no scientific basis in these studies for concluding that powerline EMFs are health hazards.

Ryan is apparently not paying attention, so there is no immediate follow-up and the last answer hangs in the air. Then, he snaps back.

RYAN: I'm sorry. I didn't hear that. Would the stenographer read the last question and answer?

The stenographer reads the last question and answer. Ryan, with his back to the witness, ponders the answer. Then he suddenly whirls around, index finger up.

Q: Is there any scientific basis to think that powerline EMFs could alter white blood cells in adults or children, such that they produce, you know, leukemia, et cetera?

A: No.

Q: *(finger still up)* Are there any scientific facts that indicate in any way that powerline EMFs will lead to the development of any disease?

A: No.

McGURK: Objection. The witness has conceded that there are such facts, even though he might disagree with them.

RYAN: *(highly irritated)* Don't look at the facts, your honor, look at the meaning of the facts.

THE COURT: Well, Mr. Ryan, what do the facts mean?

RYAN: They don't mean anything, your honor, that's the point.

THE COURT: *(looking quizzical)* Well, overruled. Proceed.

Q: *(both fingers up)* Based on your experience and training, and position, and education as a physician and a cancer researcher in the field of molecular genetics, and your work at the National Cancer Institute, is there any scientific basis for people to be concerned about exposure to powerline EMFs?

A: No. That's the honest truth.

MR. RYAN: Well, that's all I care about, the truth.

MR. McGURK: Objection, Mr. Ryan is giving a speech, your honor.

THE COURT: Sustained.

Q: Do you believe there is any scientific basis for concluding that EMFs from powerlines can trigger or propagate cancer?

McGURK: Your honor, I object: what the witness believes is immaterial.

THE COURT: *(irritated)* Sustained.

Q: Is there any scientific basis for concluding that EMFs from powerlines can trigger or propagate cancer?

A: No.

Q: Have there been any accepted observations that powerline EMFs would harm someone's health, and particularly referring to causing cancer, if the person were exposed to them?

A: No.

Q: Do we need to do more on EMFs?

A: No.

RYAN: *(triumphantly)* Pass the witness, your honor.

• • •

The days that followed brought more of the same. Margaret Tucker, a specialist in internal medicine and medical oncology at the National Cancer Institute, said under oath that EMF research "has yielded no persuasive evidence of increased risk of cancer in children or adults from exposure to EMFs." Her perspective was to require overwhelming evidence before assenting to the truth of any scientific fact, and otherwise to deny that fact. How Ryan loved that point of view!

Next came Richard Bockman. He was head of the Endocrine Service at the Hospital for Special Surgery in New York, Professor of Medicine at Cornell University Medical College, and a member of the staff at the Memorial Sloan-Kettering Cancer Center. His training combined the best of both worlds, an M.D. degree from Yale University and a Ph.D. from Rockefeller University. He told the court that he had reviewed the work of a number of scientific organizations and government agencies that had reviewed EMF research, and that it was his opinion that it was their opinion there was "no confirmed evidence that EMFs have any deleterious effects on animals or man."

Then Edward Gelmann testified. He had been trained at Yale, Stanford, the University of Chicago, and the National Cancer Institute, and had published prolifically on urologic oncology, particularly prostate cancer. He told the court he had done EMF research – by which he meant that he had studied the literature – and had concluded that subtle EMF effects had been reported, some of which might be true, but none of which related to health effects. He never explained how he or anyone else could know such a thing.

After him came Roswell Boutwell, Professor of Oncology at the University of Wisconsin, who strained to outdo the other tall tales, like a fisherman sitting around a fire. He believed that man-made chemicals didn't cause cancer, because cancer existed before man knew how to make chemicals, and he told Judge McCabe that the same thing was true for EMFs.

Herbert Terrace, a Harvard-trained psychologist and a professor at Columbia University, testified that effects of powerline EMFs were "barely detectable," and even those didn't matter because better research had shown that investigators had "failed to uncover any persuasive evidence that they are significant."

Jan Stolwijk, who had studied at Harvard and taught at Yale, said that he thought that EMFs were nothing more than a big fuss about nothing substantial. Stolwijk loved to speak in generalities and irrelevancies. Ryan usually had to school his witnesses so that their testimony did not focus on pertinent matters, but in Stolwijk's case he was already the finished product.

The last expert Ryan presented was Ken Zaner, who had a Ph.D. in physics and an M.D. degree. Ryan asked him how he would respond if one of his patients expressed concern about the issue of living or working near EMF sources, and Zaner replied as he might have when prescribing a drug for a patient to take on blind faith.

"My response," Zaner said confidently, "undoubtedly would be that he should have no concern of cancer or other effects of EMFs. If he would ask me why, if there's no reason for concern, he has nonetheless seen press reports raising the possibility of health effects, I would respond that this is the reflection of the concern expressed by a small group of scientists that receives a disproportionate amount of attention from the popular press simply because it makes a good story."

There it was. Eight experts who believed that there were EMF facts, and believed that they knew what the facts were – though they couldn't tell you *how* they knew. But that was the law. In court a scientific fact was nothing more nor less than what the expert pronounced. The law conjured up fairies, people who had no truth-seeking scheme but who knew things in mysterious ways, and then bound itself to their advice. Ryan understood that ordinary people regarded scientists with awe and respect, and that this attitude would transform the conclusions of his experts into an effective defense of his clients. In this way he could exonerate the actions of the Masters and make powerlines socially acceptable, irrespective of whether they caused cancer. What a magnificent lawyer he was!

In his closing argument, McGurk told the court that there was an honest difference of opinion between me and the experts for the defendant, and that it wasn't the court's job to decide the scientific issues. He should therefore order the Power Authority to pay damages to his clients because it had saddled them with this problem and controversy. "You must decide whether you want to let another power company get away with it," he said.

Ryan told the court that all he cared about was the truth, and that if one looked at the number and quality of the experts he had presented, it was easy to see where the truth was to be found.

Judge McCabe then made his ruling:

"It appears to the court that claimants' expert *believes* that he sees smoke and concludes that fire, health hazards, might be present. The defendants' experts don't see any smoke. Neither do I. The claimants failed to prove that the EMF of the powerlines cause any health problems."

The other five trials were mirror images of Goshen, with the same result. As the war had worn on, Ryan built up such a momentum that he didn't need all his experts. By the time we got to the case in Australia, he needed fewer than half of those he used in Goshen, although he did add Michael Repacholi from the World Health Organization, because he was Australian.

It seemed like the end of the road for me, at least as far as establishing that powerline EMFs were health risks. I had no way to fight against the interwoven opinions of Ryan's strong men. I couldn't ignore them because then nothing would change; I couldn't shout them down because there were too many of them, and I couldn't use the same tricks they did because my perception of science was so different from theirs.

• • •

The Masters had asked Patty Ryan to prove that powerline EMFs were innocent of all charges. I don't know exactly what happened, but it was probably something like this.

A group of Masters, worried, seated around a big table.

"All this crap about EMFs is coming out, and we've got a million miles of powerlines."

"Jesus Mary Joseph."

"You're talking to the wrong people. What we need is Patty Ryan."

"Who?"

"Patty Ryan. The best toxic-tort lawyer in the country. He knows about judges and science."

"Sounds like our kind of guy."

"Let's get Patty Ryan!"

"What luck, because here he is, the silver-tongued prince of the courtroom. The one and only Patty Ryan!"

Ryan dances in. He is wearing tight black pants, a red coat with tails, and a top hat, and carrying a baton.

"Gentlemen, got a problem?"

"Patty, we're facing a lot of trials, and we're scared."

Ryan, confidently:

"Don't be. I've been around a long time. Believe me, you've got nothing to worry about. It's all a circus. A three-ring circus. A trial is only show business."

Ryan glances at a contract on the table. It calls for half a million as a retainer and half a million for each case he wins. He smiles and nods.

The Masters, apprehensively:

"Judge McCabe is a tough old bird. What are you going to do?"

Ryan doffs his top hat and resumes dancing.

"The plan is simple.

It's the old razzle-dazzle.

Razzle-dazzle him,

And all he'll see are stars!

I'll give him an act with a lot of splash and dash.

The best I can get when I tap your cash.

I'll sweep him away with my top-drawer talent.

How can he see with rhinestones in his eyes?

What if you don't smell like a candy dancer?

What if your powerlines do cause cancer?

When I razzle-dazzle him,

He'll never catch on!"

Ryan dances to the other side of the room. The Masters turn their heads slowly, in unison.

"I'll do the old razzle-dazzle.

Fool and distract him with expert palaver.
Witnesses so scientific,
They'll seem beatific.
As expert after expert grows more certain,
The old boy will smile and bring down the curtain!
I'll razzle-dazzle him,
And he'll never know the score.
I'll razzle-dazzle him,
And he'll plead for more."

The Masters smile and nod to each other, but Ryan's not done.

"My boys will earn their pay.
No causal relationship, is what they'll say.
No confirmed evidence.
No ill effects.
No scientific base.
That will end the case.
How could he resist such a roar?
I'll do the razzle-dazzle and he'll plead for more."

The Masters, in a worried tone:

"What about Marino? What about Marino?"

Ryan, with a wave of his hand:

"Plexiglas!
Mr. Plexiglas,
That will be his name.
Mr. Plexiglas,
Mr. Plexiglas,
'Cause McCabe will look right through him,
As if he wasn't there.
After Harvard and Yale spout mumbo jumbo,
Who's gonna believe Marino's gumbo?"

The Masters smile but then suddenly seem to remember something unpleasant.

"We got other problems too, Australia and the rest."

Ryan smiles broadly.

"I'll give them the same razzle-dazzle,
I'll razzle-dazzle 'em all!
The Aussie judges in the powdered wigs,
Will never catch on when I do my jigs.
No persuasive evidence, no genetic damage,
I'll stuff all that in a tucker-bag.
I'll razzle-dazzle them."

The Masters, in chorus:
"Then what?"
Ryan, smiling and winking:
"I'll stay on the road with my dogs and ponies,
Who will know they're only phonies?
New Jersey, Texas, Pennsylvania, Connecticut,
I'll give 'em all the old double whammy,
Daze and dizzy 'em.
Shock and stagger 'em.
When cancer's the topic I'll just dance.
Though I'm stiffer than a girder,
I'm sure to get away with murder.
I'll give 'em the old razzle-dazzle,
And what they'll see is romance."
   Ryan does a buck-and-wing, winding up in the front of the room. The Masters nod and grin. Ryan plants his feet, stretches out his arms, leans back, and belts out his last chorus.
   "Ever since the time of old Methuselah,
Everybody's loved the big bambooz-a-la.
Give 'em the old three-ring circus.
Show 'em the first-rate sorcerer you are.
Long as you keep 'em way off balance,
How can they spot you've got no talents?
Razzle-dazzle 'em,
Razzle-dazzle 'em,
Razzle-dazzle 'em."
The Masters, purring in unison:
"Understandable, understandable,
Yes, it's perfectly understandable.
Comprehensible, comprehensible,
Not at all reprehensible.
Defensible, defensible,
It seems so sensible."
Purring more loudly:
"Isn't it grand?
Isn't it great?
Isn't it swell?
Isn't it wonderful?"
   Suddenly the Masters exhibit uncharacteristically somber expressions. In a serious tone:

"But nothing stays the same.
In 50 years, or maybe less,
Things will change, that's our guess."
Just as suddenly the Masters recover their composure and joyously proclaim:
"But this is what we want,
At least nowadays."

24
Ipse Dixit

◆———————◆

A federal judge in San Francisco held a meeting *in camera* with a group of lawyers; some represented policemen who had used radar guns and developed cancer, and others represented the companies that manufactured the radar guns. The judge told the plaintiffs' lawyers that he might not let the jury consider the causation issue involving the radar and the cancer because he had genuine doubts it could be proved, and he warned the companies that the jury was likely to be sympathetic to the plaintiffs because they had been struck down by cancer in the prime of their lives. The gist of the judge's advice was that the lawyers should settle the cases.

At the next meeting, the judge excluded the possibility that the plaintiffs could be awarded punitive damages, which took away their motivation to investigate whether the companies had conspired to hide scientific data, or whether they had known about potential risks of radar exposure but failed to evaluate them, which had been the avenues of attack that the big plaintiff-oriented law firms had used so successfully against the cigarette companies. He proposed a settlement based on each company's share of the market for radar guns, but the companies squabbled among themselves regarding who should be responsible for how much. There was also no unity of viewpoint on the plaintiffs' side; their lawyers bickered constantly about every case-management decision, large or small. In the end, the frustrated judge couldn't get a deal done. So he told the plaintiffs' lawyers to pick their most winnable case and take it to trial as a "representative plaintiff." "If you don't win that one," he said, "the whole ball game is over." At that point they contacted me, and we made arrangements to meet in Shreveport.

I met with the lawyers in Shreveport. I was given a copy of a memorandum that described the cases they had collected, and one of the lawyers summarized it for me. "One group involves melanoma or basal cell carcinoma in the head, neck, or upper back area. It occurred in situations where the transmitting antenna was mounted inside the patrol car, generally at about head level. The lymphoma and testicular cases appear to be directly related to hand-held use of radar guns. The officer usually placed the gun in his lap or alongside his thigh when he wasn't aiming it at traffic. During this time the gun would continuously emit the radar signal."

"Why didn't they turn off the radar when they weren't aiming at a car?" I asked.

"A slight warm-up time was necessary for proper operation," he replied, "so the manufacturers advised that the radar beam be left on all the time."

Then another lawyer began reading the memorandum. "Terry Rosenbalm is a 39-year-old, married police officer. He works for the U.S. National Park Service. He has a 14-month-old daughter. He had a radar antenna mounted inside his police car. He developed a melanoma just below his shoulder. He is expected to die in 6–12 months.

"Michael Vesta is an Ohio state trooper, 51 years of age presently, married, with two sons ages 24 and 27. Vesta has a disastrous experience with cancer which started on the right side of his face about an inch below the ear. He had numerous operations thereafter until half of his face and forehead have been removed, including his entire right eye and part of his brain. He was unaware of any possible connection between his radar usage and his persistent cancer, so he just continued to use the radar gun until his retirement.

"Lawrence Sudduth died last month, at age 65, after a long bout with cancer, originally a melanoma in the area of the right eye. He was a police officer in Ohio. His radar was mounted in the vehicle behind his head. He leaves a wife and three adult children.

"Dave Scarafiotti, 55, is a St. Petersburg, Florida, police officer with a history of basal-cell carcinomas on the face and neck. Because of what happened to him, the Chief of the St. Petersburg Police Department discontinued use of traffic radar.

"Robert Quarles is 29 years old and suffers from a rare form of melanoma of the eye. He is also a member of the St. Petersburg Police Department whose cancer precipitated the discontinuance of traffic radar usage by the Department. Quarles is married and has no children. His prognosis is quite poor.

"Terrance Traveler is a member of the Maine State Police; he is 28 years old and has a malignant melanoma that originated on the left tricep and metastasized to the left lung, a portion of which was excised. Another lung spot was found and Mr. Traveler is undergoing special treatment at the National Institutes of Health in Bethesda, Maryland. He used the Zapper radar by holding it in his left hand outside the driver's window directly in line with his left tricep. He has two sons, age 2 years and 1 month. His prognosis is very poor.

"Greg Koechel is a 28-year-old Indiana State Highway Patrolman with testicular cancer. As is typical of virtually all of these police officers, he was in perfect physical condition with no family cancer history whatsoever. He faces radiation therapy and a long-term medical follow-up with uncertain prognosis. He has been advised that he and his wife may not be able to have children because of this illness.

"Edward Benecke is a 32-year-old police officer for the city of Petaluma, California, Police Department. He is married, no children. He has non-Hodgkin's lymphoma, originating in the groin, after years of resting the radar gun in his lap when not aimed at traffic. He and his wife are despondent over the reality of their not being able to have children, given his constant chemotherapy, radiation, and dire life expectancy. He is an intelligent, articulate spokesman, and would make an excellent courtroom witness.

"Anthony Hutson is a retired Petaluma, California, police officer, now age 51. He used a hand-held radar gun, and his lymphoma developed in his leg. He has undergone an enormous amount of chemotherapy and radiation treatment, with a very pessimistic prognosis. His wife is a great asset to the case, and would make a very compelling witness. They have one adult son.

"Dwain Power died approximately four weeks ago of non-Hodgkin's lymphoma. He was 35 years old, and died only nine months after he first noticed a lump in the lymph glands on the left side of his neck. He had a depressingly painful course of treatment, deterioration and, ultimately, death. He is survived by his wife and their two sons. He had no family history of cancer, and was a very vigorous athlete who neither smoked nor drank.

"James Zum is a 64-year-old retired policeman, and a major league baseball scout. He estimates over 4,000 hours of radar-gun usage in the last ten years. Zum was diagnosed with non-Hodgkin's lymphoma and learned about the possible connection to his radar usage through articles in the St. Petersburg Times. He has been married to his wife for 45 years, and has three children. His prognosis is very poor.

"Edward Certain is a 35-year-old former Walnut Creek, California, police officer with Hodgkin's disease, presently in remission. He has had a dramatic struggle with the disease for more than six years, but has been symptom-free for more than a year. However, the disease is characterized by remissions and exacerbations, and the prognosis is always poor. His radar unit was mounted in the interior rear passenger-side window, and his first tumor was in the neck on the right side.

"Christopher Lindow is a police officer with the Oxford, Ohio, Police Department, age 31. He has testicular cancer. Following surgery and radiation therapy, he has had no further symptoms. Lindow is engaged to be married, and his prognosis is relatively good.

"Edward Cottom is 46 years old, and a Concord, California, police officer. He and his wife have three children. Cottom was afflicted with testicular cancer which apparently has not metastasized. The afflicted testicle was removed after which he had radiation therapy. He is only one year post-surgery, and so is still significantly at risk. He used a hand-held radar gun and rested it in his lap constantly.

"Linn Jonson is a sergeant with the Concord Police Department and was diagnosed with cervical cancer about three years ago, after five years of traffic radar usage; she rested the gun in her lap when not in use. At this point, there is no apparent metastasis. She is now 34 years old."

I agreed to become involved in the lawsuits, so the lawyers reviewed the medical records of all fifteen policemen to select the most winnable case. The prospective plaintiff needed to be tough-minded. He would be examined by medical experts, subjected to hostile depositions and physical examinations by batteries of defendants' physicians, and would undergo brutal cross-examination in court. In addition to all that, the ever-present media would dramatize the case, including intimate details of his medical condition. The lawyers eliminated the policemen who had died because the judge said that a suit by the survivors of a dead policeman would be unfair to the defendants because there would be no plaintiff to be examined by the defendants' experts or cross-examined by their lawyers. The lawyers also dropped from consideration all the policemen who had been smokers, because the defendants' lawyers were certain to blame the cancer on the cigarettes. Then the lawyers conducted mock cross-examinations of their remaining clients to assess how well they would hold up. In the end, they selected Edward Benecke as their representative client.

The policemen had used radar guns made by different manufacturers, but Benecke had used only the Zapper radar gun, so the case was brought against that company.

Initially the plaintiff's lawyers were happy with the judge because he was known as an activist, and he had a brother who was a police officer. As the case had progressed, however, his pretrial rulings didn't consistently favor one side or the other, so both sides became more uncertain of which way he leaned. One of his last rulings before the trial dealt with Zapper's motion to dismiss the case for the reason that the scientific testimony would be too complex for a lay jury, and therefore that any verdict in which the company were held liable would be arbitrary and capricious. The judge replied, "That's not the way our legal system works. The jury will decide all factual issues, regardless of the level of education of each juror. The people must be heard on this issue."

In a pre-trial memorandum Zapper claimed the freedom to do anything that wasn't illegal, which manufacturing radar guns wasn't. On the contrary, Zapper said, "most radar guns are manufactured under contract to state or federal governments. If there were something harmful about radar, the government had a responsibility to say so and to promulgate laws and rules accordingly. The government has not done so, from which it follows that radar guns are safe."

After Benecke had been chosen as the plaintiff, I had no difficulty in linking his cancer to exposure to microwave EMFs from a Zapper gun, and I agreed to describe to the jury my method of analysis and the data on which I had relied in reaching that conclusion. My plan was to explain that animal studies showed that EMFs could be carcinogens. Benecke had not been exposed to other known carcinogens, and the level of EMFs that he had experienced from the Zapper gun was far greater than the EMF levels experienced by people who didn't develop cancer. For these reasons it was probably true that he got cancer from the radar. I wrote out questions for Benecke's lawyers that would elicit the responses which would sum to this rationale.

Immediately after I took the witness stand but before I said a word to the jury, a lawyer for Zapper asked for and received permission to approach the bar. A discussion ensued between the judge and the lawyers for both sides, and when they returned to their seats the judge ordered the jury removed from the courtroom. Zapper's lawyer then read from a summary of my testimony, which the rules of evidence had required that I provide. The part he read was, "On the basis of animal studies I will testify that electromagnetic fields are probably capable of causing cancer in people."

"Your honor," he said, "the defendant has several serious problems with this proposed testimony. First, it is not generally accepted that electromagnetic fields can cause cancer. Second, even if they could, there is no evidence to indicate that microwaves such as those used in police radars could cause cancer. Third, it is not generally accepted that the ability to cause cancer in animals necessarily means that the responsible agent would also cause cancer in human beings. Finally, there are no generally accepted studies that prove that EMFs from police radar guns can cause the particular type of cancer that the plaintiff has. Dr. Marino's proposed testimony is therefore purely theoretical and speculative, and does not constitute accepted scientific knowledge. For these reasons his testimony should not be admitted into evidence."

The judge turned to me and asked, "Do other scientists accept the principle that EMFs can cause cancer?"

"Well, that's hard to say because..."

"Either they do, or they don't. Which is it?"

"The only thing I can say, your honor, is that I know some people who do, and some people who don't."

"In other words your theories aren't generally accepted by other scientists. Right?"

"Yes ... I guess so ... but."

"List for me the names of other scientists who agree with you."

"Well, your honor," I said, "I can't really speak for anybody else."

"Are you a one-man band?" he said.

I had thought of myself as part of a growing band, but its size was not part of my rationale because my inferences were new, and so couldn't be "generally accepted." The really important point, I thought, was that I had arrived at my conclusions by a *method* that was generally accepted. By my lights, it would be crazy to say that something wasn't a fact, or wasn't reliable, simply because everybody didn't already believe it. If something can't be accepted until it's a fact, and it can't be a fact until it's accepted, then the situation is hopelessly circular. I tried to tell all this to the judge, but he said, "You're a lawyer, Dr. Marino, you should know the law is that scientific evidence can't be accepted by a court unless it's first generally accepted by the scientific community. *That* is what makes it a fact," he said. The lawyers argued for a while, and in the end the judge ruled that all testimony regarding the effects of EMFs in animals was inadmissible as evidence, and that all scientific testimony would be restricted to questionnaire studies, and the opinions of physicians.

In the evening one of Benecke's lawyers told me the story of how "general acceptance" became a federal rule of evidence. "In 1922, a man named James Frye was arrested in Washington, D.C., for murder. At that time there were many amazing scientific machines on the market, one of which was a lie detector. When a subject was asked a question, either a green or a red light would flash depending on whether he told the truth or lied. Frye was asked if he had committed the murder. He replied 'No,' and the green light lit. His lawyer tried to have that result admitted into evidence to prove Frye's innocence but the judge was skeptical because of an earlier case he'd had that involved a different machine."

"What machine?" I asked.

"An oscilloclast," he said.

"I've seen an oscilloclast," I said. "It's a box with many knobs and dials. Adjusting them was supposed to transmit the precise rate of energy to a patient to counteract his disease. But the machine was a hoax."

"Yes," he said, "and it was the judge in the *Frye* case who had presided over the trial where the truth about the oscilloclast first came out."

"What happened in that trial?" I said.

"One of the parties," he said, "presented an expert who testified he had opened the machine and examined how it was wired. He said that there were no connections between what was inside the machine, and the knobs and dials on its front face, so the oscilloclast couldn't transmit anything to the patient."

"The settings of the knobs and dials didn't matter," I said.

"Exactly," he said, "and because the judge was already skeptical about the reliability of black boxes, he told Frye's lawyer to produce an expert who had examined the lie detector and could testify about how it worked."

"Did he?"

"No. The expert told the judge that the machine was hermetically sealed and that his operator's rental agreement prevented him from opening the machine. The same kind of arrangement had existed with the oscilloclast, but that expert had broken the lease, opened the machine, and the truth came out. So when Frye's expert refused to actually examine the lie detector the judge ruled that the expert's testimony was inadmissible as evidence, and the jury ultimately found Frye guilty of murder."

"Frye appealed, I suppose," I said.

"Yes. He claimed that he would not have been convicted if the results of the lie-detector test had been admitted. Judge Josiah van Orsdel was assigned to decide the appeal. He was a rancher in Wyoming who apprenticed in the law and was eventually appointed to the Court of Appeals by Teddy Roosevelt. Van Orsdel's daughter had developed cancer and a doctor treated her using an oscilloclast, but she died. Shortly afterwards the oscilloclast hoax became public."

"This story is like a Greek tragedy," I said.

"Judge van Orsdel upheld the trial judge's decision to exclude the test results."

"On what did he base that decision?" I asked.

"He invented the rule that scientific testimony can't be admitted into evidence until it's generally accepted by scientists," he said. "Since the principle of detecting liars using a machine wasn't generally accepted, he denied Frye's appeal."

"He invented the rule?"

"Yes. And for seventy years it's been respected by the courts as if it were one of the Commandments, especially so within the last few years."

"So that's the story behind the *Frye* rule?"

"Yes," he said, "according to what I've been told."

"How ironic," I replied, "that the method for identifying what constitutes reliable scientific evidence, turns out to be the *ipse dixit* of a Wyoming cowboy."

The next morning I took the witness stand to testify about the EMF questionnaire studies, which were the only kind of studies that the judge would allow me to talk about. I told the jury that it was never possible to know anything confidently on the basis of questionnaire studies, but that it was possible to use them to reach reasonably reliable conclusions when there were many positive studies, which there were in the case of EMFs. Some federal agencies reasoned exactly as I did, and I gave examples. The Environmental Protection Agency relied partly on questionnaire studies when it decided that various agents were probable carcinogens. The Food and Drug Administration routinely made its decisions on the basis of questionnaire studies when it reached decisions that a particular drug was probably safe or unsafe, or probably effective or not effective.

I ignored the federal agencies that took a different approach because I considered their policies ignorant and short-sighted. The Federal Communications Commission required absolute and certain knowledge that EMFs from TV and radio antennas caused cancer before they would take remedial steps. The Federal Trade Commission had the same attitude.

Under cross-examination, Zapper's lawyers established the following facts:

I was not a physician;

I had never treated anybody who had cancer;

I had never attended medical school;

I had never taken a course in cancer or oncology;

I had never examined the plaintiff or studied his medical records;

The EMFs in the questionnaire studies that I had described in my testimony had not come from police radar guns;

I did not know of any study that showed police radar guns caused cancer;

There were some questionnaire studies that were negative, though I did not discuss them.

After I left the witness stand, Benecke's lawyers presented two medical experts, each of whom testified that they routinely diagnosed and treated patients who had non-Hodgkin's lymphoma, that they both had examined him, and that they both believed that his cancer was caused by the microwaves emitted by the radar gun that he had used.

Then Zapper presented a medical expert named Dr. Edward Boutbad who had examined Benecke and reached the opposite conclusion. Boutbad had written a textbook on the diagnosis and treatment of non-Hodgkin's lymphoma, and was generally regarded as the world's foremost authority on the subject.

During cross-examination Benecke's lawyer asked, "Dr. Boutbad, what causes non-Hodgkin's lymphoma?"

"Bad genes, I suppose," he said. "There are some theories, but no one knows for sure."

"Well, if you don't know what causes any particular cancer, how do you know that EMFs don't cause all of them?"

"The very idea is preposterous."

"Why do you say that? On what basis?"

"On the basis of my experience. I have been treating cancer patients for almost forty years. If EMFs were carcinogens, I would know about it."

"Then your testimony is that since you don't know about it, it can't be true?"

"Objection, your honor," said one of Zapper's lawyers. "He's leading the witness."

"Sustained. Move on, counselor," the judge said.

"Are you aware of a report by the Environmental Protection Agency that labeled EMFs 'probable carcinogens?'"

"I heard about it. But it was only a draft, not an official report, and the Environmental Protection Agency backed off that claim."

"Do you know why it was not adopted?"

"No."

"Would it surprise you to learn that it was a result of political pressure?"

Before he could answer Zapper's lawyer jumped up and yelled, "Objection, your honor! He is badgering the witness."

"Sustained," the judge said to the lawyer. "Move on," which he did.

"Dr. Boutbad, you have heard the two medical experts for the plaintiff testify that microwaves caused the plaintiff's cancer."

"Yes I did, and I'm afraid it's my fault."

"Why is it your fault?"

"Because they are both former students of mine. They did their residency at my medical oncology program. The ignorance that they displayed in claiming that EMFs cause non-Hodgkin's lymphoma points to a weakness in my program, which I will correct."

The jury deliberated for two hours and then brought back a verdict in favor of Zapper, acquitting the company of all liability for causing Benecke's cancer. In an interview for television the jury foreman said, "The top man in the field said that radar didn't cause the cancer. We didn't feel that we could go against him." Another juror told a reporter outside the courtroom, "I would actually like to see these guns tested for safety."

The attorney for Zapper told the press, "There has been a lot of hysteria about microwaves, but the radar guns emit only low-level energy. It's comparable to the amount put out by a child's walkie-talkie or a nursery-room monitor. Now," he said, "police officers can breathe a little easier." He then sued Benecke to recover attorney's costs for defending against what he claimed was a frivolous lawsuit.

Soon after the Benecke trial ended, just before Christmas, someone called and said, "Dr. Marino, I'm a lawyer, my name is Jack Cordaro. I'm writing an amicus brief in a case pending before the U.S. Supreme Court, and I think you can help me."

"Powerlines or microwaves?" I asked.

"It's a dispute over a drug prescribed to pregnant women. The plaintiffs took the drug and then gave birth to malformed children. My clients, however, are scientists. They became involved as friends of the court because they're concerned with the underlying issue."

"Which is what?" I asked.

"How courts should handle scientific evidence," he said. "I know you've been involved with that question in the context of electrical energy, and even before that."

"Before EMFs?"

"The safety of BHT," he said.

"How did you know about that?" I asked.

"My father worked for the FDA. When I was growing up he would never let me or my sisters eat cereal that had BHT. Years later he told me that he had deemed it unsafe after some conversations with a scientist in Syracuse named Marino. I supposed that the BHT Marino from Syracuse and the EMF Marino from Shreveport are the same Marino. Correct?"

"Yes."

"I'll be in New Orleans on Thursday. May I drive up and see you late that afternoon?" he asked, and I agreed.

When we met he began by telling me that he had graduated from Tulane law school, after which he had clerked for a judge on the Federal Eleventh Circuit and for Justice Harry Blackmun at the U.S. Supreme Court before entering private practice. Then I said, "Tell me about the case."

"The appellants took a drug for morning sickness called Bendectin, and their children had birth defects. They sued the manufacturer, Merrill Dow, under a variety of legal theories including negligence. Merrill Dow claimed that the proposition that Bendectin could cause birth defects was not generally accepted, and therefore that scientific testimony to that effect shouldn't be allowed into evidence."

"The *Frye* rule," I said.

"Yes," he said. "The judge followed the rule and excluded all the evidence that tended to show a link between the drug and birth defects, so Merrill Dow won. But judges in other federal circuits have rejected *Frye*. The split is eight in favor and three against, and the Supreme Court finally granted cert to decide what the legal standard should be to admit scientific testimony into evidence. Many state courts use the federal rules of evidence, so whatever decision the Court makes will profoundly affect the legal system."

"Who are your clients, what's their interest?" I asked.

"They're principal investigators supported by the National Institutes of Health who are afraid that science may become marginalized in society," he said. "They think that 'general acceptance' is unrealistic because there is no community of scientists that can generally accept anything except the broadest scientific facts, like gravity, or DNA. So the effect of *Frye* is to keep science isolated in an ivory tower. That makes it seem to the public as if very little of what NIH scientists do leads somewhere or sums to something. If the public gets that idea, there will not be any grant money, at least not enough to go around."

"And your clients think that whether Bendectin can cause birth defects is not such a broad fact."

"Exactly," he said. "There probably aren't a hundred scientists in the world who care, one way or the other, and many fewer who are experts in the area."

"If the courts move beyond 'general acceptance' as a method for deciding what scientific ideas are reliable and start looking into how we actually do science," I said, "your clients may not be happy with where that leads."

He looked at me quizzically and, after a pause, said, "Tell me what you think the rule should be for admitting scientific testimony into evidence."

I said, "Anyone who claims that he knows something that can only be known by means of the scientific method should be required to explain how the method produced the knowledge."

"And if he can't?" he asked.

"Then," I said, "he has no way of knowing and therefore has no knowledge, only opinion, which is not evidence of anything except what the witness has persuaded himself is good to believe in."

"Shouldn't an expert be able to tell a judge that something in an authoritative text is reliable?"

"Only if the expert can explain how he knows it's reliable," I said, "because scientific knowledge is a product of method, not authority."

Cordaro agreed that my view was reasonable, and I continued along that same line because I thought the point was important and wanted to drive it home. "Would you accept something as factual just because someone who was rich said it was a fact?"

"No," he replied.

"What if the claim were made by a governmental official, or a preacher?"

"I would trust those claims even less than I would that of a rich man."

"The same skepticism should greet those who claim scientific knowledge. Possession of an M.D. or Ph.D. isn't necessarily a surrogate for reliable knowledge."

"That makes sense," he said, and I went further down the road. "Do you know why knowledge claims based solely on authority should always be legally insufficient?"

"Tell me."

"Because the practice of according respect to authority sprang from a conception of scientists as being free of ignorance or bias, and always motivated by the desire to do good."

"Are there such scientists?" Cordaro asked.

"No," I said, "That's precisely why the law became an ass when it conjured them up."

I could tell that Cordaro was not ready to appreciate this perspective, so I simplified my explanation by appealing to the pragmatic justification I felt he had probably already settled on.

"The problem with 'general acceptance' is that it's unworkable because no one knows what it means or how to prove it," I said.

"That's plain from all the cases that followed *Frye,*" he said. "No two courts have ever agreed on a definition."

"Isn't recognition of the practical shortcomings of 'general acceptance' sufficient for you to write a brief on behalf of your clients?" I asked.

"No, unfortunately," he replied.

"Why not?"

"When I clerked for Justice Blackmun, he always told me, 'Don't look at the legal issues. Go for the jugular. Tell me what the case is really about.' When I do that in this case I see a big problem."

"Which is?"

"The Court is increasingly conservative. One way this comes out is in its attitude toward rules that facilitate transfer of wealth from business to individuals, or that increase the cost of doing business. Toxic-tort and product liability cases do just that, so it's to be expected that the Court will be hostile to those kinds of cases. I also know it's concerned about trial lawyers bringing specious cases based on scientific evidence that bamboozles the jury."

"So you're worried that what replaces *Frye* may be even more restrictive."

"Yes," he replied, and continued, "Despite its shortcomings, the *Frye* rule is empowering for judges. It enables them to keep control of their courtroom, and they have relied on it increasingly to exclude testimony from venal experts."

"That's good," I said.

"But the price is great. Testimony can be truthful and reliable even if it's not generally accepted," he replied.

"You seem to think that the Court is poised to throw out *Frye*."

"I'm almost certain of it."

"Why?"

"Well, just look at the facts in the case. The six experts who testified for the plaintiffs were well trained in various aspects of science relating to birth defects. Their full-time jobs involved performing laboratory and animal studies on drugs, including Bendectin, and they all worked in government or university laboratories. The single expert for Merrill Dow, on the other hand, simply read the literature and concluded in an affidavit that there were problems and uncertainties and that not everyone agreed Bendectin could cause birth defects. He was the president of a consulting company that specialized in providing company-friendly advice for drug companies. So, if you intend to reject the authority of the 'general acceptance' rule, this is the perfect case for granting cert."

"Do you have a hunch about what rule the Court will put in place after it dumps *Frye*?"

"My contacts with the clerk network aren't what they used to be. When I was there the Chief was in favor of asking Congress to create a science court, with judges who were scientists."

"The Chief?"

"Chief Justice Rehnquist. One Friday, after the justices had finished discussing that week's cases, Justice Blackmun returned to our office angrier than I had ever seen him. I asked him why he was upset and he said that the justices had been discussing a particular scientific point in connection with one of the cases and the Chief had remarked, "How are we supposed to decide, nobody here knows the difference between an atom and an asshole."

"Why did the remark anger Blackmun?" I asked.

"Because he thinks of himself as a scientist," Cordaro replied.

"Blackmun, a scientist?"

"Well, sort of. He has a degree in mathematics from Harvard, and he's spent a lot of time in medical school libraries, something nobody else on the Court ever did."

"What did the other justices think about the idea of a science court?"

"They never said. They always deferred to the Chief whenever the question of the *Frye* rule was raised in a cert petition because they knew he wanted to wait for Congress to create a system of judge-scientists, so they always went along with him and denied cert. Now, that's changed. There will be a new rule."

"And it's hard for you to see what rule the Court is predisposed to adopt?"

"Whatever it is, it will be shaped by what the Court thinks is important. And what is most important to the Court is not science, it's judicial power. They want to keep control of their own courtrooms in the age of science. Also, it's a clean issue."

"What do you mean, a clean issue?" I said.

"The Supremes are political animals," he said. "They're very conscious of how they are perceived by the public. Their political standing and that of the entire judicial branch is determined by the hot-button issues that cause widespread emotional reaction and interest."

"Like Blackmun's abortion case?"

"Yes. He can't go anywhere without someone following him, waving a picket sign. At the other extreme there are cases that generate almost no emotional or political interest. Those involving scientific issues are examples, as are those involving rules of evidence. This case involves both, so there are two solid, independent reasons why the public isn't much interested in it. When the pressure is off, the Supremes always do what's best for the judiciary."

"So you need an argument that doesn't simply repeat those of other parties and that serves your client's interests, but that doesn't frighten the Supremes into thinking they will lose some power, and you think they're likely to buy such an argument because there won't be any political fallout."

"Exactly," he said.

"What are the other arguments that have been advanced thus far in the case?"

"The appellants will attack the authority of *Frye*, the respondent will claim that overruling 'general acceptance' will undermine the integrity of science, and the other amici will support one side or the other while making arguments that are flavored by their perspective regarding what is important."

"What arguments do you anticipate that the amici will make?"

"The American Association for the Advancement of Science and the National Academy of Sciences are in favor of the *Frye* rule," he said. "They claim it results in legally reliable science. I think they're worried that if there are too many controversies Congress might start poking its nose into how scientists spend money, like in the days of Proxmire's Golden Fleece Award. Evidence fuels controversy. *Frye* keeps out evidence. So, for them, it's a no-brainer."

"They won't say that in their brief, will they?"

"No, only that *Frye* is necessary to avoid outcomes in court that are what they call 'at odds with reality.'"

"Who else is in favor of *Frye*?"

"The American College of Legal Medicine," he said.

"Who are they?"

"People who think that doctors are sued too often. They want *Frye* upheld, and then extended to malpractice cases."

"Are there more amici?" I asked.

"Oh yes," he replied. "The American Insurance Association is strongly pro-*Frye*. They say it would add uncertainty to their business if the *Frye* standard were to be relaxed. Of course, uncertainty is not something they like."

I said to Cordaro, "They all argue that a particular result would be just, but in reality it's what they desire."

"Or what they realistically think they can achieve," he said. "The public position of the Chamber of Commerce is that *Frye* should not be changed. They actually favor a more pro-business rule, but their lobbyists haven't been able to generate any support for it on the hill. So I expect them to make the political decision to throw their weight behind *Frye*."

"What rule do they favor?"

"That experts shouldn't be permitted to present any scientific evidence in court unless they first present the results of a survey of all experts in the relevant field showing that at least 75% of them agree that the proffered testimony is a scientific fact."

"What other friends of the court have jumped in?"

"The Pharmaceuticals Manufacturers Association and the American Medical Association. They both say that *Frye* is far too liberal. Undoubtedly they will advise the Court to adopt a stronger test."

"What's their interest?"

"The ready availability of drugs. For them, Bendectin is the poster boy for what can happen to a drug when the legal standard for admissibility of scientific evidence is too low. Bendectin was approved by the Food and Drug Administration as safe and effective. Following the initial lawsuits that claimed it caused birth defects, the FDA reevaluated Bendectin and again concluded that it was safe. Nevertheless, the number of lawsuits continued to increase, and Merrill Dow took Bendectin off the market. Now there is what the AMA calls a "significant therapeutic gap" in the treatment of morning sickness. They're afraid that the pharmaceutical companies are going to stop pursuing all research in contraception and fertility."

"Who else supported *Frye?*"

"The American Tort Reform Association."

"Who are they?"

"An amalgam of trade associations of industries that tend to be defendants in lawsuits in which what they consider to be junk science is used against them. They would very much like to see that problem go away."

"So they want to toughen the *Frye* rule?"

"Of course," he said, "and there is more than a little basis for their concern. There have been some suits in which people claimed that exposure to an infinitesimal quantity of a substance was responsible for their illness. It's not hard to find an expert who will testify to virtually any theory of causation up to and including the fantastic."

After we agreed that every profession has its counterfeiters, Cordaro mentioned one more group that wanted to put in its two cents.

"The *New England Journal of Medicine* filed an amicus brief in support of Merrill Dow. That surprised me," he said.

"Not me," I said.

"Why?" he asked.

"It's the most famous medical journal in the world. Its editor decides what constitutes scientific knowledge, because publication there entails general acceptance. Weakening *Frye* would knock it off of its perch."

After Cordaro had brought me up to speed regarding what he thought the other amici would say, he asked, "What do you think I should urge the Court to put in place of *Frye?*"

"During a trial," I said, "suppose a lawyer proffered a witness who would testify that the traffic light was red when the plaintiff entered the intersection. Would the judge allow the witness to testify?"

"If the testimony were relevant to the case, yes."

"On what basis would the judge make that decision?"

"According to your hypothetical, the witness has knowledge about relevant events in the case."

"This personal knowledge that the witness has, how did he get it?"

"By perception. He saw what happened."

"So if the witness is asked, 'How do you know the light was red?' he can respond, 'I saw it.'"

He agreed.

"Then not only does he have knowledge, he can explain how he got it."

He agreed.

"It must be the same for an expert. He must have knowledge, and he must explain how he got it. Since his claim is that he possesses scientific knowledge, he is restricted regarding the basis upon which he can establish that he actually has such knowledge. He must have obtained it by means of the scientific method."

"That's not something judges know a lot about," he said.

"I understand," I replied, "atoms and assholes."

When he stopped laughing I said, "Judges don't have to be scientists. They only need to let the adversary system work, and if the lawyers aren't assholes it will be more or less obvious to the jury which experts are ignorant, or worse."

"I think you've got something there," he said.

As Cordaro was preparing to depart I asked him, "What's your guess regarding who will be assigned to write the opinion?"

"It will probably be my old boss," he replied.

"Blackmun?"

"Yes. The other justices usually defer to him in scientific matters, though that deferential respect irritates some of the others, particularly Justice White."

That night I found it difficult to sleep. The Supreme Court was going to decide a case that would have a profound impact on science. Their decision could push science even farther to the edges of society, making it the exclusive province of enormously specialized technocrats. Or the Court could move in the opposite direction and require scientists to explain themselves, which would bring science itself, not just technology, into the stream of ordinary life where it belonged. After all, the public owned science, the way a man owns his hand. A hand doesn't think for itself and justify itself by saying, "Leave me alone and I'll continue to do what you want if it's also what I want."

Despite the importance of the looming decision, no one seemed to care about it. A clean case, as Cordaro had said. If you had just landed from Mars and scoured the media for information, you would find almost nothing that signaled that the Court was on the verge of making such an important decision. I don't know how long I lay awake, staring at the ceiling, but eventually I fell asleep.

• • •

I saw Justice Blackmun. He was seated at his desk reading some scientific journals. He seemed very old. When he noticed me he greeted me warmly and said, "We talk to lawyers all the time, but we don't often talk to scientists."

"I would like to ask you something," I said. "Every place you go you are picketed by people who try to make your life miserable. It looks to me as if the road you've taken is not easy, and now that you are near the end I'd like to know how it looks to you. What's it like to be hated by so many people? Is it a hard time of life?"

"I'll tell you how it is, Andrew. From time to time I talk with the other justices, and most of them complain about the abuse they receive. They say it's an inevitable consequence of a life as a judge, and they look back fondly to the time in their lives when they were popular with everyone. At first I too suffered, but age has brought me peace. It is the character of a person that matters most. If someone is balanced and rational, then frenzied attacks in public are only moderately troublesome. If not, then it is hard to deal with any form of criticism. My notoriety is not a cause of pain because my decision was just."

"Why so?" I asked.

"Because it was based on scientific knowledge in journals," he replied.

I was amazed by what he said and I wanted to hear more, so I said, "Perhaps the hoi-polloi don't accept that your decision was based on scientific knowledge because they think it sprang from power. They say that justice is anything the Court says it is, like a baseball umpire calling a pitch a strike or a ball."

"There's something to that," he said, "but not so much as is generally supposed. My decision wouldn't have become law had I not been a judge, but it's not the law that makes it just, rather it is the science on which the decision was based. The man without power cannot make law, nor can the man who does not adhere to science even if he has power."

"Then your decision will be that judges should read the journals to decide questions like whether Bendectin causes birth defects, because that's where scientific knowledge is stored?"

"Yes. If that scandalizes some, then so be it." As he spoke he gently stroked a journal on his desk as one might do to the head of a small child.

"So you think that science is naturally honest, and that the information in journals is reliable?"

"That's right," he said.

"Didn't you ever hear the expression, 'He who has the gold makes the rules'?"

"Yes."

"Couldn't that sentiment account for some of the facts that appear in journals?"

"What do you mean?"

"I could tell you about published facts that are actually counterfeit. For example, if you don't want to find stress effects in rats due to electrical energy, make them live in tiny cages so that they're already stressed from being crowded. Then you can write in the journal that electrical energy doesn't cause stress. Even better, keep doing an experiment over and over until you get the results you want, and then publish that as if it were the whole truth."

"That's very sad," he said, "but I think that the truth eventually works its way out."

"All in God's time, like a ripe apple that falls from a tree," I said, but my attempt at irony failed and the old judge replied, "Well put."

"What if no one plants the tree," I said, "or fails to provide what it needs to grow, or hides the apples?"

"What has that to do with scientific knowledge?" he asked.

"I've just told you how I have often seen it obfuscated," I replied.

"The world is governed by mathematical laws," he said. "No conspiracy can stop those truths from emerging."

"A biologist has no equations. His knowledge consists only of rules derived from observations that depend on experimental conditions, and on the perspective of the observer."

"Then biological knowledge cannot be certain or absolute?" he asked.

"When I was young I thought so. Even after I had confronted the issue of health risks from electrical energy, where I saw strong differences of opinion that seemed impossible to resolve, I still believed in the existence of a canonical scientific method. After a long time, however, I understood that there was something deeper than method – there was desire, the reason why anyone makes any effort in the first place. Desire is the only absolute."

"Well, it's hard to know what to make of that," he said. "Ever since my days at Harvard I've admired the scientific method. It is the greatest machine for discovering knowledge ever invented. It's so much purer than the adversarial process."

"So you think that the scientific method and the adversarial process are different?"

"Of course," he replied. "You don't?"

"I think that they are more or less identical, as evidenced by the many things they have in common," I said.

"Such as?"

"They both seek truth."

"Yes. What else do they have in common?"

"Both activities are carried out by specialists for the benefit of others."

"That's true," he said.

"They have two other similarities. One is a process for purging error. Scientists have peer review, lawyers have cross-examination."

"What is the other similarity?"

"They both seek victory in battle."

"What do you mean?"

"What does a lawyer do when faced with evidence against his client?" I asked.

"Oppose it," he replied.

"Suppose a scientist was employed to champion nuclear power, a strong military, or the healing power of pharmaceuticals," I said, "what would he do when faced with evidence opposing his client's interests?"

"The same as the lawyer, I suppose."

He paused for a few moments during which time he resumed patting the journal, and then he commented, "But the facts in science are complex. Juries can sometimes be confused."

"So you think that judges are better than juries at determining what constitutes scientific knowledge?"

"For the most part," he said.

"I can't imagine why," I said, "because every pertinent study has concluded that juries are at least as good as judges in deciding whether scientific evidence is reliable. It's harder to razzle-dazzle twelve men than it is only one."

"Then do you think that questions like whether Bendectin can cause birth defects are exclusively within the jury's province?" he asked.

"The judge should decide whether there is any evidence produced by means of the scientific method that Bendectin can cause birth defects, like requiring someone to examine a machine to determine that it operates as advertised. Then the adversarial system should be permitted to play its role. It is for the jury to decide whether Bendectin probably can cause birth defects and whether, in the particular cases, it probably did so."

"So you would say that the question for the judge is *how* does the expert know, and the question for the jury is *what* does he know?"

"I would," I said.

"Andrew, I must leave now to sign my retirement papers and turn in my keys," he said, and he hurried away.

The *Frye* rule died on June 28, 1993, the day Justice Blackmun issued his decision. Writing for a unanimous court, he rejected "general acceptance" as the applicable rule of evidence in federal courts for admitting scientific testimony, and held that it could be admitted only if it actually was "scientific knowledge."

For a long time, Judge van Orsdel's rule, that the condition for the law to regard a scientific proposition as reliable enough to be considered by a jury was that the proposition be generally accepted by scientists, had lain dormant, like an ungerminated seed in dry soil. During that time science and law rarely encroached on each other's domain. When the law began to progressively take cognizance of scientific knowledge, it was almost always in the form of the personal opinion of an expert, the pivotal courtroom question for whom was, "What do you know?" He was never asked, "How do you know?" because the answer to that question was presumed to stem in some mysterious way from the expert's training and experience that laymen could not understand. The prototypical expert was the physician who testified in a kind of legal case that developed in the latter half of the twentieth century, medical malpractice. One expert sympathized with the patient and identified what he thought was a substandard aspect of the care provided by the defendant. Then, according to the law's formula for that cause of action, the expert would make the extraordinary statement that the defendant's substandard performance caused the plaintiff's injuries. How exactly did the expert know this? Why couldn't the cause have been the part of the care that was standard, or even the disease process itself? Don't ask! Let the experts for both sides testify, and the jury will decide. How will the jury decide? Don't ask!

When there arose the matter of whether cigarettes caused cancer, the law headed farther out into the wilderness. The physician expert testified on the basis of training and experience, his examination of the patient and, remarkably, on the basis of questionnaire studies that showed an increased probability for cancer among smokers. Statistical evidence was admitted by the courts for the first time to help establish the proposition that the plaintiff's cancer was caused by smoking. But it's one thing to say that, in general, smoking causes cancer, and it's quite another to say it did so in the plaintiff. So how does the expert know that smoking caused cancer in the plaintiff? Don't ask!

The pattern was repeated many times. Cases were brought by workers who had been exposed to asbestos which, they claimed, had caused their cancers. The proof they offered consisted of opinions of physicians and the results of questionnaire studies, and they prevailed. The Richman law firm made millions of dollars from these cases. It was less than the billions they got from winning a string of cigarette cases, but a good payday nonetheless. After that came Agent Orange, and breast implants.

While all this was going on I had arrived on the scene, full of wonder about what EMFs could do and concerns about the harm that could arise from their indiscriminate use. From the beginning I worried about how I knew what I thought I knew, and about how the experts who testified for the companies knew what they claimed to know. I became convinced that they didn't know, and that they were only telling a story in specialized language, like poets. I waited for those experts to be put to the test, but that never happened.

I didn't take seriously the proposition that a physician could look at a patient and say what caused or did not cause his disease, and I didn't think questionnaire studies were probative – and even if they were, I thought it was evil to rely on them for knowledge of what causes cancer. So I paid little attention to law cases that were founded on the opinions of physicians and the results of questionnaire studies. That kind of evidence was not relevant to what I was attempting to do. I suppose that I hoped the law would develop to the point where an expert would have to explain and defend what he claimed was scientific knowledge. I believed that evidence relevant to what had caused disease, any disease, must come from experiments involving animals, and that there was no other ethical or logical method to know such a thing. That was the course I was on, and I'd had no other choice but to ignore the misguided law of scientific evidence known as the *Frye* rule, which is what I did. During this period, *Frye* claimed even more victims. The most pathetic were Vietnam veterans who had developed cancer and birth defects after they had been sprayed with Agent Orange. The judge in those cases said that it was not generally accepted that animal studies could be used to prove whether something could cause cancer in human beings, and he declared all such testimony inadmissible as evidence. He required each veteran who was trying to prove that his cancer came from Agent Orange to submit an affidavit from a physician expressing that opinion. How absurd! How ridiculous! But how inevitable!

Then the *Daubert* decision was announced and everything changed, at least in theory. The decision did more than drive a stake through the heart of *Frye*. Every justice on the United States Supreme Court agreed that scientific knowledge was not a matter of authority or popularity, but rather the product of a method. If Supreme Court decisions were measured by a quotient consisting of the impact they eventually have on the world divided by the quantum of society's perception of the importance of the decision at the time it was issued, then the *Daubert* decision might be the Court's most important decision.

In the wake of *Daubert* a crucially important problem remained unresolved, as I saw vividly when I participated in a post-*Daubert* trial where the allegation was that powerline EMFs caused the plaintiff's leukemia. My old friend Patty Ryan ran the show for the defendant and, because of *Daubert*, he was unable to prevent me from testifying on the basis of animal studies that the plaintiff's cancer had been caused by the company's powerline. But I saw that it took more than *Frye's* death to get at the truth of the matter, it took lawyers who not only liked what I was saying because it helped their clients, but understood what I was saying and, consequently, understood why the experts who testified for the power company were advocating a vastly inferior position. Understanding requires effort, and making the jury understand requires even further effort. I saw that a prerequisite for success under the *Daubert* regime was a plaintiff's lawyer who understood scientific language and who therefore could cut through it to expose the weakness of the defendant's experts with the requisite clarity that could lead to understanding on the part of the jury. But, in my whole life, the only lawyer I ever met who knew the difference between an atom and an asshole was Patty Ryan. So from my point of view, until I found the right lawyer, further cases were a waste of time, and I turned my attention elsewhere.

Part IV

♦

Understanding: 1994-1998

Patience, iron patience, you must show;
So give it out to neither man nor woman
That you are back from wandering.

(13: 308-310)

25
Szent-Gyorgyi

◆

I was invited to speak at a symposium on bioelectricity where the famous Albert Szent-Gyorgyi, Dr. Becker's predecessor whom he most respected, was to be one of the other speakers. I quickly accepted, hoping that he might open my eyes to some possibilities concerning how EMFs affected the body. At the formal dinner for the symposium speakers I sat directly opposite him; the table was so narrow that I could have reached over and taken the meat off his plate without even fully extending my elbow. His wife sat to his right and was always vigilant to even his smallest needs. She was a beautiful, demure woman, about 50 years younger than her husband.

He told me about research he had done, and how he felt the day he heard he had won the Nobel Prize. Then he said, "I should have gotten a second Prize. All of Krebs's work was done in my lab." He talked about his theory of cancer, which had to do with how electrons moved through tissue. He believed that they had far greater mobility than was generally supposed.

The topic of our conversation changed from moment to moment depending on what he wanted to say. When he recounted his conversations with Stalin and Beria he said, "They respected scientific achievement. So, for a while after the war, things were not so bad for us. But then they got worse, so I came here to work."

Szent-Gyorgyi had asked the National Institutes of Health for money to continue his research on cancer, but he never obtained the approval of his study section. He told me that he couldn't tolerate their rules and practices for obtaining grants, and undoubtedly they returned his contempt. He said of them, "We are in the hands of pedestrian people. Robots. That's all they are. Two plus two is four, and so on. No imagination. They never make progress or learn anything, but they don't care, because they are simply robots." He alternated between passionate expressions about the importance of his theory of cancer, and lamentations because of the opposition he had experienced from the National Institutes of Health, which he blamed for blocking his road to success. But there was something discordant in what he said, like perfume that's too sweet. Although the National Institutes of Health had refused his grant requests, he had resources from private patrons who greatly respected and admired him. I thought to myself, "What's the sense of arguing about what causes cancer? Why don't you just do experiment after experiment until your proof is so strong that not even a robot could deny it? If you know how cancer comes about, you can block it. Then you'll make a lot of money, and eventually everybody will say you're right, not because you convinced them but because you were successful, and your view will be generally accepted. On the other hand, if you don't succeed because you ran out of time or money, or because it turns out you were wrong, well, you tried. What else could you have done that would have been as much fun?" Of course I didn't say anything like that to him directly, but rather probed the issue gently until I realized that he was more or less content just to lament the disrespect he had received from the scientific establishment, and was not seeking a path around controversy to victory. After all, he was a very old man.

At a point in the evening that seemed propitious, I raised the topic about which I had intended to seek his advice. I said, "In my research on electromagnetic fields I found that identical rats behaved differently when I administered the same treatment. What do you think could explain that?" "I don't know," he replied. "The principles of life are the same for each rat, and for all life, from one end of the scale to the other. The cucumber is just like the man who eats it." His tone was dismissive, so I asked no more questions about my work.

Szent-Gyorgyi had known Erwin Schrodinger and had read his little book, "What is Life?" "It's an old question," Szent-Gyorgyi said, "but a wrong-headed one. We don't have an answer because the question makes no sense. Life, in itself, does not exist. No one has ever seen it or measured it. Life is always linked to material systems. What man sees and measures are living systems of matter." I didn't interrupt with questions, so he continued to make his point. Somewhere between the level of electrons moving through proteins and that of the smiling serving girl standing beside us, he said, the property of "life" inexorably emerged, required to do so because of the motions and activities of electrons and atoms. He held out his hands, palms up, and said, "dead," "alive," as he looked respectively at an imaginary animal in each hand. "What's the difference? It's not some chemical because the chemicals in both are the same. It's electrical forces.

But robots have no imagination, so they don't study it," he said.

I asked him what the principle was that brought things together for the first time and made these systems improve themselves by building more complex structures capable of more complex functions. With an impish look more befitting a boy than an old man he said, "If I were trying to get a grant or pass a biology examination, I would say evolution was responsible, but really, there is much more to the story." I asked what he thought that was, and he replied, "Nobody knows. Perhaps some undiscovered principle in irreversible thermodynamics. Something beyond physics, something unique to biology." As he spoke I thought to myself, "What heresy. The distinction between life and non-life was not so great as to warrant thinking that there was a universal law physicists had missed entirely when they completed their book about how the world works." I considered framing a question to him about this point, but didn't because he'd already said he didn't know. So I was left to wonder why there was a Berlin wall around physics, with life on the outside.

26
Looking at EMFs I to I

◆

My meeting with Szent-Gyorgyi was the catalyst that led to my important discovery, and then to progress toward the solution of the problem that has been central in my life. These events occurred over a five-year period. I recounted them by means of letters because they best captured the development of what I learned.

May 4. Law and Life

I hope that you, Mary, and the girls are doing well. We haven't all been together for quite a while. I remember the day one of your daughters pushed my oldest son off the back porch of my house. After you left he said to me, "Daddy, don't invite them again, they play too rough!" I doubt he would say the same thing now. Raising him and his brothers has proved to be an exciting experience. Their adventures never seem to fall into a pattern, so it's hard to mitigate the damage or shape the events so that they teach some worthwhile lesson. I never say, "My son wouldn't do that," because I think they're capable of almost anything. My mother would say to me, after I had done something that drove her to a state of exasperation, "You ought to have one like you." Well, I had three. Raising a daughter, in contrast, has proved to be relatively free of stress. I suppose that has also been your experience!

William, I recently met the famous Albert Szent-Gyorgyi. He is an old man now, perhaps peculiar in some respects, but he still has a very lively mind. The meeting was really quite an event for me because Dr. Becker had drawn much of his inspiration from Szent-Gyorgyi. He too had been trained as a physician and had always been concerned with why and how people got sick. You can't study such a broad question from a narrow perspective, so he had branched into many areas of science. During our dinner, he told me, "If you want to build a house you need a strong foundation, and the bigger the house the bigger the foundation must be. I want to build modern medicine." As he said that he extended his arms from side to side as far as they would reach.

He studied anatomy first, and then progressed down the levels of organization, to organs, cells, molecules, and ultimately to atoms and electrons. He thinks that electrons moving through protein molecules are the key to understanding cancer, akin to Dr. Becker's idea that electrons moving through nerves regulate the ability of the body to regenerate limbs and organs. I never understood the details of either theory. Perhaps "theory" is too strong a word, at least in the sense of some mathematical equation derived from the laws of physics that could explain or predict why something happened. Mathematics is a foreign skill to both men, so there is nothing in the work of either that resembles an explanation of the type we were taught to give when we were graduate students, or that you now teach your students. Nevertheless, I shudder to think where the world would be without men such as them, with their ability to really see the world. Szent-Gyorgyi told an amusing story along this line. A man once said to an astronomer, "I know that you have telescopes, so I can understand how you could find all those stars, but how did you discover their names?"

You are probably wondering what the point of this letter is, so I'll tell you directly. We were joined one evening at dinner by Szent-Gyorgyi's wife. She was a lovely and gracious woman, about half a century younger than he, a fact that tells you what a vigorous man he is! Anyway, her presence pleased him very much, and his story-telling became more expansive and less technical, even though she undoubtedly would have understood such talk because she too was a trained scientist. I brought the conversation around to Schrodinger's book about "life," which you might remember I once mentioned to you. Szent-Gyorgyi told a story whose gist was that life wasn't some kind of chemical. If you had a live animal in one hand and a dead animal in the other, the thing that was responsible for the difference wasn't something a biochemist could extract and store in a bottle. Life was some form of energy. What kind of energy, and where is it to be found? Szent-Gyorgyi described a series of studies and wound up concluding that it was the motional energy of the electrons in proteins.

As he was telling this story an explosion went off in my mind and in that instant I saw that this great man, who seemed to know everything and everybody important, was completely, hopelessly, utterly wrong. I felt as if my senses had stopped working. Only two other times in my life had that happened to me. What I understood at that moment was that he had long ago stopped studying "life," and therefore could not possibly learn what it was by performing his experiments. He had moved down the scale from patients to electrons, and by the time he had arrived at that basic level he was no longer studying life because it does not exist there. Life is a feature that emerges at an organizational level far higher than that of atoms and electrons. What he was doing was like trying to study a smile by analyzing electrons in the lips. No matter what your data or how you applied the laws of physics, you could never understand why there was a smile, or predict when it would occur again. I'm not saying that the laws of physics are wrong. Far from it! The difficulty seems to be that they go in only one direction, down to the bottom of nature, never up to nature's crowning achievement, life, regarding which they are not directly relevant. The laws are incomplete.

You may recall from our many conversations in the Savoy[1] that my attitude toward science had always been rather schizophrenic. Although I found the power of science exhilarating, I could never avoid thinking that those who believed science could solve mankind's problems were confusing cause and effect. I know you will not be surprised to learn that my opinion has not changed, only intensified, because wherever I see a problem that befouls mankind I can put my finger on a development in science that generated the problem. I know you would say this is only a jaundiced evaluation by someone who sees the cup half empty rather than half full, but I do not intend to rehash that dispute and point out again why I am right and you are wrong! I do want to raise a new problem, which is that perhaps what we call "science" cannot explain everything. Do you believe that the laws of physics are complete in the sense that someday some bright person will show how they contain, in some latent way, an explanation for life and its attributes like cancer, old age, aggression, and genius? I don't! But I have spent much of my career fending off barracudas that were trying to take bites out of my hide with their sharp teeth. Perhaps your life as a teacher has afforded you the opportunity to contemplate this issue, which some bright student may have raised with you in class. I never met anyone who was more facile than you with Maxwell's equations.[2] So, if they are incomplete, who better than you to discern that fact?

June 20. Irrelevancy of Physics

I can't see where physics matters much any more, like some long-forgotten lover. Doctors study physics for the first and last time in a course such as yours, when they are barely a year out of high school. Everywhere

---

[1] A coffee house on the Syracuse University campus in the 1960's
[2] The laws of physics that govern electromagnetic fields

in biology the story is the same; if you examine that curriculum at Harvard or Yale you will see that physics has slipped below the academic waves. Nowadays students usually study physics only under duress, as we took Latin. In college, when I got on a bus, I would hold my physics book so that everyone could see it as I walked down the aisle. Today, if I ever had to appear in public carrying a physics book, I would hide it for fear of being pitied. The science that is supposed to explain everything can't explain anything about life. You say that asking whether Maxwell's equations can explain life is not a "real question." But I think you take that position only because physics has no answer. If you manipulated the equations in every conceivable way, you would never come to the conclusion that there *was* such a thing as life. But life exists, and physics has been no more helpful in understanding life than phrenology or necromancy. Monumental intellectual and practical achievements have been made in understanding the physical world since the time of Newton, but what has all that to do with life? Why should anyone believe that life could be understood in the same way that inanimate nature was mastered? Where is that written? The concept of what is "scientific" must be much broader than the explanation of the world according to physicists. What they should have done was study the aspect of their approach that was "scientific," and then develop a method to study living systems that adhered to the requirement of being "scientific." Instead, physicists simply declared that the universal laws gained from study of the motion of planets and the activity of batteries also explained how living systems worked. I suspect that the deepest part of my problem with EMFs somehow stems from this error, which was drummed into me when I was too young and inexperienced to resist.

July 1. Stink

Please allow me to clarify the issue. Suppose we had hauled Trashcan Trischka up on the top of a tall building and tossed him off. He would have accelerated at exactly the rate required by the law of gravity and would have hit the ground at a predictable speed, a little more if he tucked in his limbs, a little less if he extended them or bumped into something on the way down. Nothing about his trajectory would have depended on his being alive at the moment he left the roof and started down. A sack of sand having his shape and weight would have executed the same arc and impacted the earth with the same speed. His aliveness, which I think you must agree is a characteristic that stands out and commands attention, would be completely irrelevant to the explanation of the motion of his body. Similarly, neither an x-ray of an arm nor a photograph of the aura of a salamander depend at all on whether the subject is alive. If the explanations of physics take no account of the presence of life, then believing physics can explain life, or the reactions of living systems to stimuli like EMFs, makes no sense. I'm not out to deny anything about the laws that we learned and that you now teach, except perhaps the claim that they can explain everything that happens in a precise and conclusive way. They can't, and because of this incapacity physicists hate life. They despise its uncertainty and imprecision, so they radiate negativity like a dead animal gives off a stink. Look at the APS.[3] One of their press releases, which I read about in the *New York Times*, said, "Physicists are often asked whether EMFs are health risks," and then answered the question with a resounding "No!"

July 12. Biology Imperfect

If I were to return to EMF studies, I would continue to bump along in the humble manner of biologists, comparing reactions in the presence and absence of fields, seeking in that way to understand the impact of their presence, despite the imperfect and inconsistent knowledge that results from the use of such a method. There is no other choice, because mathematical laws are useless. Some of the spirits that guide humanity's search for knowledge escaped from Pandora's box, but the spirit for understanding life remains crouched below the lid, too timid to enter the world.

July 15. Limited Choices

Recourse to experiments with animals has proved incapable of resolving the EMF problem. On the contrary, each new experiment further confounded the situation. Whenever there was a report of a particular biological effect in animals caused by EMFs, there soon appeared a similar experiment denying that the effect occurred. I told you earlier about the experiments Richard Phillips performed in which he housed huge rats in cages so small the animals couldn't stand or even turn, and then performed measurements to ascertain whether EMFs were stressors. His rats could have more likely passed through the eye of a needle than exhibit a positive answer to the question he feigned asking. As the number of published experiments has gotten larger, the probability that there will ever be an answer to the problem of EMFs has gotten smaller.

---

[3] American Physical Society, the trade association for physicists

Some investigators have taken to the study of cells in plastic dishes to understand EMFs, but this work is doomed to fail because no one has any idea whatsoever regarding the type of cell that serves as the gateway for EMFs into the body. Light enters the body by means of cells in the eye, sound by means of cells in the ear, taste by means of cells in the tongue, sensation by means of cells in the skin. Where are the cells that allow EMFs to enter the body? Nobody knows!

Here is an example of what cell experiments are good for; at the same time it illustrates how lamentable the world of science has become. I had begun to study osteoarthritis, which develops because, for whatever reason, the cells that line the inside of the joint secrete too much of an enzyme that degrades the cartilage on the ends of the bones. I obtained some of the lining cells from biopsies of patients with osteoarthritis and grew the cells in plastic dishes to study how they produced enzymes, and I saw that the cells could not make enzymes until a change occurred in the membrane potential of the cells. I asked the Arthritis Foundation for money to study the phenomenon, but they said I must first do more work, which of course is why I had asked for the money. Still, I did more work, and discovered that the cells were all connected to one another by gap junctions, minuscule tube-like structures, like a pipe between two water tanks, and that only because of these structures could the change in membrane potential lead to the production of enzymes. When I again asked for money I was told I had not proved that the gap junctions were important. A reasonable man would think, "If they're not important, why are they there?" but the Arthritis Foundation did not see the issue in that light. So I performed additional experiments and demonstrated that in order for the cells to make enzymes calcium ions needed to pass between them through the gap junctions; when I blocked this intercellular communication the cells couldn't make the enzymes. When I again asked the Foundation for money they told me to stop sending in applications because their Louisiana chapter doesn't contribute money to the national organization, and they support research only in states where their research funds were raised. So, from this story you can see both what cell experiments are good for, and how the Arthritis Foundation works!

I did not give any serious thought to using the questionnaire method for studying EMFs. That method works well enough when the effect of the exposure is like a sledge-hammer, as with smoking, or when the victim himself is the responsible party, as with diet. With EMFs, however, as I have seen, the companies responsible for them manufacture obfuscatory evidence and then hire somebody from Harvard or Yale to swear that the exposure and disease aren't related, and that the few studies suggesting otherwise are tainted by bias. As a result few people, if any, will understand where the truth is. That's exactly what happened to poor Nancy.[4]

You can see that the options open to a lone wolf who proposes to study EMFs are rather limited.

August 17. Looking at the Brain

The dean appointed me to the committee on human research. All medical schools have such committees because, in earlier times, some horrific things had been done in the name of "research." I don't mean just by people like Mengele; sometimes the monster was the federal agency that today runs the NIH. Years ago, that agency recruited black men from Alabama who had syphilis and told them they had "bad blood" but that the government would take care of them. The men were given phony treatments so the researchers could study the consequences of untreated syphilis, which are far from pleasant. Anyway, as I learned more about the rules and thought more about what the goals of my research might be, the idea of doing human EMF experiments became attractive.

I'll explain my inner logic. A woman sits in a doctor's waiting room, soon to be told whether the tests showed she has cancer. A man is stranded on a mountain and the temperature falls below freezing. A child's body is wracked with infection. The salient common element in these stressful situations is that the nervous system forms a bridge between the external world and how the individual reacts to it. Anything capable of creating stress must first enter the body and be recognized by the brain; only then can the brain orchestrate that complex biochemical reaction. You can see that the chemical activity of the brain and hence its sign, electrical activity, *must* undergo changes when the subject is confronted with a stressor. If I proved that EMFs altered the electrical activity of the brain I could justifiably claim that the body detected the EMFs, which is a necessary prerequisite for it to be a stressor. I freely admit that observation of a change in brain electrical activity would not be clear and convincing proof that the field was a stressor, because the possibility would remain open that EMFs

---

[4] Nancy Wertheimer, whose study showing an association between childhood cancer and exposure to powerline EMFs has been attacked constantly by experts working for the Electric Power Research Institute

could change the activity of the brain but still not be stressors. I do not know of any example of such a stimulus, at least when it is applied chronically, but nevertheless I could not ignore the possibility. My critics would never allow that! Even accepting this limitation, the results, if they were as I supposed, would likely cause sober and thoughtful people to delve more deeply into the complexities of EMF biology. They would think, "Well, since EMFs can affect the electrochemistry of the brain, perhaps there is some connection between these changes and sickness. Prudence requires us to study this issue, honestly and deeply." I hope I'm not expecting too much.

September 11. Justification

You asked, "How can you contemplate exposing humans to EMFs if you believe they cause stress, and ultimately cancer?" The answer is that I plan to put only the subject's toe in the water, not drown him, because the exposure to EMFs in my experiment will last for only a few minutes. My limited goal will be to prove that the human body can detect EMFs. Admittedly such a result would be a long way from directly showing that they cause cancer, but everything has its pluses and minuses.

I had no trouble gaining approval to perform my study. I told the committee I would apply an EMF having about the same strength as one from an electric blanket or a powerline, and that I would expose the subject for only a few minutes. They accepted my conjecture that any harm from EMFs lay not in the process by which it was initially detected, but rather in the consequences of that detection when the EMFs were applied over a long time.

December 20. Inconsistent

The results of my experiments were not so clear-cut as I had hoped. I will tell you what happened. The subject sat in a dark room with electrodes attached to his scalp, in the standard manner for an electroencephalogram. Wooden walls contained the Helmholtz coils.[5] I used the common technique for measuring the electrical activity of the brain, which was to glue small electrodes to the scalp and pass the wires from the electrodes to the amplifier that recorded the voltage pattern of the brain waves. I cycled the EMF on and off, and when it was on I supposed that specialized cells somewhere in the body detected the field and sent a signal along a nerve to the brain.

I soon realized I was wasting my time trying to find an effect just by

---

[5] Loops of magnet wire arranged in such a way that they produce a magnetic field around the subject

looking. Inspection is useful for doctors when they diagnose brain tumors or epilepsy by observing abnormal patterns, but the patterns I had recorded were far too complex to yield their secrets about EMFs by that simple method. When I used spectral analysis,[6] however, the situation improved. The brain waves of some subjects were different when the EMF was present. Unfortunately, this result was not achieved in each subject. In one study I measured the brain waves in nineteen subjects but could show only that eleven subjects had actually detected the EMF. Even more disturbing, the changes that did occur differed from subject to subject.

How should I interpret my results? Tell me! If scientific data is supposed to speak for itself, what does my data say? Perhaps that only about half the population can detect the field, each in his own way. Or maybe all the subjects actually detected the EMF but spectral analysis was not sufficiently sensitive to detect detection in every subject. I know what theorists like Repacholi and Moulder and Adair[7] will say. "Inconsistent," "no robust effect," "only the results in the eight non-responders were correct," just like Herman[8] twenty years ago. To such men, the world is only a show that goes on between their ears. My findings will not rise very far above their noise, or that of Patty Ryan's servants whom I told you about previously. I feel like an insect trapped in the oozing sap of a tree. The more I struggle, the more trapped I become.

June 15. Robot

As I feared, despite four experiments and four journal articles, my results were dismissed by various blue-ribbon panels as "inconsistent" and "inconclusive" because not every subject reacted in exactly the same way, as if living things *should* react like robots. I now know that some dramatic change in thinking is needed.

August 25. Important Discovery

When I was in Syracuse last week I visited some of our old haunts. The coffee shops on Marshall Street are mostly gone, replaced by music stores and bars. The "new" physics building is now more than twenty-five years old, but if you stick your head in the laboratories what you see is much the

---

[6] A method of mathematically decomposing the brain-wave pattern into individual frequencies, thereby allowing more sensitive comparisons between two patterns

[7] Michael Repacholi, John Moulder, Robert Adair, three physicists well known for their opposition to the idea that EMFs are health hazards

[8] Herman Schwan

same as when we were in school. They moved the pole to the basement, and more plaques have been added; someone climbed it only a week before my visit.[9] My laboratory at the Veterans Administration is now only a memory, although I swear that when I walked down the hall past the rooms where I spent so many intense and exciting years I could still smell the smoke from Dr. Becker's pipe.

I visited my former landlord, you probably remember him, Mike Angelos, the blind man who fixed television sets! He asked me about my work and, since he seemed genuinely interested, I described my experiments on EMFs and brainwaves in considerable detail. I told him that I had used statistics to compare measurements made during field-on and field-off conditions, and that I found differences indicating the subjects had detected the EMF. When I finished my little speech he said only, "No equations." I asked him what he meant, and he said that the thing he remembered best about me was how much I had talked about equations, and how they seemed to contain all the answers to everything. Now, he said, even though I was still in "science," equations were conspicuously absent from my conversation. I saw that he was right in some deep way! What I had been doing was completely different from what I had been trained to do, before I came under Dr. Becker's influence. No physicist ever approaches the object of his study as I had done, but every biologist does. I came to understand that there were really two sciences in the world, one limited to the study of inanimate objects, and one devoted to the study of life, *each with its own method for saying what was or was not a fact*. I think I have made an important discovery!

September 2. Great Goethe

I thank you, William, for your excellent exposition, but if you look in any journal to see how biological experiments are actually done, you will find that the practice is at war with your perspective. Biologists understand life in terms of law, but not physical law imposed on living systems from the outside. Instead, each living system follows its own inner law. You would explain an observation in terms of a general law and a set of initial conditions particular to the system. That is how you can say where Mars will be next week, next year, or anytime. In my experiments, however, each subject was the source of his own unique law which governed his unique "becom-

---

[9] Climbing a wooden pole is a traditional ceremony in the Physics Department at Syracuse University for students who have passed the defense of their doctoral dissertation

ing." The law and the subject were inseparable. This is always the case in biology. I don't care whether the object of study is a human brain or a single cell, or whether the stressor is an EMF or something else. In all cases, biologists explain observations in terms of an archetype that comprises all possible particular forms of whatever is being studied. Each individual is a self-enclosed whole that changes from one form to another by a process of continuous dynamic development. We can learn something about the nature of the law that governs this process only by studying the individual behavior of a fair number of individuals and searching for a pattern. We learn much less, if anything, when we confine our study to a single isolated object, as in physics. Suppose someone came here from another galaxy and sought to understand what human beings were. Do you think that intimate knowledge of one soul would suffice? No! That limited approach wouldn't lead to understanding of how people act, and it won't lead to understanding of how they react to EMFs. In both instances, the law is particular to each subject, not universal as in physics. So I think the so-called inconsistency in my experiments did not stem from any inconspicuous fault in what I did or any unreality in the causal nature of the effects of EMFs, but from the *nature of lawfulness in biology*. Squeezing blood from a stone would be easier than extracting precision or certainty from biological study. I agree that EMF studies are "inconsistent," but to criticize them on that basis is nonsensical, as the APS did when it demanded demonstration of universality and consistency of causal connections in living systems. They say that the whole world should be approached according to the stereotype of physics! I say "no" because life forbids it. Such an approach is antithetical to the behavior of every living thing from the tiniest virus to man himself – the struggle for existence. No solid or liquid or gas, nor any object made by man struggles for existence, but every living thing struggles to remain in being. One could never show that this struggle was a causal result of some particular preconditions. Preconditions surely have their influence, but we will never be able to say that they were the sole cause. We can say that a virus or a man reacted in a particular way in certain circumstances, but we can never say that what we observed was required to occur; it's always the case that the behavior could have been otherwise.

I think you must agree with me that physical laws are not the only laws that exist, but rather are only a special case of lawfulness involving a particular class of objects. The use of only physical laws to study living systems amounts to impressing upon them an alien lawfulness, which renders almost impossible any understanding of how they react to EMFs. Before me, there was someone else who had recognized biology for what it is. Shall I say who? Goethe! You can begin to understand what a great man he was when you realize that writing *Faust* was only the second most important thing he did.

October 15. Contempt

I wrote to two professors of philosophy, one at the University of California and the other at Oxford, and told them of my important discovery. They replied in buttery jargon that required the use of a specialized dictionary to decipher, although the gist of what each man said was clear enough. The man from Oxford wrote that the laws of physics speak the truth about nature, otherwise they could not have been so successful. And since life is part of nature, life must obey the laws. From this he deduced that the laws must explain life. I argued with him for a while. "If you look you will see that biologists carry out their work as if physics doesn't exist! Look at the curriculum at Harvard or Yale and you will see that physics is not taught to biologists. These facts are evidence that physics is useless to biologists. If you have some other explanation, please tell me." Do you know what he said to me? "You are probably not right about your wild charges regarding the curriculum. Even if you were right, that would show only that physics *should* be taught there, not that physics was superfluous in biology." The California professor was wildly opposite. He told me that the laws of physics are lies and referred me to a book by his former student. I soon found that the author made her case by using the age-old trick of the rationalist, the method of pre-selecting favorable evidence. He was even less hospitable to biological thought, which he dismissed as "aphilosophical." Never in my life have I witnessed such intense devotion to the content of one's own mind. I respect the old philosophers. Today's breed is contemptible.

Next week I will go to a conference on experimental biology. I am sure there will be some there who will grasp the significance of my discovery, or at least point out the error of my ways if indeed they are in error.

October 29. Unavoidably Uncertain Method

I told the audience that some of the subjects in my experiments had detected the EMF, and that further similar studies would "probably" give the same results, more or less, because my subjects were typical. In biology, this claim means that the investigator hopes or expects or thinks or believes that would be the case. He doesn't "know" it because a biologist never "knows" anything with the certitude that a physicist "knows" something. One or the other specialist should use a different word. Anyway, my "probable" assertion raised no hackles, nor should it have because my reasoning was standard according to their lights. Then I went further and began talking about the method I had used, in comparison with that of physics. In my mind they were distinctly different sciences, and I spoke from this perspective. I said explicitly what was obviously true, that a biologist could never be certain of any "fact," or that any observation occurred for the reason he supposed, or that any particular event would actually happen. Why raise false hopes or set impossible standards of exactitude? Biology is true science, appropriately matched to its objects of study and, to tell the truth, far more important to the world than anything physicists do (and I am not forgetting technology). But I had failed to understand that most of my audience regarded physicists as heroes, and had great respect for the success of their project to mathematize nature in terms of a small number of basic laws. Many in the audience seemed embarrassed by the notion that their method was unavoidably uncertain. This reaction was particularly galling to me because biologists had, in practice, actually accepted the uncertain character and subjective tinge of biology. At an operational level they *knew* that thinking in terms of pure "facts" was meaningless and self-delusional. Evidence of this acceptance could be found everywhere in the rich variation and equivocation of the language that they themselves used to describe the link between their studies and the putative generalized fact they imagined by listing some of the euphemisms I had heard during the four days of the meeting. "Suggests"; "indicate…"; "…may have been instrumental…"; "…results in…"; "…reasonable to say…"; "…consistent with…"; "…provide direct evidence for…"; "…is the most likely…"; "…is involved in"; "…raised the possibility…"; "…believed that…"; "…may underlie…"; "…provide insight into…"; "…support a determining role…"; "…orchestrated…"; "…readily accounts for…"; "…showed…"; "…confirmed the role of…." The vagueness inherent in these words contrasts profoundly with the precision of the mathematical statements of physicists. Nevertheless, the euphemisms contained sublime truth because they expressed reality in the clearest way it was susceptible of being expressed. I made my point as persuasively as I could. Some in the audience smiled and nodded approvingly, but others got hot and claimed they were true scientists, like physicists. Sometimes I think my career is nothing but a campaign to make other scientists despise me! Despite appearances to the contrary, that is not what I'm trying to do.

December 7. Why Did Raymond Get Cancer?

My friend Raymond was diagnosed with brain cancer. I talked about this sad news with an oncologist named John. I couldn't seem to verbalize the questions that were really in my heart. I wanted to know what he thought Raymond did wrong, or what wrong thing was done to him that had caused the tumor to start growing. There had to be a rational explanation. But how do you even talk about that? Initially, I asked, "How did Raymond get the tumor?" even though I knew that wasn't the right question because John hadn't shadowed Raymond and taken notes regarding everything he did. Even if he had, so what? No one knows what "causing cancer" looks like, so John wouldn't have known that's what he was seeing even if he saw it. John, however, had an expedient answer: "The p53 tumor suppressor gene was down-regulated and the ras gene was over-expressed." He had programmed his mind so that when he heard, "How did so-and-so get cancer?" he did not give the question the meaning, "What made the tumor start to grow?" but rather, "What is the genetic machinery of the cancer?" "No, no," I said, "Do you think that one day his p53 and ras genes decided of their own free will to stop behaving normally and begin acting malignantly? Genes don't have free will. Something must have forced them to misbehave. What are the items on the list of agents that have the power to do so?" But I soon saw that John had made a Faustian bargain in which he agreed to never seriously entertain such a question. Fifteen years earlier, while he was still at Harvard, the Air Force had decided to shoot its PAVE PAWS[10] beam right over his coastline property on the Cape, and he sought my advice about EMF health hazards. Now that he was a richly funded and well-connected investigator, he was careful to never ask questions that upset his corporate and governmental sponsors. He has unfortunately forgotten what it's like to be on the receiving end of risk for the benefit of another. He devotes his energies to tackling only the machinery of the disease. He teaches that "almost all" cancers are caused by smoking, poor diet, and bad genes, and he beards me by bringing in speakers who spin stories for the medical and graduate students about how all links between EMFs and cancer are nonsense. Still, I must deal with him and

---

[10] A radar system intended to provide early warning of nuclear attack

people like him. I am like a traveler ascending a mountain. If there were no mountain the road would be shorter and easier, but the mountain is there and the traveler must get over it.

March 2. Mother-Worship

Much progress has been made since biology began blazing its own path. But what was the price of this progress? Did it allow us to see some things while blinding us to things of even greater importance? I fear so. There has been precious little progress in understanding what causes the behaviors exhibited by living things. What, for example, causes cancer? Note that I don't ask "how?", but "what?", so don't talk to me about genes. Every living thing on earth has this in common: it exhibits behaviors and responses that more or less parallel those previously exhibited by others of its species. Never exactly so, but always approximately so. But the lack of progress in identifying the agents that cause disease is absolutely breathtaking. Excepting infection and trauma, in almost no instance can we say with any reliability what the cause of a person's sickness is. I will tell you why I think this is so. The biologist looks up to the physicist, as a child to its mother. And mother is such a simple soul! Whenever she sees a leaf fall she thinks some force must have pushed it. Biologists also go around saying "caused this" or "caused that," but they don't think about what they mean. So they don't realize "cause" can't have the simple mechanical meaning invoked by physicists, because never is only one thing completely responsible for the way something alive behaves.

Mother-worship has engendered many errors. First and foremost, when biologists adopted her "machine" metaphor they denied chaotic nature. Think about it! What more effective way to misunderstand experiments than to have inferences based on averages? Then, if an agent causes one person's cancer but prevents another's, we must falsely conclude the agent had no effect on cancer.

As a class, biologists seem to believe that the laws of physics will someday explain life, so they do not trouble themselves to think seriously about biological lawfulness, which actually is internal. This is the far-reaching principle about the nature of the law that governs living things which mother's concept of "law from the outside" has blocked from consciousness. Her dearest belief, that there is an invariable proportionality between a cause – that is to say, "force" – and its effect thwarts biological thinking, like sand in the gears of a machine.

As a result they believe in "consistency" as the natural order of things, and this religion forces the biologist to deny with his mind what he sees with his eyes. I once heard a presentation from a biologist who studied the effect of a drug on the ability of rats to negotiate a maze. He hypothesized that the drug would impair their motor efficiency with the result that it would take longer for them to negotiate the maze. That was what he observed in twelve rats in his study group; three other rats, however, were like dynamos and solved the maze three times faster than the others, and two other rats were never able to complete the task even though he waited all day. Do you know what the investigator did when he evaluated the data? He said, "Well, there was clearly something aberrant about these five rats, so I excluded them from the study." Such silly interpretations come about for one reason – false belief. If you ask nature a question and she answers it, then you must accept that answer even if it is against your will.

April 15. Biology in Servitude

The worst error of the biologists, one that directly mimics another of mother's errors, is their willing acceptance of a state of servitude to government in return for grants or contracts. This cultivated obeisance arrests scientific development because the government regards most new frontiers and directions with suspicion and contempt; in the EMF area government has made a caricature of biology by requiring biologists to deny reality. Government has that kind of power; if the president said there was green cheese on the far side of the moon, 70% of the population would believe him.

May 25. Internal Law

There was a time when people hoped that weather could someday be so well understood that you could set your watch by your thermometer. We now know that is impossible. Tiny changes in temperature can grow to a hurricane or a tornado, and each time it happens we are surprised. Isolated parts of the atmosphere follow well-known laws and are susceptible to understanding using the old method. Changes in temperature occur exactly as required by the laws of thermodynamics, and every small volume of atmospheric gas moves exactly as it should. The global phenomenon of weather, however, follows its own law, as if the weather were a living being. I think the effects caused by EMFs are like the weather in the sense that the tiny changes produced by fields when they interact with the body can have profound consequences. Ask Richard Phillips about that! He could not imagine that the growth rate of his animals could go up or down when an EMF was present. Effects of that sort are enough to make a man who is trying to please his client pull out his hair in frustration. Did you know I found good evidence that EMFs were linked to suicide? I think EMFs produce opposite responses in different people, regardless of the response the investigator happens to be measuring. Nothing has been as consistent as the inconsistency of the EMF studies involving animals. This consistent inconsistency means EMF-induced bioeffects are real, despite what old Chauncey Starr[11] wants the truth to be. There was a Spaniard named Delgado who reported that EMFs caused deformed skeletons in chick embryos. Morton Miller, my old friend from New York, repeated the experiment and, of course, found that Delgado was completely wrong. But everyone who studied EMFs knew that Miller was a fraud and that his work was not intended for the consumption of serious scientists. So the Environmental Protection Agency set out to answer the question with precision and finality. It awarded funds to six different investigators at six different locations, all of whom promised to do exactly the same experiment. The people at EPA in charge of the project were shocked by what happened. The investigator in Las Vegas found no evidence whatsoever that the EMF produced deformed chicks. In Maryland, however, the results were exactly opposite, to a statistical certainty. Wait! There's more! The investigator in North Carolina agreed precisely with the investigator in Las Vegas, but the investigator in Ontario confirmed the results of the Maryland investigator because she too found that the EMFs increased the number of skeletally abnormal chicks. The story would not be complete if I did not tell you that the investigators in Sweden and Madrid agreed with the investigator in Las Vegas. If the investigators were a jury the vote would be 4-2 in favor of convicting EMFs of deforming chick embryos. Isn't it apparent that whether the effect of the EMFs will occur in particular cases is beyond our power to understand? There is no law that descends on an EMF-exposed organism and requires it to act in a completely specific manner, as gravity requires an object to fall and thermodynamics requires heated air to rise. The method for studying the effects of EMFs on people and animals must be based on the conception that living things are not ruled only by an external law. Even some nonliving systems are partly governed by unpredictable inner laws, the weather for example. The government has thousands of computers and

---

[11] The president of the Electric Power Research Institute

a fleet of airplanes all for the purpose of trying to guess how the animal that surrounds us will react to tomorrow's sunlight. Appreciation of the role of inner law is no dark mysticism, no teleology, no form of spiritualism. Rather, it is real, honest, science appropriately matched to the object of study. That is what I must learn to do with EMFs.

June 14. Congress Steps In

Sadly, some organizations are incapable of appreciating how much people value their health. I'm confident you remember our discussions about the Electric Power Research Institute in Palo Alto, which paid for many inglorious EMF experiments that now contaminate the journals with wide-ranging myths. "He who has the gold makes the rules" was Chauncey Starr's motto but in at least one instance things haven't gone his way. Congressmen have been besieged by mothers who don't want their children to get cancer from powerlines. Someone on the staff of a congressman from California told me of a new law that will require the government to pay $40 million for experiments to get at the bottom of the matter. Starr will have difficulty controlling that agenda, but I wouldn't be surprised were he to succeed. His shtick has been that cigarettes cause far more cancer than EMFs, that the evidence against EMFs is not conclusive, and so on. Now the well-meaning Congress has appropriated money to pay for scientific studies of the link between EMFs and disease. Perhaps we will learn whether powerlines are safe, as they stoutly maintain in Palo Alto, or unsafe, as I claim. At least one thing has happened well.

July 5. What Bright Minds?

In New York, fifteen years ago, the $4 million that was squeezed out of the power companies at the price of Dr. Becker's laboratory and my career at the VA was wasted because the companies retained enough control to insure that whatever they didn't like wouldn't happen. Did Congress do any better? I fear not. When I was contacted by the staff of the congressman who wrote the bill, I argued with all my energy that the money must be controlled by a group with balanced views that owed no allegiance to any permanent bureaucracy. But I learned yesterday that Congress simply gave its money to the director of the National Institute of Environmental Health Sciences.[12] He freely admitted he knew nothing about the parameters of the problem, but promised to use the brightest minds in the government

---

[12] One of numerous "Institutes" in the National Institutes of Health

to manage the project and interpret the results. What bright minds? There are none because the cornerstone for building a bright mind is intellectual freedom which, by law, is something no non-elected government mind possesses. What other organization could Congress entrust with the task? Alas, none that I know of. Every group of scientists I know cares primarily for its own interests. Whatever disinterestedness exists in science can be found only in individuals.

April 1. My Plan

I decided to ask the NIH for half a million dollars to help breathe life into my theory that EMFs cause cancer. You would expect that I had no chance for a favorable response, but you would be wrong! I'll tell you the game I played to get their money, although I think my success was probably due in larger part to some benevolent spirit whom I have been unable to identify, rather than anything clever that I did. Such patronage has graced me previously at the NIH.

I've focused my attention on cancer because of its popularity, not because I believe it is the most likely disease manifested by those exposed to EMFs, which I think predispose to every disease under the sun. According to the rules for getting a piece of the $40 million, the proposal had to involve plans to study possible mechanisms by which EMFs could cause cancer. That was the information NIH said Congress had said it wanted. I never argue with a judge, so I brought my brain-wave studies to a screeching halt and began thinking about "mechanisms." I found the activity repugnant but I had to do it because I needed the money. I learned about "natural killer cells" that circulate through the blood. When they come upon cancer cells, the killers somehow recognize the deviation from normality and promptly destroy it. According to some immunologists (by no means all!) the reason the cancer rate is as low as it is stems from this immunosurveillance function of the killer cells, which destroy cancers at their point of beginning, before they grow to a palpable lump. I realized that if EMFs impaired the efficiency of immunosurveillance they could thereby be a "cause" of cancer – not of the initial oncogenic transformation, but rather of the lump a woman felt in her breast, because the original aberrant cell that produced the lump might otherwise have been destroyed. Why would I hypothesize EMFs would degrade the efficiency of killer cells? Because it is well known that any stressor does so.

The National Institutes of Health are like sausage makers. Each produces a popular product using a hidden process and a formula that abjures novelty. The basic requirement for obtaining funding is presentation of evidence that the results of the proposed experiment will be as predicted. I therefore had to show that EMFs would alter killer cells in animals before I could receive money to do the experiment using large numbers of animals and thereby prove my point absolutely. Occurrence of a change using ten animals, called "preliminary evidence," would entitle me to ask for money to find the effect in five hundred animals, which is called "proof!" So, I measured killer cells in ten mice treated for two weeks with an EMF of the same strength as the one produced by grandmother's electric blanket. God smiled on me because the results showed the killer cells were far less robust than those from ten sibling mice that had no EMFs. Had I repeated the experiment, my thirty years' experience with EMFs suggests to me that the killer cells in the exposed mice and their controls would not have differed, and had I done the experiment a third time I might well have found that the cells were more robust in the exposed mice. This is reality. This is what EMFs do to living organisms. Had I told NIH about my perspective, the radiation study section would have given me a 5.0[13] because, to them, nature is a robot. So I did not perform the second and third experiment, but simply submitted the results of my first experiment, and asked for money to prove that EMFs degrade killer cells. I knew that kind of hypothesis would be to their liking, to the extent anything I could write would be to their liking. If the killer cells go up, that's good. If they go down, that's bad. Everything quite neat. That's the fantasy the radiation study section makes of the world.

    I submitted the grant application, hoping for the best, and thinking perhaps I had a little edge because NIH would probably want to avoid the charge that they were completely biased against EMF effects on health, which could easily be alleged because essentially all of their grants had been awarded to professional grant-writers who move to wherever the money is, young assistant professors who need to get grants to earn tenure, and investigators from the herd of the Electric Power Research Institute. You can see that the inclusion of someone experienced and with my views might add "balance."

    My proposal had to pass between the Scylla of the project officers at the

---

[13] The rating scale used by NIH is 1 to 5, with 5 the worst

National Institute of Environmental Health Sciences and the Charybdis of the radiation study section. Charybdis hates EMFs, and drowns any proposal not based on a plan to compare averages; so, to win a 1.0, an investigator must express doubt that EMFs will have any effect, and he must promise to put out his eyes so that he cannot see chaotic nature. Scylla swoops down and swallows all proposals having any propensity to produce controversial inferences regarding hazardous factors in the environment. So effective is the Institute in this regard, their funded investigators have never found a hazardous agent in the environment. What a shock if EMFs were the first. Bearing all this in mind, I put things in the proposal that I hoped would help me avoid both dangers. I promised that I would follow the law of averages; I surmised that tack would be unsuccessful, so I made plans to use forbidden methods to analyze my data. To deal with the other danger, I dredged up the ongoing controversy among immunologists concerning whether killer cells actually carried out an immunosurveillance function. The gist of what I said was, "Don't be too concerned about the political fallout if the data should show that EMFs impaired killer cells. After all, immunologists themselves cannot agree whether any such effect, however produced, has any direct implications for whether or not someone will develop cancer." Because of all this scheming and planning, and with the help of a guardian angel, I got the money! Now I shall see what I can discover.

February 14. Real Immune Effects

The people running RAPID[14] continue to dole out money, but the rumor is that they have not received much positive evidence, which is not surprising. I'll tell you how I spent their money and what results I achieved and sent to them. I cannot yet tell you about their reaction. They are crafty people, adept at not betraying their true thoughts. As for me, I believe I have the correct conception of things, and I am excited.

My co-workers and I took out the blood and the blood-forming organs in mice, and performed a battery of measurements of parameters that reflected the activity of the immune system. I measured killer cells, but also various sub-classes of T and B cells,[15] all according to the most modern methods. When I compared the average values of these parameters in fifteen mice that had been treated with EMFs with fifteen other mice that had received

---

[14] The acronym for the name given by the director of the National Institute of Environmental Health Sciences to the congressionally mandated program to assess whether powerline EMFs were health risks

[15] Two kinds of blood cells

no EMFs, the two groups were identical. I was not the least bit discouraged because that was exactly the result I would expect if each animal followed its own law, any particular parameter increasing in one animal, decreasing in another, and not changing at all in a third. The individual responses are meaningful but they sum to nothing. I had a plan for dealing with this situation which, as I told you, I anticipated. The story began when an article about an improved method for analyzing data was published in a statistical journal. The method was not the answer to my problem but the author, a professor at Indiana University, struck me as someone who could help find a way to evaluate my data so that its message was not averaged away. I agreed to pay his fee, which I could afford thanks to my RAPID money, and after some back-and-forth communication between us he produced a method capable of recognizing a causal influence of EMFs, by which I mean I could be more than 95% certain regarding whatever the truth was.

I did my experiment over and over, with small variations on the basic theme. I exposed animals for 10 days, 25 days, 100 days, and 175 days. I exposed them to 1 gauss, and to 5 gauss. I used males and females. The result was always the same. The immune system of the mice that spent time in an EMF consistently differed from the immune system of the controls to more than a 95% certainty. I can now say without any reasonable fear of error that the effect of EMFs on the immune system is real, and can be realized by anyone who, when he designs his experiment, keeps in mind what biological lawfulness is.

Only a poet could put into words how I felt when I finished the studies and realized what the results meant, so I'll tell you in a way that I'm better suited for. My hypothalamus sent electrical impulses to my pituitary which then sent hormones through my bloodstream to my endocrines. They, in turn, secreted other hormones that were carried to receptors on nerve cells throughout my body, and those cells sent impulses back to my brain, producing a feeling of intense joy and satisfaction that approached the way I felt when our children were born and I saw that they were perfect in every anatomic detail.

May 17. Ballistic

I informed the radiation study section of how I had evaluated the data, and they went ballistic. I'm pretty sure I'll never get another NIH grant.

June 1. Christopher

The power behind the throne of the RAPID program is a small athletic-looking man named Christopher, a statistician by profession, who seemed fit enough to run a marathon. My initial impression of his understanding of the subtleties of the EMF problem was favorable. During a speech to the RAPID investigators he said, "Data does not speak for itself, it must be analyzed properly so that its meaning can be determined." He also said, "I have come here to San Antonio to ask your opinion and seek your advice about EMFs so that my decision concerning whether they are health risks will be based on the best available science." After he finished, I waited in a line for an opportunity to speak with him, as if he were a mafia don; I felt uneasy because I had no reason to think he would give any credence to what I had to say. When my turn came I said, "It is not enough to simply listen to someone's opinion about why he thinks something is this way or that way. The opinion should be written so that others might determine exactly what it is, and whether it was arrived at using consistent and proper methods. If there is no written record, people can say essentially anything, because there is no price to be paid for misstatements, faulty reasoning, or vague language." When I finished he remained silent for a few moments. He knew I was a lawyer. Usually, that knowledge serves only to make people wary, but sometimes it serves as the predicate for a sardonic remark. I sensed that Christopher was preparing to make such a remark, and he did: "You sound as if you want to cross-examine people." I knew instantly that the complete appropriateness of his comment was entirely unintended, but I answered as if the opposite were true. "Precisely. Cross-examination is the greatest vehicle ever invented for ferreting out what is truthful and reliable. If you base your decision on opinions that have been tested by cross-examination, you will have a worthy basis for the making of public policy." He listened but said nothing. The director of his Institute had gone before Congress where, concerned mainly with form and ceremony, he spoke in generalities about RAPID which, he probably thought, he headed. I believe he is governed by Christopher, who knows the scientific details. Who, therefore, is really the chief? It seems to me it is the person who knows the subject and possesses the strength or skill to insure that his designs are executed. What Christopher's designs were I couldn't tell.

July 1. Happy

There is no reason for anyone to take note of my work. It is a solution to a problem that is perceived by too few. Most people accept the tradition that EMFs are harmless, so they have no reason to consider my work. As for my perceived need for a new way to study life, the cure-mongers have been busy, so almost no one sees any reason to look for a cause in order to prevent diseases in the first place. To most people, life is nothing more than a complex machine governed by outside laws, so the old way is good enough! The more I learn, the more I am forced to part company with the old way. I must try to plan experiments that will provide the confrontation I desire so desperately, so that if I am wrong in some particular or general way I can learn to see how or why. If my isolation continues, fine. I am happy. I cannot conceive that there is another soul on this earth that is luckier than me, or who is having more fun.

July 10. Better than Snapshots

All effects that EMFs cause in the body must begin with transduction,[16] immediately followed by changes in brain waves, because the brain is how we know about the world around us. The changes in the immune system I found surely occurred, far, far subsequent to transduction of the field, thereby allowing time for myriad factors to exert their influence, thus obscuring the initial causal role of the applied energy. If we had drowned old Trashcan in the headwaters of the Mississippi, sooner or later his DNA would have drifted past New Orleans, its concentration and state of preserve no doubt influenced by the myriad forces that shape and govern the flow of the river. Any measurements that could be made there on his genes would be a poor indicator of what he had been when he was deep-sixed. That was what the immune measurements were like, far downstream so that I could not say much more than that the EMF had been there.

Here's another problem with my immune study, or more exactly, with the design behind it. When I killed each mouse, took out its immune system, and measured twenty things, it was as if I took twenty snapshots of what that animal had been at the moment I killed it. What if the snapshots had been made a minute earlier? An hour earlier? A day earlier? Would the results have been the same? Of course not! Every living thing follows its own law and is forever changing. Every measurement of every kind that it is possible to make reflects that law, so any number that results from a measurement must be in a constant state of flux. Not without limits, of course, but forever changing within limits. Poor me! I was trying to understand a dynamic process by looking at only one time point. Taking snapshots is no proper way to study dynamism. I must do better. I need a way to descend more deeply into the nature of the

---

[16] Biological jargon for the conversion of a stimulus into the electrical and chemical language of the body

impact of EMFs on life. I have known this for some time.

July 15. Szent-Gyorgyi

Brain waves may be the key to understanding EMFs. Perhaps I should return to them. Their infinitely variable pattern of changes reflect exactly and precisely from moment to moment what is transpiring in the internal and external world of the subject. To what can they be likened? A canvas on which an artist paints what he feels or thinks or sees? The undulating surface of the ocean, pushed and pulled by forces from above and below? These are pale metaphors to explain what brain waves are in relation to what produces them. I can think of no better way to describe brain waves than to say that they are a book in which is written the time-dependent vector sum of every actual influence on the subject. Pure dynamism, the essence of life, as Szent-Gyorgyi said. But it is a book that no one has ever succeeded in deciphering because its bewildering complexity renders meaningless attempts to interpret the effect of particular influences, except, of course, for influences that are so obvious that their effect can be seen at a glance by the use of averages, and the hoary rule-of-three.[17]

September 15. New Developments

As I was sitting here on my island with my magic bow and foul-smelling foot, waiting for visitors, Christopher invited me to participate in a grand meeting he had organized where all of the world's experts on EMFs were to be assembled for the purpose of reasoning together to establish what the true facts were regarding EMFs. Then I was contacted by an immensely powerful lawyer who could easily match the infinite resources that Patty Ryan brings to a lawsuit when he represents his EMF clients. I may have opportunities to argue my case.

---

[17] A rule first formulated by a 17th-century English merchant, thereafter used in commerce and science to identify so-called "true causes." The rule says that true causes always produce proportional effects. Mathematically, if a and c are different levels of a cause, say an EMF, and b and d are different levels of an effect, say a change in natural-killer cells, then $a \div b = c \div d$.

Part V

♦

Return: 1998-2010

Then we shall make our plans how all may come out best for us.

(13: 365)

27
Christopher

♦

The EMF problem had continued to heat up. Scientists at the Environmental Protection Agency had written a report saying EMFs were carcinogens, and someone leaked a draft copy. The president's Science Advisor had rejected the report, nevertheless people still complained about EMFs to their representatives in Washington, some of whom contacted me. A senate staffer who was investigating the matter had told me, "Cancer is a terrible thing, but I don't see how anybody can link it with powerlines. Every time there's a study that says one thing, there's another study that says something different." I had learned from an assistant to another congressman that he had looked into the possibility of burying powerlines and learned it would cost billions. The assistant said his boss didn't want to be known in his district as the man who was responsible for doubling everyone's electric bill. A policy advisor for a California congressman had told me he considered holding public hearings but decided against it after he received complaints about the possible impact on property values near powerlines. I had heard directly from a Massachusetts congressman about a delegation of homeowners that visited him. They were concerned about a cluster of cancers in children in their neighborhood, and told him they didn't think the cluster just happened because everything happens for a reason; they suspected the powerline that passed through their neighborhood. All these inquiries had led to much harrumphing in Congress concerning the EMF issue, particularly among the Democrats. A committee chairman had said in a speech, "The occurrence of even a single case of cancer is a great tragedy. We are faced with a public-health hazard of unknown proportions. The people have a right to know the facts. If something they live with or that is present in their neighborhood is killing them, they ought to be told. We have to find the facts."

Shortly thereafter Congress had passed a law that gave $40 million to Kenneth Olden, the director of one of the institutes of NIH, and directed him to find the facts about EMFs and cancer. The program, which he called RAPID, had sent shock waves through the NIH because, consistent with its motto "science by scientists for the sake of science," NIH despised the involvement of laymen in its activities, and was especially fearful of open politics. Olden surely had felt revulsion at the idea the power companies might be causing cancer simply because not burying the powerlines was economical. Just as surely, as a black physician he must have been sickened by the failure of physicians at his agency to treat the poor black syphilitics from Tuskegee in a manner consistent with their Hippocratic oath, and this must have motivated him to insure that he did not commit the same sin, neglect of one's highest duty. Olden had said: "The best scientific minds in the world will resolve the EMF issue, and the process will be open, transparent, objective, scholarly, and timely under Congress's mandate." Only later did I learn how Olden had coped with the impossible task Congress had given him; his strategy had been the same as that Andy Bassett had used with Bernard Baruch, and I had used when I advised Dennis Thibodeaux concerning his legal problems.

Olden had appointed John Christopher to run RAPID. When I had first met him he told me, "I promised Dr. Olden I would solve the problem with no lingering concerns, and that is what I intend to do." Three years later, after spending all of the money he got from Congress on experiments he thought meritorious, including mine on the immune system of mice, Christopher was ready to fulfill his promise. He hand-picked a panel of experts and organized a meeting to debate the meaning of all the results that had been obtained in the RAPID studies. I was one of those whom he invited to participate.

When Erica, my graduate student, learned of the meeting, she was anxious to attend. Some of the other invitees told me they planned to bring students or other guests to observe the debate, so I agreed to bring her.

We went down to Dallas for the meeting. A little before dawn the phone rang and woke me up. It was Erica. I expected bad news, but she was excited because the circus was in town and asked me to go with her to watch the unloading of the animals.

First we saw tigers skulking in their cages as a boom-crane lifted them off flatcars and onto wagons. Then a troop of chattering chimpanzees were led down a gangplank to join the parade. Next we saw the elephants lumber out of their boxcars and organize themselves into a line, all in response to sharp raps from the long slender sticks of their handlers. Each elephant wrapped its trunk around the tail of the elephant it followed, except for the first elephant which raised its trunk high over its head, as if the air there smelled better. The crowd stared, but the animals seemed to take no notice of them.

"It seems like a sad life," Erica said as we watched the spectacle.

"Perhaps not," I said. "The animals are well fed and all their needs are met."

We walked back to the hotel slowly, to pass the time until it got light and the meeting started. To test Erica's mettle I began to question her about her ambitions.

"You could be back in the laboratory doing experiments, but instead you came here. What do you expect to learn?"

"Well, Professor Marino, I suppose that the best science is done at the NIH, and I would like to meet the people who work there, and learn what I can from them."

"Do you mean that you would like to make connections so that when you apply for a grant the people at the NIH will know who you are and perhaps look favorably on your application?"

As a streak of daylight betrayed her blush, she said, "What interests me most is learning how science can be used to prevent disease."

"Is that what you think the NIH does?" I asked.

"That's what they say," she replied.

"So you think that 'preventing disease' is a precise subject that can be taught?" I asked.

"Yes, I have no reason to think otherwise."

"Before you began working in my laboratory, we corresponded for almost a year. Then we had several long discussions. After all that you decided to devote yourself to conducting EMF experiments because you had an understanding of the person from whom you expected to learn. But you don't know anyone at the NIH, so why do you think you can believe what they say?" I asked.

"I just presume that what they say is trustworthy," she said.

"But how do you know that you can trust what you hear at the meeting?"

She had no answer so I suggested that she give the matter some thought, and I promised to introduce her to Christopher at the first opportunity.

At the hotel where the meeting was to be held, our path was blocked by picketers who were protesting cruelty to the circus animals. I told them we were scientists and had no connection with the circus. Reluctantly they stood aside and let us pass.

When we entered the meeting room we saw Christopher. He was walking in the front, and walking with him in two lines were many experts whom I recognized. On one side I saw Neil Chernoff of the Environmental Protection Agency, Imre Gyuk of the Department of Energy, and Phu Phant of the Food and Drug Administration. On the other side I saw John Morrison of the Electric Power Research Institute and Russel Reiter of the University of Texas. Those who followed behind listening to their conversation seemed mostly to be some of the foreigners that Christopher had drawn into his EMF program, but I also saw some Americans in the troop. When Christopher turned around, the troop divided perfectly and circled to the rear.

After that I recognized Martin Ruhig, who was Christopher's deputy; he was sitting at the opposite end of the room, and seated around him were men and women not much older than Erica. I supposed they were asking him questions on strategies for obtaining grants from the NIH.

Then I spied Don Justesen, or at least someone who looked like him, sitting on his suitcase and looking even more glum than usual; he had arrived too late the previous evening, and so had lost his room. Sitting on a couch was Charles Pick, a physicist from Rhode Island; he appeared to be with the woman who sat next to him. There was Asher Sheppard, owner of the Asher Sheppard Consulting Company, talking with Monica Lugner and her sister Eva Stum, both of the Electric Power Research Institute. I would have loved to know what they were discussing. James Barnes of Wyoming sipped coffee and stared into space, looking very much like the pensive college professor he was. He asked a young man, whom he took to be a caterer, for a doughnut, but the man turned out to be Paul Afelis from the Department of Energy, who had become involved with EMFs so recently that Barnes didn't know him.

Just after we had come in, Nancy Wertheimer, the woman from Colorado who wrote the first papers about EMFs and childhood cancer, entered immediately behind us, accompanied by David Savitz from the University of North Carolina. They spoke cordially and I saw no signs of resentment, even though she had paid for her studies out of her own pocket and Savitz had received millions from the Electric Power Research Institute and the

NIH to repeat her work.

In the hallway outside the room I saw the enigmatic Ross Adey, from California, pacing back and forth along a straight line taking care not to veer to either side, like a funambulist.

After taking in the scene we went up to Christopher and I said, "Dr. Christopher, this is Erica. She has the feeling that you can teach her something, though she doesn't know exactly what that might be."

He said to her, "Each day you're here you will learn more about science, and when you go home, you'll be better than when you came."

"Better at what?" I asked.

"Not in measuring enzyme levels, cloning a gene, or designing an experiment" – here he looked directly at Erica – "but in your ability to evaluate scientific evidence to predict the circumstances and conditions that will or will not cause disease. We will determine whether powerline EMFs threaten human health, but the principles that will guide our decision are applicable to any agent that is thought to cause disease."

"Isn't a threat something immediate," I asked, "and therefore shouldn't the answer be obvious?"

"Special skill is needed to assess the truth of the claim that EMFs cause cancer," he replied.

"Some questions can be answered exactly, some approximately, and some not at all. What kind of a question is the safety of powerlines?" I asked.

He replied quickly, "The kind that can be answered exactly, because it's a scientific question."

I did not want to impose any further on Christopher, who I could see was anxious to start the meeting, so I ended our conversation by thanking him for his kindness in speaking to Erica. As she and I walked away, she told me how excited she was at the prospect of learning what Christopher had promised he would teach her.

"It's the most important thing anyone could teach," I said, "because everyone wants to avoid getting sick. The question of whether anyone can know or teach that is something I have thought about a lot."

She had no idea what I meant, so I explained.

"Many well-known persons advise the public about health matters. For instance, Pat Boone says that a particular drug will cure acne, or Larry King says that garlic will promote health. People who accept their opinions haven't really been taught anything because Boone and King have no special expertise. If we consider the fate of experts, we reach the conclusion that they did not know what made people sick. No one knew more about biochemistry than Philip Handler or more about medicine than Andrew Bassett, yet both died of cancer. I could mention many other experts who were also unable to save themselves. So I don't believe that the health or disease of human beings can be predicted with certainty. There are no equations for that, and I don't imagine that there are fortune-tellers who really possess such knowledge. Still, I suppose we must wait and listen to Christopher. If he can demonstrate that the ability to predict disease is something that can be known and taught, you and I will both be wiser."

We then walked to our seats, Erica to the place designated for observers, me to a chair at the square of tables designated for the experts. Christopher began the meeting by explaining that he had been trained as a statistician and immunotoxicologist, but subsequently had become a specialist in the subject of discovering the overall meaning of a group of scientific studies. Then he explained what was at issue. "Our task is to determine with scientific certainty whether EMFs from high-voltage powerlines cause leukemia or brain cancer. There have been thousands of scientific reports, and they have confused the public and the decision-makers. Our government needs to know the truth, which we can find by ascertaining whether laboratory studies have convincingly shown that EMFs can damage cells in a way that is harmful to human health, or whether EMFs common in homes can cause health problems."

The experts murmured when they heard that weighty responsibility described with such clarity; even the foreigners looked somber. Christopher continued: "Foremost among the principles that must guide your judgment is that of the plausibility of a relationship between EMFs and either damaged cells or health problems. Does it make sense? Is there a mechanism by which it can occur? If we understand how an agent affects the body, we can more readily accept that it might lead to disease. For example, we know from laboratory studies that chemicals in cigarette smoke can form DNA adducts; adducts, of course, lead to altered gene activity, and ultimately cancer. So we can see the means by which smoking leads to cancer. Do laboratory studies similarly disclose plausible mechanisms that might explain how EMFs could cause cancer?

"A related principle is that of proportionality between dose and effect. The reality of a causal relationship between an environmental agent and its effect is more readily acceptable if more serious consequences follow from higher levels of the agent. For example, the cancer rate is far greater among those who smoked for many years compared with those who smoked for only a few years. Are the effects attributed to EMFs proportional to EMF dose?

"A third principle is that of coherence. If similar experiments produce opposite results, then we can be confident that the effects of EMFs are not real. All properly done studies of the link between smoking and cancer point to a link between them. Do the EMF reports similarly agree with one another?

"Your successful competition for NIH grants is evidence that you have mastered these three great principles. My role is to guide you in their application to the EMF studies during open scholarly debate."

When he finished, Christopher asked Rose Mandel of the University of Toronto to give her opinion of the EMF immunology studies, and to state whether they showed that EMFs had the power to alter the immune system. She described her studies, those of Jacob Juko from the University of Kupio, James Morrison from Battelle Institute, Meike Mevissen from the University of Berlin, and Thomas Tenforde of the Bonneville Power Authority, along with the reports from about a dozen other investigators. She did not mention my studies except to say that my methodology was novel and had not yet been generally accepted. She concluded, "The studies evaluated a wide variety of immune-function endpoints in mice, rats, baboons, sheep, and humans, including immune-system structure, cell- and humoral-mediated immunity, and innate immunity. Limitations of the studies have included a lack of consistency in study design and exposure parameters, and a failure to repeat the experiments to insure that the results were correct. For these reasons the studies do not sum to anything, and give no indication of a possible mechanism by which EMFs could alter the immune system. It is therefore not possible to draw firm conclusions regarding the potential effects of EMF exposure on the immune system, and further studies should be conducted."

Christopher expressed satisfaction with Mandel's evaluation, and others echoed his view, especially Pick and Morrison, who each made strong speeches. Pick said, "Unsubstantiated claims have generated fears of powerlines in some communities, leading to expensive mitigation efforts, and, in some cases, to lengthy and divisive court proceedings. The costs of mitigation and litigation relative to the powerline–cancer connection have risen into the billions of dollars and threaten to go much higher. The diversion of these resources to eliminate a threat which has no persuasive scientific basis is disturbing."

Morrison played the same tune: "More serious environmental problems are neglected for lack of funding and public attention, and the burden of cost placed on the American public is incommensurate with the risk, if any."

When the views from that side had been completely aired I spoke for the first time, directly addressing Mandel. "Suppose you had come to the opposite conclusion, namely that a particular mechanism was the means by which EMFs alter the immune system. Could that be a complete explanation, or would it still be necessary to understand the mechanism of that mechanism?" I asked.

"I don't understand your question."

"For example, accepting that DNA adducts are part of the mechanism for lung cancer, would you agree there must be a mechanism by which adducts play their specific role?"

"Of course."

Christopher interrupted the discussion and said, "I don't see where this is leading."

I acted as if I hadn't heard him and continued to address Mandel. "Then there is something that happens after the formation of the adduct but before the development of the cancer, and whatever that process is, it can more accurately be called the true mechanism because it is more basic."

"Yes," she said.

"Wouldn't there always be an even deeper mechanism, so that it is impossible to conceive of any specific structure that could truly be called the mechanism by which the EMF altered the immune system?"

"The hope," she said, "would be that we might come sufficiently close to understanding what mediates the effects of EMFs that we could, for all practical purposes, declare an end to our quest."

"Well, if even in principle there is no unique mechanism, the requirement that a mechanism must be identified before the causal impact of EMFs on cancer can be accepted would seem to make acceptance impossible."

That remark was too much for Pick who confidently observed that "the NIH obviously doesn't agree that the process is an infinite regress as you seem to suggest. It spends billions of dollars every year to find biological mechanisms."

"Why do they seek knowledge of mechanisms?" I asked.

"Because it is good science," he replied.

"But the pursuit of mechanisms is not the hallmark of good science. What, for example, is the mechanism by which gravity works? Or, if you prefer, explain the mechanism of Ohm's law."

"They are forces," he replied. "It is meaningless to ask what the mechanism is of a force. Forces are governed by laws, so knowledge of mechanisms is unnecessary. It is precisely because there are no such laws in biology that it is necessary to identify mechanisms before we can believe what we see, and accept it as real."

"If there are no laws in biology," I said, "why do living things show patterns of behavior and response? Are these patterns chance events?"

"Of course not," he replied. "Living things obey laws, it's simply the case they don't obey mathematical laws, as in physics."

"I find that shocking! Do you say that living things don't obey the laws of physics?"

"Insofar as living things are objects they obey the laws," he said. "A living thing will be attracted by gravity in exactly the same way as a non-living thing. If I apply a voltage to something that is alive, the current that flows will be exactly the same as when the voltage is applied to something that is not alive but has the same electrical resistance. The point is that living things don't obey any other general law that we know about."

"Do you think they obey general laws that we don't know about?" I asked.

"I didn't say that. You're putting words in my mouth," he replied.

I told Pick that was not my intention, and he settled down. Then I looked away from Christopher and asked no one in particular, "Why is it necessary to identify a mechanism before the ability of EMFs to alter the immune system is accepted as plausible? Shouldn't any attempt to identify mechanisms, if that is possible, be taken after the ability to affect the immune system has been established, for the reason that only real effects can have mechanisms?"

The room became silent, as if I had been rude to our host, but Christopher's thoughtful reply put me at ease. "Then you suggest that the principle of biological plausibility puts the cart before the horse?"

"It seems that way to me," I replied.

At this point, several committee members who had previously been silent spoke up and expressed one degree or another of criticism of this or that aspect of the point I had raised, and tempers flared. When this happened the observers sitting with Erica perked up, as if a show was about to begin. The situation was skillfully rescued by Christopher, however, who calmed the committee by thanking its members for their deep interest in the issues. I took advantage of the calm to ask Mandel, "If the laboratory evidence explains how EMFs could cause cancer, would it then follow that powerlines are unsafe?"

"Not necessarily. Just as it does not follow that low levels of mercury in drinking water are unsafe even though mercury is toxic."

"Because toxicity always depends on dose?" I asked.

She agreed.

"For example," I said, "someone might need to drink as much water as an elephant before the mercury levels could cause harm."

She agreed.

"Suppose we conclude that the laboratory evidence does not explain how EMFs cause leukemia or brain cancer. Would it then follow that powerlines are safe?"

"It would certainly seem so. From a scientific point of view, if a thing is not unsafe, it is safe."

"Then the principle is the same as in algebra: a negative multiplied by a negative is a positive."

She agreed.

At this point several committee members called for a vote because, they said, the discussion had become tiresome and it was time to move on to a new topic. Christopher acquiesced and put to the vote the question, "Do the laboratory studies of the effects of EMFs on the immune system of animals convincingly demonstrate that EMFs can affect the immune system?" The final vote was no, 16, yes, 8, and 3 abstentions.

After lunch, when a sense of comity had returned, Christopher asked Juko to summarize the studies of the effects of EMFs on the brain, to assess whether there were such effects and, if so, whether they could be responsible for physiological changes leading to brain cancer.

In his speech Juko said that the studies involving the effects of EMFs on the brain had reported every kind of result imaginable. To the amusement of many on the committee, he said, "Whatever your favorite result is, I can point to someone who found it and someone else who didn't," and he proceeded to give examples.

"There are reports that EMFs affected the electrical activity of the brain," he said, "but other reports concluded that no such effects existed. Among those who report effects, some said that the energy in the brain waves increased but others said that it decreased. The story is the same for studies involving intracellular levels of calcium ions. Some said it went up, others that it went down, and still others that it never changed."

He said that because the studies were "incoherent" he had to conclude that EMFs alone can't affect the brain, at least not to the extent that they could cause cancer. He said his own research, though, had shown that EMFs "may act as a co-carcinogen."

Barnes, recognizing that Juko was a foreigner and perhaps not knowledgeable regarding the nuances of English, asked him, "In English, 'may' could be 'I just don't know,' or 'Definitely yes, but only in particular circumstances.' Which is your meaning?" Juko replied that he had both meanings in mind.

Justesen asked Juko whether he had meant to say that the EMF research was "incoherent" or "inconsistent." When pressed to explain what he understood the difference to be, Justesen said that "incoherent" meant that two things did not rise and fall together in a fixed relationship, like two waves spreading on a pond, and that "inconsistent" meant that two things could not both be true, because the truth of one thing opposed the truth of the other thing. For example, it would be inconsistent to claim that a clown was happy and sad because being one precluded being the other. Juko confessed to not recognizing any distinction between the words, and after a long discussion a consensus developed that while a distinction might exist between the terms in some contexts, there was none in the case of the EMF studies.

I then asked Juko, "Is the evidence 'incoherent' or 'inconsistent' because, for example, one study claims that EMFs increased the energy of brain waves in people whereas a similar study claims EMFs decreased it and a third study claims there was no effect?"

"Yes, that's a fair summary of the pattern of results," he replied.

"Then the evidence is incoherent because, according to what you believe, if a phenomenon were real it would be verifiable."

"Not only according to what I believe," he said, "but according to science."

"And verification was absent because in one case the energy increased, in the second case it decreased, and in the third case it was unchanged. Is that right?" I said.

"Yes," he said.

"What exactly must be verified?" I asked.

"If the reality were that EMFs increased the energy, then that phenomenon would have to be verified before it could be accepted. If EMFs actually decreased the energy, then that would have to be verified."

"By 'reality' you mean what is objectively true?" I asked.

"Of course," he replied.

"Couldn't it be the case that the phenomenon to be verified was exactly what was observed – that sometimes the energy increased, sometimes it decreased, and sometimes it was unaffected?"

"That would be a strange state of affairs," he said. "There must be a law that determines whether the energy is to increase or decrease because, according to science, everything that happens does so by necessity. It is law that requires one alternative or the other. If we observe both kinds of changes, and also neither, the plain meaning is that there is no law. Wouldn't you agree that is a proper scientific explanation?"

"Undoubtedly, at least for gravity and alchemy," I said, "because that explanation makes it clear why one is useful and the other isn't. Still, I wonder whether your explanation is no more than an opinion, or perhaps only a special rule. Would it not be permissible for a scientist to believe that the rule applied in only some cases, and not in others, and that EMFs were an example of the kinds of cases in which the rule did not apply?"

"I don't know what you mean," he said.

"Suppose two theories claimed to explain something but one of them was more in tune with the observations. Would you agree that it would be more preferable?"

"Of course."

"Doesn't my theory – which is that EMFs can affect the energy in human brain waves, either increasing it or decreasing it, but that the occurrence and the direction of the effect in specific human beings is unknowable – fit the data better than your theory that the body of research has no meaning because it resembles a sequence of positive and negative numbers that add up to zero?"

"Why on earth would you say that the occurrence of an effect in a particular case, or its direction, is unknowable?" he asked.

"Why would you say it must always be otherwise?" I said.

"All of our scientific laws require consistency. When the conditions are the same, the results must be the same."

"I would not deny that our laws require consistency. But isn't that because they were made that way?"

"Wait. That's another error on your part. Our general laws were not made by man – they were made by nature and discovered by man."

"But as Dr. Pick pointed out this morning, living things don't obey general laws. There is no force that descends on human beings and requires a particular response in each individual. This being true, I think there is no reason to suppose that the consistency built into the laws that describe nonliving things should be imputed to the behavior of living things, for which there are only particular laws."

As I suspected would happen, Pick chimed in.

"Dr. Marino, is it necessary for you to use such vague language?" Before I could reply Christopher said, "Andy, could you make your point using clearer language?"

I asked Christopher whether I should do so by continuing the argument, or by means of a story, and he replied that the choice was mine. Several dozing committee members suddenly awoke and said it would be more interesting if I were to tell a story, and so that's what I did.

"In the beginning, God made an infinite number of equations that governed everything that happened in the world. But He was dissatisfied because the world was only a machine, and therefore had no capacity to love and honor Him, so He created human beings and gave each one a unique equation that was not controlled by an outside force. To the humblest and most self-effacing human beings God gave superior intellects so that they could eventually come to understand His handiwork. Those so favored, who became known as physicists, eventually learned to understand the equations that controlled the part of the world that was a machine, and progressively distilled them into smaller and fewer classes of greater generality. After a long time the physicists finally intuited the perfect equation, the one that completely governed the machine.

"It had been God's plan that the humble physicists would also learn that there was no single universal equation that governed life, because He had intended that human beings would not be machines, and hence not perfectly predictable. Now there was an angel named Grun, who was one of those who had disobeyed God and was banished from heaven. In retribution, Grun went among human beings attempting to implant in their minds the belief that they were only machines, like the rest of the world. He hoped to thwart God's plan and drive people to despair, because machines have no purpose.

"Grun was rebuffed by most human beings who immediately saw that his message was preposterous on its face. The physicists, however, who were always receptive to ideas that benefited all human beings, became victims of their own open-mindedness; they embraced Grun's message and became its proselytizers among other scientists, who profoundly respected what physicists had accomplished and therefore believed in the truth of everything they said. As a consequence, scientists came to believe they could know everything, as God knows it. This explains why Juko believes as he does."

When I finished, animated discussions broke out among the committee members. On one side, there were those who were anxious to show their support for Dr. Pick, who had announced that he hadn't come to listen to myths aimed at rehabilitating worthless studies, and that it was now time to vote and move on. There was also a group that seemed to think my theory made sense. Most of the action was in the center ring, where a shouting match occurred regarding the basis for choosing between Juko's theory and mine.

Christopher saw that no one was winning any converts so he said, "Let us put aside for now the discussion of what the proper underlying view is. We can return to it later." Then he began to question me directly, something he had not done to anyone else on the committee. "If EMFs affect brain waves and lead to brain cancer in some cases, as you say, then the evidence that we have on the large scale would seemingly be inexplicable. Since high-voltage powerlines were first built in the United States, consumption of electrical energy has increased each year, but death from brain cancer has not risen at the same rate. This shows that powerline EMFs pose no significant hazard to the average person."

I thought about Christopher's argument for a few moments and then said: "Even if one believes that the cancer rate has lagged behind the rate of powerline growth, the difference might have meanings other than the one you suggested."

"What are you getting at?" he said.

"Suppose we asked what it means when an animal wags its tail," I said. For a cat the act is a manifestation of discontent. For a dog, tail-wagging indicates health because a sick dog never wags its tail. Some say that a cow wagging its tail manifests contentment, and the same may be true for a deer, which is known to indicate fear when it freezes its tail in an upright position. So the fact of tail-wagging can have many different meanings."

"You'll have to do better," Christopher said, "wagging tails are different than brain cancer."

"Dr. Philip Handler," I said, "when he was president of the National Academy of Sciences, appointed three different blue-ribbon committees to study the question of health risks due to powerline EMFs. These experts examined all sides of the issue, yet none concluded that EMFs were safe because of a mismatch between the rates of powerline construction and cancer death. The possibility should therefore be considered that your argument has a hidden defect that is not presently apparent to you."

Christopher had no ready reply, so I continued. "It is unprofitable for us to spend our time debating the meaning of expert committees whom we cannot question directly. Rather we should try to get at the truth by the direct interplay of our own thoughts."

He nodded in agreement.

"I would therefore like to suggest an alternative to your interpretation of the facts."

"Please do," he said.

"EMFs affect the brain waves in each exposed subject," I said, "but the consequences of that detection process are different in different people, like two adjacent raindrops released from a cloud and striking the earth at widely separated points. The claim is not that EMFs can cause only brain cancer, but rather that they can cause all cancers, indeed all diseases, in the sense that their presence always raises the probability for disease in an individual, compared with their absence. This capability is not unique to EMFs – it is probably manifested by all stressors. One reason, therefore, that the incidence of brain cancer has not increased in proportion to the levels of EMF in the environment could be that EMFs are killing people in other ways."

The fractious nature of the issues under discussion again led the committee to boil over. Christopher responded by adjourning the meeting for the day, but not until he promised me that we would return to the discussion of the two theories.

As Erica and I were passing through the lobby on our way to dinner, we came across a young girl and her father who were arguing with a woman who wore bright red cheek-rouge. Her name-tag identified her as the superintendent of the local school district. We listened, and learned that the girl had been a contestant in a high-school science fair being held at the hotel, and that her project had been judged the best but the superintendent had refused to award her the trophy. When the girl's father pressed for an explanation the superintendent told him, "The project basically encourages sex, and our philosophy is abstinence." When Erica and I went over to where the projects were on display, we saw that the girl had tested six different brands of condoms and found that one brand was stronger than the others.

After dinner, on our way back to our hotel we were delayed because traffic was stopped to allow rescue vehicles and police cars to come and go. Later that evening I learned that an aerial performer at the circus had been twirling from a long piece of chiffon, which had snapped. There was no safety net and she was not wearing a harness, so she fell thirty feet onto the concrete floor and died.

The next morning, Christopher and Ruhig joined Erica and me at breakfast. Soon after he sat down Christopher said, "If, as you speculated yesterday, the consequences of EMFs differ from person to person, how is it possible to decide whether EMFs cause cancer? Wouldn't we be left with only mysticism or some kind of teleology?"

"Would you agree," I said, "that what makes something scientific is that it can be described by a rule that allows predictions?"

"Yes."

"And that it would still be scientific even if the predictions were not perfect, like predictions about whether a medicine or a treatment for a disease will be successful?"

"Yes," he replied. Then he asked, "What method is there to make reliable predictions?"

"The same method that was used to discover the general laws of physics," I replied.

"Induction?"

"Yes. The method of induction. It's not possible to answer our question about EMFs using the method of proof, because there is no underlying law that governs everything by necessity. Instead, every human being is a self-enclosed whole, governed by a unique law. But it would be reasonable to search for a pattern in how human beings react to living in a sea of EMFs. That pattern would be the answer to the question whether people who live or work in high EMFs have a higher rate of brain cancer or leukemia compared with people who live in low EMFs or no EMFs."

Christopher nodded, and then left the table to prepare for the next session of the committee, which was to be devoted to the human studies. When he was well out of earshot, Erica said, "He seems troubled."

"He has good reason," I said, "because he has a heavy responsibility."

In his opening remarks to the committee on the last day of the meeting, Christopher said only, "Today we will consider whether the EMFs in homes or industry can cause health problems. Do people who live or work in EMFs have a higher than expected cancer rate?" Before he called on anybody to summarize the evidence, I said, "Who has the burden of proof, those who would say 'yes,' or those who would say 'no,' and by what standard of certainty?"

Pick couldn't let this pass. He said, "There is only one standard in science, truth, so it makes no difference who has the burden of proof. In science, at least, the search for truth is not an adversarial process. We are not lawyers, at least not most of us."

When the buzzing at the table and in the audience seated behind us had ceased, I said, "But we are human beings and therefore prone to error. If you think about the consequences of error, you can see why the choice of a standard is crucial. We could err by accepting that EMFs don't cause cancer when in reality they do, or by accepting that they do when they don't. The two errors do not have equally bad consequences. Either the people who live near powerlines will be injured by loss of health, or the power companies will be forced to needlessly bear the costs of undergrounding the powerlines. If you believe the people have at stake the interest of greater importance, you will not conclude there is no risk when there is a reasonable doubt about the truth of that claim."

Pick's reply illuminated his value system. "It would be far better for the overall stability of society," he said, "to advise the public that powerlines were safe, at least until we were certain that this is not the case, because a definite answer is preferable to vague allusions of potential problems."

Barnes then gave his point of view, saying, "I think it is better to be safe than sorry. So it would be prudent to make worst-case assumptions about the human studies and build in a margin of safety to avoid risk." Pick then accused Barnes of denying that there was an identifiable standard for assessing what is or is not a fact in science.

I said to Pick, "Regarding the health risks of EMFs at least, it looks as if the facts are whatever those in authority say they are by the expedient of choosing and assigning the burden of proof, like setting the bar so high only a few jumpers can clear it. So the facts are the choices they make."

"Who are 'they?'" he asked.

"We should have a discussion about that," I replied.

Christopher diverted attention from that suggestion by turning to Monica Lugner and Eva Stum, two sisters who represented the largest stakeholder in the EMF dispute, the Electric Power Research Institute. They had formerly worked as experts regarding the health implications of depleted uranium and the hole in the ozone layer, and had recently moved into the EMF area and begun to opine that high-voltage powerlines were completely and absolutely safe, with no ifs, ands, or buts. Many businessmen and housewives had been swayed by the sisters' view. He said to them, "Obviously you cannot discuss comprehensively the human studies in the short time available, but tell us this. Besides satisfying those who are already convinced that the human studies show powerlines are safe, do you think you can appeal to someone who is not yet convinced?"

"Definitely," Lugner answered. "But rather than simply giving a lecture for which we might be criticized for picking and choosing the evidence that we considered, I would like to ask for someone on the committee to volunteer to respond to our questions concerning the basic principles of human studies. If you have no objection." Christopher nodded his approval.

At this point Afelis spoke up and said, "I would be happy to respond to questions, so that we can quickly get to the bottom of the matter."

"Tell me, do powerlines cause cancer?" Lugner asked.

"If we believed the Wertheimer study, and several others, we would say yes. But if we accepted your studies we would say no," Afelis replied.

"Did Wertheimer claim that children who lived beside powerlines got cancer more often than children who lived elsewhere?"

"Yes."

"Because of EMFs?"

"Yes."

"But the children who didn't live near powerlines also got EMFs from other sources. Isn't that true?" Lugner asked.

"Surely," Afelis replied.

"Then if EMFs were present in both groups, isn't it illogical to claim that they caused more cancer in one group than the other?"

"Perhaps there were more EMFs along the powerline?" he said.

"Did Wertheimer measure the EMFs along the powerline?" she asked.

"No," he replied.

"Then she would have no way of knowing whether there were more of them at that location, correct?"

"Well, I suppose it's reasonable to assume that EMFs are higher near powerlines," Afelis replied nervously.

"But it's not established as a fact in her study, is it? It's just speculation."

"Yes."

"Suppose we studied cancer among men who worked as stockbrokers and insurance salesmen," she said. "We would not expect to find exactly the same number of cancers in the two groups, but rather that the number would be greater in one of them, correct?"

"It would be quite extraordinary if the numbers were exactly the same," he said.

"Suppose it was higher among the stockbrokers. Could we conclude that selling stocks caused cancer?"

"That would be foolish."

"Suppose we conducted the same study among women who were either nurses or teachers, and we found that the number of cancers was higher among the teachers. Could we conclude that teaching causes cancer?"

"No."

"If children who drank white milk got more cancers than children who drank chocolate milk, would it be fair to say that white milk causes cancer in children?"

"Of course not."

"Isn't comparing children who live beside powerlines with those who don't the same as these cases, so that there is no reason to claim that powerlines cause cancer."

Before Afelis could answer, Stum piped up and said, "Actually, powerlines cure cancer."

"I never heard anything like that," Afelis said, with a surprised look on his face.

The pair's admirers on the committee cheered, while others were speechless with amazement. I suppose to astound us even more, Lugner kept on relentlessly questioning Afelis with the same method.

"You know, don't you," she said, "that EMFs are used by physicians to treat disease?"

"Yes."

"And that the government has said they are effective?"

"Yes, for treating bone diseases."

"Do you agree that a thing cannot at the same time be itself and its opposite?"

"I don't know what you mean."

"If something is hot, it cannot at the same time be cold. If a thing is good, it cannot at the same time be bad."

"Yes," he said, "but…"

"Then if EMFs cure disease," Stum interrupted, "they cannot cause disease, and since cancer is a disease, EMFs cannot cause cancer."

"I can't argue with you," Afelis said.

The sisters were winding up to throw Afelis still another curve ball, but I had had enough, so I jumped in and said to him, "It would have been appropriate to make a distinction between the conditions under which EMFs are applied. Their successful use under the careful control of a physician does not imply that good results would occur when EMFs are applied to everyone by the power company. Things are not good or bad in themselves. And earlier, when the probative weakness of questionnaire studies was pointed out, you didn't seem to appreciate that the trick being played was to deny the usefulness of questionnaire studies simply because they do not yield clear and unambiguous results, like making the perfect the enemy of the good."

Then I said to the sisters, "Enough of these games. Any more would be superfluous. Please begin to try to convince us that EMFs are safe, as you say."

I watched to see if Christopher would take up my theme and encourage the sisters to present more substantial arguments. He said nothing, however, and they launched into more, similar arguments, which neither I nor anyone else on the committee was willing to address. Near the end of their performance, Stum addressed the committee as a whole and said, "The studies that purport to link powerlines with cancer of the brain or blood all suffer from statistical uncertainties, so it is a fallacy to suggest that the results mean anyone's chances of getting cancer are increased by EMFs."

Christopher shifted the focus of the conversation by asking, "Dr. Lugner, some experts have written about what has been called 'the precautionary principle.' Would you tell us how you think it ought to be applied in evaluating the human studies about EMFs?"

"There is no evidence in the human studies that can rationalize the need for any precautions insofar as EMFs are concerned," she replied.

That was too much for some of the members of the committee and shouting matches developed, leading to a loss of decorum that prevented Christopher from taking a vote regarding the significance of the human studies on EMFs. The next morning he folded his tent and took his show on the road to two other cities.

On the way home I asked Erica whether Christopher had lived up to her expectations. "Not in the way I expected," she said. "I think preventing disease is a subject that can be learned and taught, but not by means of his three great principles. His approach reminds me of the way someone might train a dog."

"Then how?" I asked.

"I think your way is better."

Christopher finally returned to the NIH and began to ponder his decision. What happened during those crucial months, according to a source within the NIH who was present at almost all the important meetings, shaped both Christopher's fate and the fate of the public. The story began at the Dallas meeting, which had been where Christopher came to believe that there was a realistic possibility that EMFs from powerlines caused cancer, and even that EMFs might have other effects on health that were largely unknown because they were largely unstudied. He had expected the representatives of the Electric Power Research Institute to act like barkers at a carnival, and had included them in the meetings only because he was required to do so by Olden. But Christopher had believed he still could orchestrate a give-and-take among the other participants that would lead inexorably to the identification of the presence or absence of good science capable of resolving the EMF issue. That idea died in Dallas, and his view of science fell into disequilibrium. Ultimately, he concluded that the best answer to the question posed by Congress was that the issue of EMF safety was very complicated and could not be resolved on purely scientific grounds in the absence of rules regarding the meaning of uncertainty.

Christopher sought Ruhig's support and found him sympathetic. He too felt that Olden's order to "resolve the issue" couldn't be implemented in light of what had taken place at the meetings. Nevertheless, Ruhig reminded Christopher that they both worked for Olden, who was attempting to obey a direct congressional mandate, and that he had told them that he would accept one of only two possible outcomes: either there is strong, certain evidence that EMFs cause cancer, or EMFs are safe. Ruhig told Christopher there was nothing they could do about Olden's order, as much as they would like to, and he advised Christopher to obey it.

Christopher, however, resolved to follow his conscience. He told Olden that the question put by Congress could not be answered on the basis of scientific principles alone, and that the moral force of science, and the respect and confidence that people have in it should not be weakened by asserting scientific certitude where none existed. Olden again reminded Christopher that the only permissible outcomes of the inquiry were that there is strong, certain evidence that EMFs cause cancer, or that EMFs are safe. And that if the NIH were to proclaim that EMFs caused cancer, Christopher needed to be as certain of that fact as a human being could be. When Christopher held his ground Olden removed him from the adjudicatory process and ordered Ruhig to draft a report that concluded EMFs were safe.

Ruhig asked his staff to prepare a statistical estimate of the number of deaths due to brain cancer and leukemia that could be attributed to powerline EMFs. He had thought that the number would be minuscule, and thus that the estimate could serve to support Olden's decision. According to the analysis, between six and six hundred cancer deaths would occur annually for each one million people in the population. When Olden learned of this estimate, he began to have doubts about his course of action, and particularly about whether the oath he had sworn as a physician to do no harm would permit him to issue a judgment that could result in such severe harm to so many people. He knew that the EMF research program would be ended if he concluded that EMFs were safe, and thus that his conclusion would not be challenged. His doubts deepened; he became depressed, and he began to complain about not feeling well.

In the meantime, Ruhig produced a draft final report that concluded EMFs were safe, as he had been instructed to do; the statistical estimate was not included because it pointed in the wrong direction.

As Olden agonized about what he would do, the whole affair rapidly came to a conclusion by means of forces he could not control. The end began when someone leaked the draft final report to congressmen who had been instrumental in the creation of the EMF program. They immediately went on CNN and took credit for solving the problem of EMFs, and the story appeared in all the major newspapers. In the face of this publicity, Olden decided that the only course open to him was to release the report, which he did. It was lavishly praised by the Electric Power Research Institute, which said that the process NIH had followed to evaluate powerline EMFs should also be used to evaluate the supposed problems with nuclear waste.

28
Young Richman
◆

I was invited to give a talk at an international conference on environmental science in Washington, DC. The topics presented included global warming, the safety of food additives, the side-effects of drugs, the health risks of environmental electromagnetic fields, breast implants, Agent Orange, tampons and toxic shock, the hazards of lead paint, asbestos, the effects of pesticides and herbicides, the link between vaccines and autism, mercury poisoning, the dangers of x-rays, indoor spraying of DDT, Gulf War Syndrome, and benzene in drinking water. Experts of various stripes attended the conference, many of whom were towed around by lawyers. I was interested in what all these issues had in common.

Immediately after my presentation, while I was still pumped up, some men approached me. The oldest appeared to be in his mid-70s; he identified himself as a lawyer named Peter Richman, and introduced me to the others in the group, which included his son who was a lawyer and an engineer. Young Richman reminded me of Bob Simpson, and as I was reflecting on their similarities in appearance and manner, Peter Richman asked me to meet with him to discuss some cell-phone cases. As he spoke I suddenly placed him – he was the famous trial lawyer who had become a billionaire by winning some big cases against tobacco and asbestos companies.

Before I could respond to the invitation he departed, saying only that he was late for an appointment to discuss the cases with someone else. When I recovered from my surprise that he had not waited for my reply I turned to young Richman and told him that I would not attend the meeting. He was not in a mood to accept no for an answer, and urged me to attend. I declined again, and explained that I did not approve of his father's legal approach to toxic-tort cases because I thought it was based on a mistaken understanding of environmental science. At that moment one of the others in the group, a young man named Dr. Faul, said to me that he thought Peter Richman actually had a good idea of what environmental science was because "I myself have worked with him on many different cases and have always taken the trouble to point out what it was."

"I'd like to hear what you think it is," I said.

"Easy enough," he replied. "Take, for example, the presentation we heard today on global warming. Various kinds of scientific facts were mentioned; they included temperature measurements, data concerning wind velocity, mathematical models for predicting the weather, and so forth. All this information adds up to the environmental science of global warming. Similarly with the environmental science of breast implants. We know how many women received the implants, how many developed problems with their immune system, and how many who had such problems never had breast implants. Everything that affects people has a different environmental science because the facts differ from issue to issue."

"All you've done is give me a gang of instances," I said. "Suppose I had asked you what a cow was. Would you answer by pointing to each cow you saw, or would you tell me what they all had in common?"

"One cow is pretty much like another," he said.

"That's what I'm after – what do all of the topics at this conference have in common?"

"I don't know, exactly," he said.

"If we can't identify it, there is no such thing as environmental science," I said.

"Well, I certainly didn't mean to deny that," he replied.

At this point Dr. Stein, a friend of Faul's, said to him, "Why don't you just tell him what environmental science is and end this charade?"

"I think he tried," I said to Stein.

"And failed," he replied.

"If you know that, you must know what it is. Why don't you tell me?" I asked.

"Environmental science is health information of the highest quality, objectivity, utility, and integrity," he barked.

"What do those words mean?" I asked.

"It's a matter of judgment by experts," he replied.

"How do they know?"

"Because of their training and experience."

"When experts disagree, whose information is highest?"

"The consensus."

"A consensus is never highest because it is a harmony of different opinions," I replied. "So, on your account, there is no such thing as environmental science."

One of the others in the group jumped into the conversation. After telling me that his name was Dr. Milkin and he worked for an electronics company, he said, "If you just look around, I think it's plain what environmental science is. People hear all kinds of stories about what's going to happen unless they change their ways. If they use too much energy the glaciers will melt. If they eat tuna, they'll get mercury poisoning. If a woman gets a breast implant, it will destroy her immune system. On and on. The businessman knows that these are not problems, but opportunities. When a need arises in a free market, someone comes along to fill that need. In this way we get nuclear power, which doesn't lead to global warming, methods for growing tuna in underwater farms where there is no mercury, and breast implants that don't activate the immune system. Cigarettes present another opportunity. Somebody needs to figure out which of the thousands of chemicals in cigarette smoke are actually responsible for cancer and remove them. It's all simply a matter of technology. Environmental science is technology that promotes health."

"Suppose the thing impacted by the technology had no relation to health," I suggested.

"What do you mean?" he said.

"Just this. If man's activities didn't cause global warming, or breast implants didn't compromise the immune system, or mercury didn't cause brain damage, or smoking didn't cause cancer, then the technology that changed or removed these factors wouldn't be environmental science because it wouldn't promote health," I said.

"I suppose so," he replied.

"So environmental science must be prior to technology, rather than technology itself."

At this point Dr. Morris, whose name tag identified him as being from the Harvard Center for Risk Analysis, raised his hand as if he were asking for permission to speak. Milkin nodded, and Morris said, "Suppose we say that environmental science is technology whether or not there is a true threat."

"No, no, no," Milkin said. "Environmental science is based on truth, not assumptions."

"If so, there must be knowers who are recognized as such," I said.

"Of course, there are scientific experts," Milkin replied as he waved his hand in the general direction of the meeting participants who were milling about.

I said, "There are scientists here from the National Academy of Sciences, the World Health Organization, and the Mobile Phone Manufacturers Forum. Would you say that they are experts?"

"They could be, I just don't know," he replied.

"In other words, they are not necessarily experts, even though they are scientists."

"Yes."

"So it seems that you don't know what environmental science is because you don't know who is truly an expert. I think you were closer to some truth when you said that technology was the work of businessmen."

"Everybody knows what environmental science is," he replied, and then turned away.

Young Richman finally broke his silence. "So environmental science cannot be explained by means of examples, and it's not quality data or technology. Instead of telling us what it is not, can you tell us what it is?"

"I think so," I replied.

"Would you like to really persuade us, or don't you care what we think?" he asked.

"I'd like to persuade you," I replied.

"Suppose I tell you the kind of thing people say environmental science is, and where it came from, and then explain why they don't think it's any good. If you intend to make any headway with us you will need to overcome these points."

"That's fine with me," I said, "please proceed."

"When powerful organizations like governments or companies want something they pursue it. If they get what they want they are happy, and they don't think about the suffering they caused. The people who are harmed resent this behavior, but they recognize that governments are necessary and that companies are the source of their livelihoods. Moreover the weak admire the strong and regard them as naturally superior. So the weak don't insist that the strong entirely avoid causing harm, only that the harm not be excessive. Since the strong are relatively few in a democracy but the weak are plentiful, they can force the strong to establish such rules. This is the genesis of environmental science."

I could see young Richman had not yet finished so I waited, and he continued. "Environmental science restricts the powerful. They follow the rules out of practical necessity while at the same time working to weaken their effects and cloud public perception of the nature and extent of the harm that the rules address. That's only natural. Anyone would do exactly the same thing. There's a story about a professor who had often testified in court against tobacco companies, where he swore that smoking was addictive, and caused cancer and heart disease. He criticized the companies for suppressing evidence and for developing new strains of super-addicting tobacco. One day he found some stock certificates in a trunk in the attic of an old house he had recently inherited from his grandfather. The certificates indicated that the bearer was a part owner in a new company whose business was manufacturing cigarettes. Over time the company had grown into one of the tobacco companies whose activities he had attacked. Because of stock splits, mergers, and acquisitions, and a healthy financial environment for tobacco, the value of the stock had grown into a great fortune. Thereafter, he had only good things to say about the research published by the companies, which he called true lovers of science."

"I heard about him," I said. "He taught at Harvard, didn't he?"

"I don't remember," young Richman replied, "but the case is a good proof that people are not willing to accept evidence when it is against their interests. Indeed, if the professor had continued his public pronouncements that smoking causes cancer, everyone would have said that was exceedingly strange behavior for an owner of a tobacco company, whether or not what he said were true."

"I agree," I said, "that when people make decisions for themselves about what is or is not true, they do so with regard to their own interests. The point was in the front of my mind when I came to this conference."

Young Richman continued, "The more a company opposes environmental science the more scientific the company will seem to be in the eyes of ordinary people, so the best course for any company is to totally oppose environmental science."

This assertion seemed contradictory on its face, so I asked for an explanation and he replied, "Consider two companies that follow starkly contrasting approaches to the issue of whether their products have harmful side-effects. One company diligently seeks to avoid inadvertently causing harm. When hazardous situations are uncovered, this company immediately warns the people so that they can protect themselves. The other company never discloses even a hint of potential risk, but rather manipulates research data and calls the data of others 'junk science.' Let's consider the fate of these companies, each of which is perfect at what it does, either discovering or obfuscating practical scientific knowledge. The first of the two companies will be seen to be error-prone, while the other will be seen to be perfectly able to guarantee that its customers and the public are completely safe. So the company that knows the least will seem to be the one that knows the most.

"I'm saying all this," young Richman said, "to explain why companies always deny that they are bringing about unintended health consequences, not because I think what they do is right."

"Not only companies," I said. "Some government agencies do the same thing."

"Are you conceding the argument?" he asked.

"Quite the contrary," I responded.

"Well, persuade me about what environmental science is and why it's a good thing," he replied.

"All right. First imagine a vertical line," I said. "The region on the left side represents the information that we have about things in the world, and the region on the right represents the corresponding mental state concerning how confident we are that the information is true. Can you visualize what I'm talking about?"

"I can," he said.

"Now suppose we divide the line into three segments," I said. "The region on the left side of the bottom segment corresponds to having no information. The same region on the right side would then correspond to a guess, because if we had no information about something we could always guess at what it might be. Do you agree?"

"I do."

"The region on the left side of the top segment corresponds to having perfect information. For example, a mathematical system. In this case, on the right side we could say that we are certain we understand what the truth is, because every mathematical statement is either certainly true or certainly not true," I said.

"Not exactly," Morris interjected. "You are forgetting Godel's Incompleteness Theorem."

"I would like to neglect that for now," I said. No one objected so I continued. "The middle segment corresponds to imperfect information on the left side and some level of uncertainty on the right side."

"Yes. I see that," young Richman said.

"If we were to represent the regions using colors, we could imagine that the region of certainty is red, that of uncertainty is green, and that of a guess is yellow," I said.

"If you like," he replied.

"The green region, unlike the other two, would need to be graded continuously from light to dark," I said, "to represent progressively more certainty and less uncertainty."

"Yes," he said.

"Within the region of graded green, which we have agreed corresponds to imperfect knowledge, what types of such knowledge do you recognize?" I asked.

"I'm not sure what you mean," he replied.

"Would you agree that common sense is a kind of knowledge?"

"Yes."

"And that on the basis of common sense, sometimes we are confident that we understand what is true and other times less confident, but that we never say we are absolutely certain based on common sense."

"Yes, because, as you have said, our information is always less than perfect."

"Do you recognize other kinds of imperfect knowledge?" I asked.

"I would say that skill is another kind," he said. "For example, the skill to steer a supertanker or try a lawsuit or hit a baseball."

"And in these cases some people are better than others, which is another way of saying that they know more about how to do the thing."

"Yes."

"But in matters of skill, no one is perfect."

"Of course not."

"Science is a third kind of imperfect knowledge, isn't it?" I asked.

Everyone agreed except Stein, who only grunted.

"And so the different kinds of science have different shades of green."

"Do you mean because they differ regarding how perfect their knowledge is or can be?" young Richman asked.

"That's exactly what I mean," I replied.

Then Faul asked, "What color do you think should be assigned to the various parts of science?"

"I would put some parts in the yellow region," I replied. "For example, astrology, or phrenology."

"I suppose that creation science would also be yellow," he said.

"Not necessarily," I replied. "It is not science, but it might be a part of common sense, or some other kind of imperfect knowledge. Let's discuss that point some other day." He nodded.

I turned back to young Richman and said, "Mathematics belongs in the red region, at least if we continue to ignore Godel's Theorem."

He agreed.

"In the empirical realm, gravity belongs at the top in the green region," I said.

"Our knowledge about that subject is certainly impressive," he commented.

"Next," I said, "we could list technology. We know how to build reliable machines. Just imagine, you can use any telephone to contact any other telephone in the world, and when you punch in the same number you reach the same phone. Not 100% of the time, because the system is not perfect, but with a degree of success that is truly impressive," I said.

"Very true," he replied, after which he asked, "What sciences would you say are a paler shade of green than technology?"

"Medicine is one," I replied. "We have wonderful drugs, surgeries, and forms of therapy. Sometimes they work and sometimes they don't, but no one can predict with certainty what will occur in any particular case, or tell why."

"You seem to have a place for everything," he said, "but what's all this to do with defining environmental science and explaining why it's a good thing?"

"Well, the first important point is that environmental science is a form of imperfect scientific knowledge."

"What shade of green?" he asked.

"The same as medicine," I replied.

"How so?"

"Because its methodology permits the formulation of statements whose truth values are like those of medicine, and because both sciences aim to promote health."

Stein said, "Many of the claims of environmental science are yellow."

"As in medicine," I replied. "Nevertheless, on average, they are both green."

"Still," young Richman said, "I'm not persuaded that you have adequately explained what environmental science is."

"I haven't finished," I said. "I've told you why it's like medicine, but I haven't explained how the two differ. You can't judge me until I've finished."

"Sorry. How do they differ?"

"Environmental science is based on justice," I said.

That remark seemed to rub Milkin the wrong way, and in an irritated tone of voice he said, "You're going to have to explain what you mean. I always believed that science was based on laws and principles, like gravity and DNA."

"Different parts of science are based on different laws or principles," I replied. "What distinguishes environmental science is that it is partly based on the principle of justice."

"What relationship do you see between environmental science and justice?" young Richman asked.

"If we consider a particular case we will be able to understand," I said.

"Please proceed," he replied.

"Let's take cell phones," I said. "I know they interest you and your father."

"Fine," he said.

"Companies that manufacture them expect to make a profit."

"Indeed."

"And the people who buy cell phones do so because they are convenient and useful."

"Correct."

"A cell phone works by means of EMFs that pass from the phone to a base station and from the base station to the phone, like two people shaking hands."

"Yes, that's the technology of cell phones," he said.

"It is also true," I said, "that part of the EMF from a cell phone does not travel to the base station but goes in the opposite direction and enters the user's head and is absorbed by the user's brain."

"There's no denying that fact," he said.

"It also cannot be denied that the brain is an electrical organ that sends electrical messages to all parts of the body and also receives messages from those parts."

"Yes."

"And some of the information remains in the brain in the form of a mysterious code that allows us, for example, to recognize the faces of our children, recall the beauty of a sunrise, acquire expertise, and examine our lives so that we understand who we are."

"Yes."

"It is therefore prudent," I said, "to inquire into whether the EMFs produced by cell phones interfere with what brains do."

"Yes," he said. "People would demand it."

"Some people who use cell phones might not do so if it were true that the radiation altered the function of the brain, or injured it, for example by causing the brain to develop cancer, and if they *knew* that the radiation was capable of producing these consequences."

"Obviously."

"Every company that makes cell phones," I said, "also publishes promotional literature saying that the phones are completely safe."

"Yes," he said. "Every company says that."

"If cell phones caused disease," I said, "it would not be true that they were completely safe."

He agreed.

"Suppose there was evidence that cell phones could cause brain cancer. Would the company's claim be untrue?"

"From the company's perspective, only if the company agreed that's what the evidence showed."

"A company can't accept or reject evidence," I said. "Only *people* at the company can do that."

"Of course," he said.

"Which people?" I said. "Would it be the experts who work there?"

"Surely they would have something to say, but I think that the decider would be the company president because the buck always stops there."

"But regardless of the evidence, a president of a cell phone company is unlikely to say publicly that the cell phones manufactured by his company can cause cancer," I said. "Do you agree?"

"Yes," he replied with a smile.

"So, under my hypothesis the company's claim would be untrue, whether or not the president accepted the evidence."

"Yes."

"And if he privately accepted the evidence, the claim would be a lie."

"Yes."

"Now, let's suppose that some experts outside his company disagree with his expert risk assessors regarding whether cell phones are safe."

"Yes."

"Isn't it only fair," I said, "for the company president to tell the public that the experts are in disagreement regarding whether cell-phone EMFs can cause cancer?"

"In principle, yes," young Richman replied. "In practice, that's not going to happen."

"I'm trying to explain the connection between justice and the environmental science of cell phones," I said, "so it's the principle that's important."

"Then I think the even-handed approach would be for the company president to tell both sides of the story, regardless of which story he adopted on behalf of his company."

"Suppose that evidence of the proper sort..."

"What do you mean by that?" he interrupted.

"Peer-reviewed publications in archival scientific journals," I said.

"Excuse me for interrupting," he said, "but I just wanted to be completely clear regarding what kind of evidence you believed was necessary. Please continue."

"Suppose that the proper kind of evidence showed that radiation from cell phones could alter the function of the brain, but that there was no evidence directly showing a link between cell-phone EMFs and disease. Would it be misleading to say that cell phones were safe?"

"Not if, according to his experts, the alteration had no health consequences."

"Otherwise it would be unjust?"

"Yes," he said.

"Would his experts need to know that there were no health consequences," I said, "or would it be enough to know that there was no direct evidence of such consequences?"

"There can't be evidence of nothing," he said, "unless you call the absence of evidence of something evidence of nothing."

"I don't," I said.

"Then the strongest statement the company experts could make is that there is no direct evidence of health consequences."

"Would that mean that cell phones are safe?" I asked.

"Only if you believe that direct evidence is needed," he replied.

"I don't," I said.

"Then evidence of effects of cell-phone EMFs on brain function would falsify the claim that cell-phone EMFs are safe."

"In that case," I said, "continuing to make the claim would be unjust."

"Yes," he said.

"How should we determine whether cell-phone EMFs affect the function of the brain," I asked, "or cause brain cancer?"

"By experiments, I think," he said. "There is no other way."

"Who would do these experiments? Scientists employed by the companies?"

"I suppose that would be putting the saddle on the wrong horse," he said.

"We know from experience," I said, "that studies controlled by companies always conclude that their products are safe, whether or not that is the reality."

He agreed.

"We must also consider the kinds of studies that we will accept as evidence regarding whether EMFs can alter what takes place in the brain. For example, suppose a Harvard professor used physics and mathematics to show that EMFs had no effect on the brain," I said. "Should we consider his evidence?"

"Of course," he said, "but if that were possible, there would be no serious dispute about the propensity of EMFs to cause health problems. Since there is a dispute it's safe to say that we won't encounter a professor who offers that kind of evidence, at least not in court."

"Yes – some do offer it in the media where they can't be effectively challenged. So, next question. Should we allow questionnaire studies?" I asked.

"My friend Judge Weinstein would become apoplectic if he heard you ask that," he said. "Why would we even consider not allowing them?"

"If questionnaire studies were permitted," I said, "people would have cell phones before anyone knew that they were hazardous to health because cell phones would need to be in general use before anyone could do a questionnaire study."

"Yes," he said.

"In the beginning, when there were no questionnaire studies, the companies would claim that cell phones were safe."

"Yes," he said. "That's how everything new gets started, it's a kind of fundamental principle of technological progress."

"Then if it should turn out that brain cancer actually did occur more frequently among users of cell phones, the news would be greeted with skepticism by those who had been initially deceived into believing that cell phones were harmless."

"I suppose that reliance on questionnaire studies would give the companies an illegitimate advantage," he said.

"That's not the half of it," I said. "An *intention* to rely on questionnaire studies amounts to a conscious decision to use human beings as guinea pigs, which is wrong."

"It does seem unethical," he said. "On the other hand, the principle of trial and error was exactly how our forefathers determined what was safe to eat, or what was poison."

"Their experimentation was voluntary. If it were done by force or deceit, it would have been wrong."

"All right," he said.

"Even a scientist who had no moral scruples about involuntary human experimentation would object to using questionnaire studies to determine if cell phones were safe."

"What objection do you imagine he would make?"

"That epidemiology has no conceptual bedrock because there are no basic concepts or agreed-upon methods and that the investigators have no control over their 'experiment,' so every study leads to controversy, even those involving cigarettes."

"The point is that epidemiology, at least insofar as cell phones are concerned, is the palest possible shade of green, perhaps even yellow."

"Where does this leave us, then? What are good experiments?"

"The morally and scientifically correct thing," I said, "would be to rely on experiments with mice and rats and other animals. That is the standard practice."

"By whom?" he asked.

"The Environmental Protection Agency," I said, "relies on animal studies for setting guidelines for cancer risk."

"Yes."

"And also for determining risks due to pesticides and herbicides."

"Yes."

"And the Occupational Safety and Health Administration relies on animal studies when it sets occupational exposure limits, as for example to ethylene oxide."

"Yes."

"And the Food and Drug Administration relies on animal studies when it evaluates the safety of drugs and medical devices."

"Yes."

"And even though animal studies are far from perfect in predicting human risk, they are better than questionnaire studies which are not even real experiments. It's wrong to trick people into believing something is truly beneficial when it's just cheap and easy."

He agreed and said he thought laboratory studies on human subjects who had given their consent to participate were also good experiments, and I quickly agreed with him. "So now," I said, "we must face the problem of how to connect the results of laboratory experiments with the possibility of health consequences in people. This task requires an expert.

"Experts are necessary," I said, "but they present a special problem."

"What kind of problem?"

"The kind President Truman had with General MacArthur. After carefully studying the military situation MacArthur decided that the best course would be to launch a nuclear attack against enemy sanctuaries. The President, whose military knowledge was far less than that of MacArthur's, disagreed profoundly. Should the President have deferred to the general, who was the expert?"

"The President is the decider," he said, "but what does choosing a nuclear option have to do with environmental science?"

"Just this," I said. "Experts are like well-trained dogs. It is good to breed them and use them for our protection and convenience, but we must always remember who is the dog and who is the master."

"What type of opinion should we demand from our dogs?" he asked.

"What type would you say?" I answered.

"I can only give the traditional answer," he said, "the elucidation of safe levels."

"Shall I tell you what safe levels are?"

"Please do."

"They are rules," I said, "made by experts for the purpose of gaining freedom of action for their principals."

"That's what I said environmental science was," he said.

"What you actually described was what are popularly known as safety levels."

"Then what judgments should we demand of experts?" he asked.

"That all possibilities be investigated, that the evidence be interpreted honestly, and that all reasonable inferences be disclosed in their stark significance. Then every person could make his own choice, like deciding whether or not to smoke. The importance of values must be recognized. Different people value pleasure and health in different proportions. For those who attach great value to health, environmental science is a good thing."

"But people think that only experts can say what is or is not true in science," he said.

"Yes, many people think that, but they are mistaken in what they believe environmental science is like."

"What is environmental science like?"

"Not like a diamond waiting to be discovered, but like the student who had a revelation and saw the truth, which was not something in his texts or the words of his teacher, all of which only alluded obliquely to truth, but rather something he had added to the world."

"You are like a sculptor, Dr. Marino," he said. "You have made a science that is fair and good."

"Only for those who share my values," I said. "Don't suppose that what I have said applies to those for whom the health of the body is not a paramount concern."

"Now what?" he said. "Is your description of environmental science just theory?"

"It has legal and thus social consequences. *Daubert* made that possible."

"How would you construct a lawsuit?" he asked.

As I was preparing to respond, Stein grunted and walked off. The other experts also departed because, they said, many years' experience testifying for the Richmans in tobacco and asbestos cases had taught them to stick to science and leave legal matters to the lawyers. When all the experts were well out of earshot I said to young Richman, "Is what you have in mind a method for punishing companies who abuse environmental science?"

"That's what I want," he said.

"Because punishing companies is what you do for a living?"

He grinned.

"Let's begin by considering what kind of case we should bring," I said. "Perhaps the case should be for assault and battery, or even murder, if the victim dies. If a few company presidents were electrocuted there would probably be a groundswell of corporate support for the principles of environmental science."

"But we agreed that environmental science is a type of imperfect scientific knowledge, and the standard of proof in a criminal case is 'beyond reasonable doubt.' If we took your suggestion we would never win."

"Then an intentional tort would be a better choice," I said. "A 'preponderance' standard fits the nature of scientific evidence better than any strict criterion. A company that assaults a victim should have to pay up. Besides, lawyers can make more money in civil cases. But I suppose I don't have to tell you that."

He grinned again and said, "We shouldn't be troubling ourselves with the difficulties of proving intention. The defendant should pay if he did the act, regardless of his intention."

"Agreed," I said. "We should prosecute an unintentional tort."

"My father and I have a series of cases involving cell phones and brain cancer. Why don't you come down to our law offices so that we can discuss them. We will meet in the boardroom, which can easily accommodate the firm's attorneys who are working on the cases, and you can then tell us in concrete terms about the method you have in mind."

"When someone is old and set in his ways," I said, "he is not likely to welcome new ideas or novel methods, particularly when he has enjoyed great success. Such a man has neither the motivation nor the capacity for change."

"You should come and present your point of view, and defend it – perhaps you will win converts. Without the backing of a law firm such as ours, your environmental science will remain only an abstraction without any real presence in the society, and will therefore be unable to promote justice. So if you are serious, you will come."

"I will do so," I said, "under the condition that I may begin the discussion by explaining how environmental science should properly be used by lawyers in court cases."

He agreed.

Formal Articulation of Environmental Science

Kinds of Knowledge	Mental State	Knowledge Hierarchy
Perfect Knowledge	Certainty	Mathematics
Imperfect Knowledge	Uncertainty	Gravity
		Technology
		Medicine & Environmental Science
		Questionnaire Studies
No Knowledge	Guess	Astrology
		Phrenology

29
Measure for Measure
―――――◆―――――

It was raining when I went to the Richman building. "Do you know who last sat where you are now?" the taxi driver said. "No."

"Henry Kissinger."

A receptionist in the lobby told me that if I took the express elevator to the twenty-second floor and then walked down a flight of steps, I would arrive at the floor where the Richmans had their offices. I met young Richman at his office, and together we went into the law firm's boardroom. I saw an elegant mahogany table surrounded by plush armchairs that could be raised or lowered by pressing a button on the armrest. The outside wall was made of glass, which afforded good views of the Washington Monument and the Supreme Court building. The other walls held framed newspaper articles about huge judgments the law firm had won in litigation involving smoking and asbestos.

After a few minutes Peter Richman and some others entered the room and we all introduced ourselves. Most of those in the entourage were lawyers but it also included Faul, Morris, Milkin, and Phillip Keine, who had apparently changed sides. Just before we got down to business an older man walked slowly through the boardroom. He never said a word, but I recognized that he was Allan Frey, whom I had not seen since the meeting I had attended at the Barbizon Plaza Hotel, almost thirty years earlier. I had heard rumors he was advising plaintiffs' attorneys concerning the health risks of EMFs, but I had no idea he had bagged such a big client as Peter Richman.

The purpose of the meeting, Richman told me, was to discuss the cellphone cases but, without missing a beat, he began recounting some of his past cases, pointing to the stories on the walls as he spoke. It wasn't long before I said, "I know about your successes, but in my opinion they weren't achieved in the right way." The room fell silent except for the hum from the vibrating window glass and remained so until he said, "Dr. Marino, how many cases have you tried?"

"None," I replied.

"Then from where does your knowledge regarding how to try lawsuits come?"

"From an understanding of environmental science," I said. "Would you like me to tell you how victory should be pursued?"

Before he could respond young Richman said, "Perhaps if I give you some details regarding one of our cases, you could use them as a framework." I agreed, so he opened a file that was in front of him on the table, and glancing down at it from time to time he said, "Patrick Keogh is a forty-six-year-old brain surgeon who lives and works in Washington. He is married, with four children, including a set of twins; his oldest child is ten years old. About a year after he started using a cell phone he began experiencing headaches and blurred vision. A year later he was diagnosed with brain cancer."

"To what extent did he use the phone?" I asked.

"About a thousand minutes a month," he replied.

"Who manufactured the phone?"

"Motorola, a Delaware corporation that has its principal place of business in Illinois and is doing business in Maryland, which is where we got long-arm jurisdiction."

"So it will be Motorola you sue?" I asked.

"Yes, and they deserve it," he replied. "I'll tell you what Motorola did. Its ads deceived users of cell telephones by relaying testimonials from satisfied users, and by manipulating statistics to suggest that cell phones were safe, while downplaying, understating, and not stating their serious health effects. Its promotional literature fraudulently kept relevant information from cell-phone users, and minimized user concern regarding the safety of its cell phones. It began funding a huge research program that it said would prove cell phones were safe. But after the initial studies reported results that supported the opposite conclusion, Motorola discontinued funding the program. It also manipulated the results of testing by others, and concealed evidence that cell-phone radiation was harmful. This information would have caused Keogh and other prudent consumers to stop using cell phones, at least with the antenna held directly against their head. Instead, Keogh and the public were grossly misled and misinformed regarding the biological risks associated with usage of cell telephones precisely because Motorola purposely downplayed, understated, and did not state the health hazards and risks associated with cell phones."

"I take it that your goal is to make Motorola pay for causing Keogh's cancer," I said, "measure for measure."

He nodded emphatically.

"Then we should think like lawyers and concentrate on the things that matter."

"Like what?" he asked.

"First, Motorola knew that cell-phone EMFs could cause brain cancer."

"Wait. Stop right there," Morris interjected. "The president of Motorola is not stupid. Of course he will say that he did not know that." As Morris said "know" he flicked his two index fingers in the air as if he were adding quotation marks.

"He had a duty to know that cell phones can cause brain cancer because it is true," I said. "So either he knew it or he should have known it, and the distinction is irrelevant in the eyes of the law." All of the lawyers nodded approval, so I proceeded by asking no one in particular, "What should Motorola have done?"

When one of the lawyers said, "Motorola should have made a safe cell phone," I asked him, "What is a safe cell phone?"

"One that does not kill or harm the user," he replied.

"Would you say that a safe automobile should be defined in the same way?"

"No, not exactly," he said.

"Would it be fair to say that the objectionable aspect of Motorola's behavior was not that they made a cell phone that could cause cancer, but rather that they were aware of the possibility and yet failed to inform Keogh?"

When he agreed, I pointed out, "That's what negligence is." Then I asked young Richman, "Is it quite certain that Keogh developed a brain cancer?"

"He didn't undergo five craniotomies, radiation therapy, and chemotherapy for nothing."

"Was he exposed to anything else that causes brain cancer?"

"No, as far as we know."

"There you have it," I said. "All you need to do is prove that cell-phone EMFs probably can cause cancer, and that they probably did cause Keogh's cancer, and Motorola will be forced to pay for its negligence, although the measure he receives may be little comfort to him."

"Well, Dr. Marino," Richman said, "according to my son, you had something to say that could help us. If so, I haven't heard it yet."

Before I could respond, he answered a page and left the room. That was just about enough for me, so I lowered my chair and prepared to leave. Young Richman held me down, however, and respectfully asked me to proceed as we had planned.

"That's not what your father wants," I said. "We would all be wasting our time."

"My father built this law firm," he said, "but times change and sometimes men do not change with them, even when there is an earthquake."

"I suppose that a great man can misread the times just as easily as a small man," I replied, and he said, "Let's proceed and see where the discussion goes." He was a hard man to resist. Besides, I knew that the enormous resources of the Richman law firm would give me the best possible chance to show how the law ought to utilize environmental science, which was something that I very much wanted to do. So I raised my chair and began to say what I had gone to the boardroom to say.

"In a toxic-tort lawsuit the plaintiff says to the defendant, 'You caused my disease,' and the defendant replies, 'I did not.' The MacGuffin can be a drug like Bendectin, a contaminant like dioxin, a medical device like breast implants, or a product side-effect as with cigarettes, powerlines, and cell phones. The common factor in the cases is the question, 'Did the toxic agent cause what happened to the plaintiff?'"

At this point I paused. I had no pipe to fiddle with, so I just sat still for a few moments to give the impression that something significant would follow. Then I said, "We must first determine what *cause* means at law." I knew from past experience that if we failed to do so the discussion would inexorably descend into confusion and ambiguity.

"Are you talking about some kind of philosophical definition of cause?" one of the lawyers said.

"Philosophers deal only in opinions, which anyone is free to reject or accept," I said.

"That's why we hire experts," young Richman said. "They deal in facts, and are trained to look at nature and say whether or not there was a cause-and-effect."

"We are not talking about how to identify a cause-and-effect, but rather about what 'cause' *means*. The opinions of experts regarding that meaning are no more binding on the law than are the opinions of philosophers."

Young Richman frowned as if he were meditating. He must have realized the absurdity of the notion of philosophers or experts mandating to the law what the law meant by "cause," because he then said, "The people define for themselves, through their law, what 'cause' means. *Palsgraf,* for example."

"*Palsgraf,*" I repeated slowly. "The case says in black letters that 'cause' in torts means a particular, foreseeable *but-for* cause of the plaintiff's harm, namely a but-for cause that for policy reasons the law allows as a basis for assigning liability."

"I haven't heard that said explicitly since my first class in torts," one of the lawyers said. The others nodded in agreement, but the experts looked at each other as if I had spoken in Greek.

"The meaning of 'cause' pertinent to what Motorola did," I said, "is not a philosophical concept or a scientific fact, it is a social construction. The law's social policy is the primary determinant of what 'cause' means. That meaning is only *shaded* by philosophy, and by what scientists mean by that term when they talk to one another."

"*Palsgraf* is an old case, and it didn't involve expert testimony," one of the lawyers pointed out.

"No court ever limited the applicability of *Palsgraf* or its progeny to matters outside science. The rule of that case therefore applies in every kind of tort, including toxic torts."

After I said that I looked specifically at each of the lawyers in the room, one at a time, waiting to see if anyone would challenge what I had said, but no lawyer spoke. Next, I asked no one in particular, "What would you say is the social policy of the tort law as regards the dangers of commercial products?"

"That when the manufacturer places a product in the stream of commerce that product should be safe," one of the lawyers responded.

"Yes," I said, "and I suppose you'll agree that the way the law implements social policy is to create legal duties."

No lawyer objected.

"In our situation," I said, "Motorola had two legal duties: One was to investigate the safety of its cell phones and disclose whether they can cause cancer. If it failed in this duty and the failure caused someone to get cancer, Motorola's second duty was to pay damages."

All the lawyers nodded except for one who was staring out the window and probably didn't hear what I said. "Taking first things first," I continued, "we must consider whether cell-phone EMFs *can* cause cancer."

"Which brings us back to the experts," young Richman said, "because when the matter in issue involves knowledge not normally possessed by the ordinary citizen, only experts are competent to give opinions that can be counted as facts."

"That is the traditional jumping-off point," I said. "Let me tell you a story about how this tradition got started."

"All right," he said.

"Back when the English common law was first forming, there was a case in which a woman was accused of having given birth to a sheep. Some priests said the event proved that the devil was at work in the world, but others said that God would not have permitted such a thing, and that therefore the purported event never really happened. The king appointed a committee of scholars to investigate the matter, and they ultimately concluded that the woman could not have given birth to a sheep. The people were afraid to ask the learned men how they knew that, and the king did not do so because he wanted to end the controversy as soon as possible. From then on, whenever experts gave opinions, no one presumed to ask the experts how they had come by their knowledge of the causes of what takes place in the world.

"Of course from time to time there were those who suspected that what experts said was no more reliable than what priests said. More typically, however, the people believed what either priests or experts said, and for the same reason."

"You think this attitude continued even into the twentieth century?" young Richman asked.

"Not only continued but grew stronger, because the success of technology was taken to mean that experts *always* knew what they were talking about."

"Then?"

"We reached a turning point. The faith that people had in the word of authority figures began to erode."

"What was the turning point?"

"I can only give my opinion," I said.

"Please do," he answered.

"When President Eisenhower lied about the U2 spy flights over Russia. People thought, 'If you can't trust Ike who can you trust?' and their willingness to believe anyone in authority began to decrease. It wasn't long before this attitude was transferred to experts. Distrust increased with each lie that the experts told, even though some lies were arguably helpful in certain respects."

"For example?"

"They said that the high cancer rates in the soldiers who witnessed the hydrogen bomb tests weren't caused by the radiation, and that the convenience of x-raying children to insure that their shoes fit properly outweighed any risk that the children might get cancer. They told black men at Tuskegee that there was no treatment for their syphilis, and they said that DDT didn't hurt the environment. They said that thalidomide, Agent Orange, and cigarettes were safe products. They told an infinite number of smaller lies that you never hear about because they were never publicized, as for example that artificial ligaments made from carbon fibers are unsafe."

"What has changed as a result of the lies?" he asked.

"The breaking of the chain of inference by which an advanced academic degree had been taken to imply expertise which, in turn, had been taken as a sign of knowledge. This break ended the time of the law's uncritical faith in experts."

"What broke the chain?" he asked.

"The Supreme Court," I said. "In *Daubert* it rejected the antediluvian practice of blindly trusting experts and laid down three requirements for identifying knowledge."

"Which were?" young Richman asked.

"First, that there is a method for producing knowledge. Second, that the method is a commonplace. Third, that the expert followed the method."

"Do you mean the scientific method?" young Richman asked.

"Are you asking me what I suppose the Court meant by 'method?'"

"Yes," he said.

"Some particular form of thinking is needed to connect evidence with knowledge, because evidence doesn't speak for itself. The thinking that does this job is what I suppose the Court meant by 'method.' I wouldn't call that method the 'scientific method' because the term has become a cliché."

"People seem to agree on what it means," one of the lawyers said.

"Only because a person who hears the phrase assumes he knows what was meant. That's how its vagueness is perpetuated."

"What's the way out of that problem?" asked one of the lawyers.

"To say directly what the method is, not simply call it 'scientific,'"

I replied.

He asked me to say what I thought the method was, but before I could do so another lawyer piped up and said, "It is a specialized form of thinking used only by scientists."

"I suppose that you know what form of reasoning is used by lawyers," I said, "but how would you know what form is used by scientists?"

"I don't know," he said.

"Then how would you know that the reasoning they use is specialized?" I said.

"All right," he said, "you tell me, is it?"

"The thinking that scientists do is not unique to scientists. It is the same as the thinking of laymen. So we should use the same name we use when we describe how laymen think."

"What name?" young Richman said.

"'Common sense' is as good as any," I said.

"Then why do we need experts?" he asked.

"Because the evidence on which the common-sense inferences of scientists are based is of a special kind," I replied.

"If science is just common sense, why do experts cost so much?" one of the lawyers said, much to the amusement of all the other lawyers except young Richman.

"Scientists are specialists in what they do," I said, "like boat-builders or lawyers. Everyone is a specialist in something, and when he is consulted in that capacity he is entitled to be paid, and his opinions deserve respect."

"What's the special evidence you mentioned?" he asked.

"One kind consists of the laws of physics, which is where all our scientific deductions come from. Another kind consists of careful observations and measurements. Bona fide experts usually find these kinds of evidence clear and direct, so they rarely result in disputes."

"But disputes do occur," Morris interjected. "Remember cold fusion."

"What's another kind of evidence?" young Richman said.

"A work product containing observations and measurements commingled with opinions of the author that invest meaning in the data. These products are typically called 'studies,' but they may also be called 'analyses,' 'reports,' 'judgments,' 'results,' 'assessments,' 'reviews,' 'decisions,' 'determinations,' or 'findings.' They are the primary means by which scientists exchange with each other what they regard as knowledge. Studies are also the portals for communication between scientists and laymen. Unfortunately, studies can easily be manipulated."

"Can you expand on that?" young Richman asked.

"Like the ingredients in bread, the ingredients in studies disappear. Laymen can judge opinions but not data, so companies that despise environmental science make the kind of bread that best nourishes their business. And they always do it quickly."

"Why quickly?" he said.

"They want their point of view publicized as early as possible because they know that the tradition of reverence for experts has conditioned the layman's mind to be like the jaws of a crocodile – the muscles that close it are strong but those that open it are weak. That's why Motorola hired Mays Swicord; I would be more delicate were I in court, but here I am among friends."

"What does he do for Motorola?" young Richman asked.

"Arranges for the publication of allegorical studies whose moral dovetails with Motorola's interests. Motorola is like the company that had been making food additives for a long time. One day someone asked, 'Is your additive safe?' The company president realized that the question could alarm the public who might expect that there was an answer, so he hired experts who produced data showing that, as far as they could tell, consumption of the additive had no effect on growth in mice. The experts then opined in their report that the food additive was safe for human consumption, which vindicated the status quo. That's what Swicord's studies are designed to do, and which is possible because all biomedical studies consist of commingled data and opinion."

"I don't know what you're talking about," said one of the lawyers. "Experts are supposed to tell us what the data means."

"I disagree. You pay a boat-builder for his expertise in building boats," I said, "you don't ask him about the purpose or meaning of the boat. Similarly, you hire a lawyer for his expertise in winning cases, not because you want him to find the truth. All lawyers know that's not their jobs."

"We do?" the lawyer said.

"Suppose you represented a stone quarry," I said, "that was sued because its operations were too loud, and you hired an expert who made measurements of the sound level and agreed with the complainants. What would you do?"

"I'd fire that expert and hire another one," he said.

When the laughter stopped I said, "When the putative side-effects of a commercial product are under consideration, the strangest thing in the world would be for all the experts to agree on the meaning of the pertinent studies."

"So are you saying the question whether EMFs *can cause* cancer is to be resolved by a jury when the opinions of experts are in conflict?" one of the other lawyers asked.

"I think so," I said, "but let's not jump the gun. Common sense often fails when people attach different meanings to the same words. We must first ask each kind of expert to tell us what he understands to be the meaning of 'can cause.'"

Faul said, "For me the meaning derives from experiments. I won't attempt a grand definition but I'll give you an example. Suppose I am investigating whether cell-phone EMFs can cause changes in the amount of neurotransmitters in the brain. I would form two groups of rats, only one of which would be exposed to the EMFs. After a predetermined time period I would kill the rats and measure the neurotransmitter levels in each brain. The range of the values in each of the two groups will overlap, so I would use the t test to compare the averages and assess the odds that the two averages were really different. If the odds were greater than about 20:1, I would say that the EMF exposure caused a difference in the average level of brain neurotransmitters."

"Suppose the odds were less than 20:1," I asked.

"Then I would conclude that the EMF had no effect on the neurotransmitters," he replied.

"It's possible that an effect occurs only after a different predetermined time," I said.

"Yes."

"Or with a stronger or weaker EMF."

"Certainly."

"Or with an EMF of a different frequency."

"Obviously."

"Or if you used a different animal species."

"Clearly."

"You wouldn't propose that the experiment be done with human beings, would you?" I asked.

"Of course not," he replied.

"Suppose that the odds were greater than 20:1," I said. "How could you say whether or not cell-phone EMFs can cause brain cancer in human beings?"

"I would form an opinion based on the number, quality, and kind of all relevant animal studies."

"What number, quality, and kind would you require?"

"I can't say for sure, but I usually know it when I see it."

"We've heard from a biologist," I said, "now let us hear from a physicist." I looked at Dr. Stein and said, "Please tell us what 'can cause' means to you."

He stood up, said, "Cause means force, therefore 'can cause' means *can force,*" and sat down.

From the expressions on their faces, the lawyers in the room found that explanation unedifying, so I asked, "Would you kindly explain in slightly more detail?"

"Maxwell's equations govern the interaction of EMFs and tissue. So 'can cause' means that the effect is predicted by these equations." He swiped his hands against one another and sat down.

One of the lawyers said, "When an expert says 'can cause' perhaps he should be required to wear a colored hat, green for the biological meaning and red if he uses the term like a physicist."

"We might need a third color, perhaps yellow, because we have not yet heard from Dr. Keine." As the old man rose, his bones creaked like a set of castanets. When he was standing as tall as he could he said, "Suppose that everyone in Maryland who got cancer filled out a questionnaire and disclosed every possible fact about his life, including the extent of his usage of cell phones. To explore whether cell phones can cause cancer, I would find a person who was a doppelganger for each person who got cancer, except that the persons chosen would not have cancer. I would then compute the percentage of cellphone users among the doppelgangers. If it were less than the percentage in the cancer group, I would say that cell phones were associated with cancer, and perhaps caused it."

"What would it take for you to drop the 'perhaps?'" I asked.

"At least fifteen studies in all of which there was a higher percentage of cancer in the cell-phone group."

"Would fourteen studies be enough?"

"No."

"Dr. Keine," I said, "have you ever seen an instance where it was possible to identify a doppelganger?"

"Not really," he said. "Everybody is different."

As he sat down, his bones played another tune.

"If different kinds of scientists disagree over what 'can cause' means," young Richman said, "how can we expect a jury to sort out the mess?"

"There are indeed problems," I said. "I can think of two. But they don't include a lack of intelligence or incisiveness on the part of juries. Given half a chance, they will find the best truth."

"I wish I had your confidence," he said.

"I think you do," I said, "at least you act as if you do."

"What do you mean?"

"You already agreed," I said, "that any liability Motorola might have stems not from the fact that cell-phone EMFs cause cancer, but rather from Motorola's failure to warn Keogh of that possibility."

"Yes."

"That failure to warn is called negligence, which is nothing more than acting unreasonably," I said.

"Yes."

"In a particular set of circumstances, what is unreasonable is determined by the consensus of laymen," I said.

"Yes."

"You wouldn't say that was hopeless."

"No," he said. "Juries do it all the time. That's our system. They evaluate the credibility of witnesses, and make judgments regarding what is reasonable."

"It's the same with expert witnesses," I said, "so even a jury of high school graduates can tell which experts make the most sense, assuming, of course, that the lawyers are competent."

"Yes," he said. "The longer I practice law the more I think that they should teach the art of cross-examination in law schools. But you mentioned two problems. What are they?"

"People expect too much certainty and too much honesty in scientific matters," I said. "This mythic background can be overcome only by hard work and proper preparation by the plaintiff's lawyer. A top lawyer conducts a proper cross-examination, a lazy lawyer doesn't make the effort."

"Let's assume that we have proved that cell-phone EMFs can cause cancer," young Richman said. "How do you decide whether Keogh's cancer was caused by EMFs?"

"First I would prove that he had cancer. Then I would show how much EMF exposure he had and how long, and that he had no substantial exposure to other known cancer-causing agents. The conclusion follows that the cancer was probably caused by the EMFs."

"Why couldn't it have been caused by something else?" he asked.

"It could," I replied. "No one can ever be certain. But look at the facts. Keogh has cancer, and he experienced extensive EMF exposure. Now, something caused his cancer. Something made it happen. Something such that, if you took it away, you would take away the cancer. We know EMFs can cause cancer. It's possible something else that we don't know about caused Keogh's cancer, but that's unlikely because the other things that we know can cause cancer weren't experienced by Keogh. The choice, therefore, must be made between the two alternatives. Either we say that the EMF caused the cancer, or that it was caused by some unknown or unsuspected agent. It's more likely that EMFs were responsible, because we know that they can cause cancer and that they were extensively present in the case."

Just as I finished speaking Peter Richman returned to the boardroom and took the vacant seat at the head of the table. After apologizing perfunctorily for his absence he said, "Dr. Marino, would you summarize what you have told my colleagues?"

"I said that the key question was whether cell-phone EMFs caused Keogh's cancer, by which I meant that they were a but-for cause. If so, Motorola will have to pay up. Experts who truly know from laboratory studies that EMFs can cause cancer are needed. The surmise that EMFs caused Keogh's cancer is more or less obvious, given his extensive EMF exposure and the absence of other putative causes. Victory will depend on your skill in showing the jury that Ryan's experts are mistaken."

Richman laughed for what seemed a long time and then said, "Experts are never mistaken; they just disagree with one another. If Ryan presents an expert from Harvard, I present one from Yale. An expert from Yale is not like one from LSU."

"No," I said, "Yale is more famous."

"Juries don't understand experts," he said, "and when juries guess I win at least half the time. As you can see," he said as he pointed to the walls, "my approach has enabled me to provide well for my family."

"If I could make you see what you have become," I said, "you would kill yourself."

"What!?"

"Your aim is to make Motorola pay whether or not cell-phone EMFs can really cause cancer. It's a loathsome goal, and by chasing it you have weakened society."

Richman slammed his left hand on the boardroom table, said, "Meeting over!" and left the room. I never saw him again.

I lowered my chair and waited for everyone to leave, which they did, except for young Richman who, like me, remained seated and silent. My inability to connect with Peter Richman was a giant blow to me. I had engaged in honest dialogue and meant everything I said for the reasons I gave. But my sincerity was not returned, and I was not able to draw him into a discussion and convince him of the error of his ways. He was rich, so money would not have been a barrier to winning the case. The victory would have been earned in the right way, and would therefore have begun the process of establishing environmental science as the evidentiary frame of reference for scientifically related issues. With my knowledge and his money we could have done this, and then gone on to other similar cases. Together we would have been like Eisenhower and Montgomery invading Europe to free it from tyranny. What a legacy we could have left to our children!

Young Richman and I went up to his office on the twenty-second floor, and as we waited there for the limousine that would take me to the airport we had our last conversation. He said he would press forward as best he could to try and prove that the EMFs from Motorola's cell phone had caused Keogh's brain cancer.

On the way home from Washington, while my failure to team up with the Richmans was still vivid in my mind, I sat next to a professor from the University of Rochester on the airplane.

"I know someone from your university," I said. "Morton Miller. I haven't seen him in many years. Do you know him?"

"Yes indeed," he replied. "Morton is doing quite well. He was recently notified by NIH that his grant will be renewed again. He has been continuously funded by NIH for the last 28 years."

30
Mind War

♦

After I returned home Richman hired some experts to represent his point of view in the Keogh case; young Richman had told me his father would not proceed unless he were "90% sure of victory." When Motorola was informed of the identity of the experts, the company took their depositions and brought on a *Daubert* hearing to argue that their testimony was unreliable. In their normal element they were confident scientists, but in the depositions they were hapless victims of circumstance who stumbled into every trap Ryan had set. There was little Richman could do to stanch the damage to his case because he did not understand the scientific mumbo-jumbo his experts used. Even if he did, his normal strategy was not to defend his experts but to attack those of his opponents, as in the tobacco cases where he usually won. Motorola's experts were adroit; Ryan had recruited them from Harvard and Yale and schooled them on how to duck and weave during Richman's cross-examination. In the end, the judge denied Richman the right to present his experts in court because, she said, their testimony was not generally accepted. He should have swung the *Daubert* sword against Motorola's experts, as I proposed. The world is still waiting for the bold lawyer who will do that; he is certain to become rich.

I went to England to talk at Oxford about EMFs. I hoped the holiday would help me see the big picture, and decide what I should do. I had a secure job, intellectual freedom, tremendous financial support from my boss and friend Jim Albright, a fully furnished electrophysiology laboratory staffed by four wonderful scientists, and graduate students who were interested in my interest. Although I was 62, six years older than Dr. Becker when he was forced out, retirement was far from my mind. I wanted to find the best way for me to advance the environmental science of EMFs.

In the dining room of Christ Church College I met an Australian man whose outlook was similar to mine. I told him that the Electric Power Research Institute had won, at least temporarily, by being unjust, and that the Mobile Manufacturers Forum was now busy doing the same thing. Thinking to cheer me up, he told me about his experience.

"The Aspirin Foundation," he said, "opposed a labeling requirement notifying users that aspirin could cause Reye's syndrome. It took years, but the Foundation was finally beaten."

"How?" I asked.

"By sound science," he said.

When I returned home I told our secretary, a devout Baptist, about the dinner. "We all stood and toasted the Queen, and as I did I looked around at the portraits on the walls of the famous men that had graduated," I said. "I saw Lewis Carroll, William Penn, and Charles Darwin."

"Ah," she said, "the evil Mr. Darwin."

My situation was something I couldn't have imagined in high school when I had first realized that science was the thing that most fascinated me. I had learned a lot during my long journey and gotten, as far as I could tell, closer to truth about EMFs than anyone else. But that wasn't enough. Although my knowledge was something real, so was the will of those who opposed progress in the environmental science of EMFs; their strategy had been as effective as it was unjust.

The injustices I had encountered each contained two elements. There was something that someone didn't know, and hence hadn't taken into consideration when resolving a particular question. There was also a default heuristic for deciding, in the context of that ignorance, whether it was worse to allow a naïve person to get cancer or to require an innocent company to spend money for prevention or compensation. How one answered this question depended on his value system, and what he thought was fair. In my philosophy the worst sin was injustice toward others, and intentionally injuring an innocent person was an unjust act. The Masters at Motorola knew or should have known that cell phones probably caused disease. But they hid that knowledge from an unsuspecting population, and then went farther and said cell phones were almost perfectly safe. Their cruel perspective led them to ignore objective morality and morality rooted in humanism in favor of morality that was entirely self-referential.

There was nothing I could do about the concept of justice that others adopted. Making that choice was something each person did alone, like dreaming. Knowledge is different. It's not any particular person's decision, but rather the fruit of a method. That was how Justice Blackmun had defined knowledge, and he was correct. Proceeding along this line, I decided that knowledge was more fundamental than justice and that my goal should be to concentrate on my lifelong goal of learning more about EMFs.

For me, the brutality of relying on questionnaire studies had ruled them out of bounds as a proper method. And even neglecting their brutal nature, I really had no choice because the meaning of every questionnaire study is always already deferred. So I had turned to a consideration of high-level scientific studies, that is, careful, prospective, controlled inquiries governed by hypothesis. I had the simple idea that if there were enough such studies focused in a particular area and as clear-cut in their results as scientific studies could be, people might then say, "It looks as if there truly is a problem, and therefore we should consider doing X," and there would be a civil debate regarding which X was best. In this way knowledge would lead to justice.

Ever since my first halting attempts at doing animal research, I had understood the logic behind the link between laboratory research and judgments about human health risk. The ultimate goal was to reliably answer the question whether environmental EMFs caused or contributed to human disease. The best method would be to form a group of human beings who were absolutely identical in all respects, randomly divide them in half, and require them to live their lives in exactly the same way except that only one half would be exposed to environmental EMFs. Whether EMF exposure caused disease could then be directly determined.

The experiment would be impossible, of course, which left only two alternatives. Animals could be used. On the practical plane however, this approach led immediately to problems involving judgment, opinion, and perspective. Choices had to be made regarding which animal species to use, the duration of exposure, and the endpoints to be measured. Moreover there was the reality that no two animals of any species were ever truly identical. Another complication was that many phenomena were species specific. Occurrence of brain cancer in rats, for example, would not necessarily imply that similarly exposed human beings would also get brain cancer. Despite all these difficulties, we still desire an answer to the question and the use of animals is a rational approach, assuming one is willing to accept the uncertainty inherent in any possible answer.

The other alternative was human laboratory research, the kind that could be done ethically. Normally, when society must decide the question whether it is worse to allow a naïve person to risk developing disease or to require an innocent company to spend money for prevention, the company is expected to furnish evidence that, on balance, any inherent risk triggered by their product should be regarded as acceptable. In the old days the evidence was often arrived at by deeming. Subsequently, a philosophy evolved under which at least some experimental data was required. The Environmental Protection Agency, for example, answers questions regarding toxicity of pesticides by requiring the manufacturer to measure the dose required to kill 50% of a test group of animals; the pesticide is deemed safe for people when the dose they receive is lower. The approach was somewhat of an improvement over deeming, but EMFs don't kill enough animals quickly, so the EPA method doesn't work with EMFs. There were those who interpreted this to mean EMFs were "safe," but the more enlightened view was that the lethality method was too insensitive to measure EMF-induced risk. Unfortunately, no consensus developed regarding what other paradigm for animal studies should be used to answer the question of risk. Consequently, no matter what EMF animal study was done and no matter what the result, it was always possible for Motorola to hire experts from Harvard and Yale who would say, "So what?" and "The results have no meaning with regard to human health risks," and "When I tried to repeat the study I got different results."

Animal studies were no answer, at least not for me, so I decided to concentrate on doing human laboratory research because I had no other choice. My earlier studies on brain waves had not left a perceptible imprint on the EMF story, but I felt I had matured as a scientist. I understood the primacy of internal law, and the necessity to develop methods capable of revealing its operation. My hope was to generate knowledge about EMF effects on the brain that would lead to environmental justice, perhaps even compel it.

One evening I disclosed to a friend what I intended to do. He asked why I had chosen that goal, his implication being that it was likely to be a long and arduous effort, not ideally suited to someone of my age. He offered some possible reasons thinking, I supposed, that I would choose one. Was it something I thought God wanted me to do? Was it the search for truth that is frequently imagined to be the motivating factor for scientists? Was it a desire for fame or recognition so that I would be remembered after I died? I denied each suggested reason, but had no tidy answer to his question. Then something happened a week later which made me realize that my personality, and nothing more, was the explanation.

I was the director of a Musculoskeletal Medicine course for the medical students. After I had sent out the final grades, a student who received a B realized he would have an A if only one answer on his final examination were changed, so he challenged a question he had missed. The question had referred to the "posterior tibial nerve," the official name for which had been changed to "tibial nerve" in 1976. He argued there was no such thing as a "posterior tibial nerve," and hence no correct answer to the question, and asked the dean to order me to give him credit for the question. I opposed the student's argument on the basis that three orthopaedic surgeons had vetted the question and found it to be clear and appropriate, and because more than 85% of the class had answered it correctly, which clearly showed there was a correct answer.

I expected the dean to reject the challenge and take the student behind the woodshed for failing to adhere to the ethical standards expected of doctors. Instead the dean sided with the student and ordered me to change his grade. I felt the decision was wildly wrong, so I refused. Some of my colleagues thought my decision was irrational. "In the scheme of things, who cares?" one of them said. "Do what he wants."

For me, the issue was whose concept of right and wrong was I going to follow, and if it were mine, at what price? The dean was powerful. He couldn't have fired me because I had tenure, but he could have made my life miserable in innumerable small ways. At that time, for example, a document sat on his desk awaiting his signature that would have authorized a salary raise for my Italian post-doctoral student to the level required by federal immigration authorities. I would have lost her had he refused to sign, and the research program I had been busy planning would have been thwarted. In the end, the dean acquiesced; he sent me an unkind note saying that I had mistreated the student, but he denied the challenge. And he signed the document.

Most who knew about the incident thought I had risked too much, but that's the way I had always been. Mr. Schmidt, Andrew Bassett, Trashcan Trischka, Dr. Franklin Hart, Herman Schwan, Cedric Minkin, Joe Berndt, Edward Berlin, Governor Carey, Asher Sheppard, Hans Tauschen, Thomas Matias, Frank Wallace, Robert Flugum, Bennett Miller, Dwight Spieler, Chauncey Starr, Richard Phillips, Philip Handler, Woodland Hastings, Marguerite Hays, Don Justesen, John Mattill, Nirmal Mishra, Anthony Erdgas, Karl Gustavson, Phillip Keine, Ken Olden. An automatic mechanism inside me was triggered whenever someone tried to push me around.

I understood what kind of a thing scientific knowledge about EMFs was. Some inferences might be sufficiently grounded to be regarded as true, and others sufficiently ungrounded to be regarded as false. None, however, could be true or false. I had pity for those who thought otherwise, the devout and the nihilistic. I also had a sober idea of how knowledge about the effects of EMFs would be received. First it would be ignored: by the electric-power and cell-phone industries because they would perceive it to be against their interests; by the government because politics trumps knowledge; by other scientists because their main interest was in making their own discoveries. Eventually, a newspaper article, court case, or other publicity-generating event would prompt the industry to hire a passel of experts to package and present its story. This time, however, I thought the end result might be different because I had a plan.

My original jumping-off place for the connection between EMFs and brain waves had been the idea I described in *Electromagnetism & Life* in 1980, that EMFs entered the body by means of specialized cells, just like light, sound, taste, or touch. Stimuli have no meaning prior to uniting with something in the biosphere – to be, is to be detected. I drew strong parallels between the processes that mediated prosaic stimuli, and the processes involved in the body's response to EMFs. A scene in the movie "10" had crystallized my thinking. Bo Derek and Dudley Moore raced toward each other with their arms extended in anticipation of an embrace. If Moore had no sensory cells in his eyes, he would not have known she was on the beach. But a marvelous detection system in his head told him otherwise. Light had entered his eye, passed into his retinal cells, and was absorbed by the protein rhodopsin. An immediate consequence was that another protein, essentially a door to the cell, opened and allowed sodium ions outside the cell to enter. The presence of the ions inside the cell changed its electrical characteristics, resulting in the propagation of an electrical signal to the brain. Moore saw Derek only after that signal had arrived at his brain. The sequence of events must be the same with EMFs, except that the brain's processing of the EMF-induced signal did not involve connections up to the top of the brain, where consciousness occurred. Although Moore wouldn't have been aware of an EMF, his brain would have known it had been there.

When I had designed my first brain-wave experiments, I started in the right place, with Goethe, not Newton. If I wanted to think about why a steam generator in a nuclear submarine wouldn't break free of its supports when the sub made a crash dive, or about how much thrust was needed to insure that two orbiting space ships would rendezvous as planned, then I would think the way I had been taught by my teachers. They could trace their knowledge back to Newton, who apparently got it from God. I would write equations, put in the values of some predetermined parameters, like the strength of the steel in the generator's support or the initial separation of the two space ships, and then I could be confident about what would happen. In matters involving the biosphere, however, I had learned to think like Johann Wolfgang von Goethe. God had told him that there were no universal governing equations for life, but that each living thing followed its own law. Newton wasn't wrong, but there were some things he didn't know. The most important thing he missed was life. Dr. Becker and Szent-Gyorgyi, the two greatest men I ever personally knew, also missed it. They spent their lives looking for answers at the level of electrons, which they presumed were governed as in Newton's world view. But their grandest dreams were incapable of fulfillment because they set their sights too low. Life didn't exist down where they aimed, but rather was an emergent property of an organized unitary organism whose behavior was governed by its own law. They hadn't understood what they were studying. They weren't alone. Everyone since Goethe had failed to understand life, so they hadn't designed their experiments properly. Instead, they had followed Boltzmann.

Ludwig Boltzmann, one of Newton's heirs, had concerned himself with the problem of explaining how the million billion billion atoms of air in an ordinary hat-box–sized container generated the pressure that could be measured on the sides of the box. The atoms moved incessantly, colliding with one another and with the walls. Even though the motion of each atom was governed by exactly the same law, the atoms moved with different speeds and consequently had different energies. Listing the million billion billion atomic energies would have been impractical, so Boltzmann expressed them expediently by taking their average, and used it to explain pressure. His justification for averaging was that the atoms were identical and were governed by the same law. In the biosphere, however, no two individuals in any species are identical or governed by the same law. So there was no reason to suppose that averaging the data was also the canonical way to study how living things react or behave.

Unlike Ludwig Boltzmann with atoms, and biologists with animals and human beings, I had avoided averaging my results across all the subjects exposed to the EMF. Instead, I had treated each subject as a separate experiment by measuring the EEG in the presence and absence of an EMF. Although I had realized that the focus needed to be on the response of the individual – so I had started in the right place – I went wrong when I used the technique called spectral analysis to reduce the EEG to numbers. As it turned out, spectral analysis was still too much like what Boltzmann did. Eventually I had published a series of articles in neuroscience journals in which I reported that EMFs altered brain activity in about half the subjects. The articles had generated only mild interest because they were nothing special – sometimes yes, sometimes no, which was always the case with EMF studies. My Italian post-doc said to me, *"Diranno che hai scoperto l'acqua calda."*

Looking back at those studies, I began to consider the possibility that perhaps the subjects I couldn't prove had detected the EMF had actually done so, but spectral analysis was too insensitive to detect the effect. What had prompted me to return to this idea was a report about an experiment using light. From an analysis of the EEG the investigators had been able to show objectively that only about half the subjects had detected the light, even though all of them said they saw it. This reinforced my suspicion that the thing standing in the way of my achieving my goal was my analytical method, and therefore that something better than spectral analysis was needed.

During my studies of the immune system in mice, I had consulted a friendly mathematician from Indiana University, and together we worked out a method for analyzing my data without averaging away the result I sought, which was to prove that EMFs had affected the immune system. In those experiments, my data had consisted of fifteen different measurements of the immune system of each animal, all of which I obtained after I had killed the animal, which was necessary to obtain the immune cells. Consequently all the data had applied to one time point, the instant of death. In the brain experiments, in contrast, the EEG was measured 300 times a second, typically for 1,000 seconds, resulting in a huge series of time points stored in a computer. Thus my data-analysis problem was far more complex, but unfortunately, like Rosalind Russell's lover in "Auntie Mame," my mathematician friend had fallen off a mountain in Switzerland

and was not available to help me solve the problem.

The key that had led to the solution had been my reformulation of the experimental goal. I had not tried to measure average differences in specific immune parameters between exposed and control animals, or to identify the mechanisms that might have been responsible. Instead, I had addressed the seminal question: Did the EMFs affect the immune system? The realization that the same kind of approach was needed in the brain-wave studies, plus my luck in finding the work of two obscure but insightful physiology professors from Chicago, led to my eventual success in finding a recipe for demonstrating the effect of EMFs on human brain waves that made the phenomenon as reproducible as gravity.

Everything began to fall into place when I noticed the similarity between the EEG and the fluctuating signal I had seen many years ago, during a lecture by a man named Edward Lorenz; put side by side with an EEG, I could hardly tell the two signals apart. His signal, which had come from a calculation of the temperature at a point in the atmosphere, looked like a random series of peaks and valleys; but the opposite was true because it had been produced by a mathematical equation. A law had produced perfect lawlessness. A paradox? No – simply a case of not looking at the situation from a good perspective; when the signal was transformed into phase space, a mathematical construct useful for analyzing certain forms of data, the famous "butterfly wings" pattern appeared. I thought I too might see a hidden pattern after I transformed my brain waves into phase space. I was right! My pattern wasn't as beautiful as Lorenz's, but considering that my butterfly came from the real world where nothing is perfect, I was quite pleased.

Naturally my next step was to try to attach a number to the pattern. That was a big problem. Mathematicians had devised many ways of quantifying the phase-space pattern of a series of points in time that came from an equation, but each method required the time points to meet certain mathematical conditions that time points from real-world measurements simply could not meet. Some other form of what had come to be called "nonlinear analysis" was needed. One day I googled that term and found the work of Zbilut and Webber.

Joe Zbilut was a professor in the Department of Molecular Biophsyics and Physiology at Rush University, and Chuck Webber was a professor of physiology at the Loyola University School of Medicine in Chicago. They had developed a way to quantify signals that was free of the limitations which encumbered all the other methods. Their approach was based on earlier work by a German mathematician who had devised a technique called the recurrence plot, which revealed the order that otherwise was hidden in series of time points generated by mathematical equations. The plots had a geometrically artistic mien; still, they were only pictures. If someone wanted to compare two different plots, the only possibility was to examine two pictures, the psychological impact of which was certain to vary from observer to observer.

Zbilut and Webber had invented a way to attach numbers to a recurrence plot, based on the arrangement of its points, and they named their quantification scheme "recurrence quantification analysis" or RQA. Then they took the momentous step of recognizing that RQA could be applied to a series of time points that came from the world, not simply those from mathematical equations. In their first publications they applied RQA to uncover the hidden patterns in signals derived from heart activity and muscle action. Unlike every other form of nonlinear analysis that had been devised, RQA did not depend for its validity on the mathematical properties of the time series. Like Roentgen when he had discovered x-rays, the Chicago professors made their discovery freely available by openly distributed software they had written that permitted anyone to do RQA. I obtained the software and began using it to analyze EEG signals.

Patterns come about as the result of the operation of a governing law, known or unknown, which is the mathematical way of expressing a cause-and-effect relationship. A change in a pattern entails the idea of a change in the law. If a subject's EEG consistently differed between the presence and the absence of the field, that consistent difference would be manifested in the RQA results because RQA detects something more basic than the direction or magnitude of a change, it detects *change itself*. For that reason the method was perfectly matched to my task, which was to detect changes in EEG energy, even if it went down in some trials and up in others.

Using RQA took me into the counter-intuitive world of nonlinearity where novelty was normal and predictable phenomena like the rising and setting of the sun were the exception. I was reminded of a trip I had taken to Hungary. I didn't know how much anything cost because the official exchange rate was wildly ignored, and varied in mysterious ways; when I made a phone call to America the charge was almost as much as my airline ticket. Navigating the cities was a problem because the language was so strange that a posted sign contained not a hint of its message to an English speaker. I used RQA on a mathematical model system whose properties I knew with certainty, because I had built the model from known parts, and found that the method worked perfectly. That eased my mind about RQA and convinced me that its potential benefits outweighed any concerns about its strangeness and novelty.

As the former head of my school's committee for regulating human research, I had applied the federal rules to the research projects of other faculty members. Now, for the second time regarding studies of brain waves, I appeared before the committee as an applicant, to answer its questions.

"You plan to expose subjects to EMFs, is that correct?"

"Yes."

"What is your purpose?"

"To establish the fact that EMFs such as those from cell phones alter the electrical activity of the human brain."

"What exactly will you do?"

"I'll measure brain electrical activity in each subject during EMF exposure and compare it with that same subject's brain activity when the EMF is not applied. I expect to prove that the EMF stimulus caused a change in the subject's brain activity, which would mean that the subject actually detected the EMF."

"Is what you are proposing safe?"

"I plan to apply the EMF for two seconds and turn it off for at least two seconds, and then repeat the process a hundred times. So each subject will be exposed for only 200 seconds. That's equivalent to talking on a cell phone for less than four minutes, which is nothing out of the ordinary."

I answered several other questions of the type I had asked innumerable times when I headed the committee, and then was asked to leave the room while they discussed my application.

As I waited, my mind drifted and I began to imagine how my interview would have gone had I sought permission to perform a study that could actually lead to results about cell-phone safety that were as certain as science could be.

"Do you plan to expose children to EMFs?"

"Yes."

"Why?"

"To see what happens. Children are more vulnerable than adults to drugs and toxins, so if cell-phone EMFs can cause cancer, the effects should be easier to detect in children."

"What exactly will you do?"

"Expose 1,000 newborns from poor families to cell-phone EMFs while the babies are sleeping. Their cribs will be designed so that the phone begins to radiate when the baby is placed on the mattress. I expect to prove that the subjects get leukemia and other diseases more often than normal."

"Their parents would never agree to such a proposal."

"I think I'll have no difficulty convincing them to allow their children to participate."

"I don't see why."

"Motorola says that cell phones are perfectly safe, so it would appear to even the most conscientious parents that they were doing their children no harm because almost everyone believes Motorola."

"The whole idea sounds revolting."

"Why? If that's what Motorola says is true then I think you have no reason to be upset. If you do, what is it?"

"The idea is preposterous. That's my reason. But tell me why you would include only babies whose parents are poor."

"I'll pay $5 a day. That's a lot of money to poor people. There is also another reason. If the results of the study went against Motorola, they would hire experts from Harvard or Yale to argue that not EMFs but some unknown sociological factor accounted for why the babies got sick. But if all the families were poor, they likely would be homogeneous with respect to any sociological factors the experts might latch onto."

"You seem to have thought of everything. Have you thought about how much this study would cost?"

"At $5 a day for 360 days, 1,000 babies could be studied for only $1.8 million. The American Cancer Society has agreed to fund the study. We all know how interested they are in finding the cause of cancer. The director of the society told me that he is willing to pay, wanting to pay, and waiting to pay."

When I was called back into the committee room I learned I had received permission to begin my experiments, provided that the subjects signed a written informed consent.

In my first experiment, using RQA, I found that the subject's brain electrical activity was different during EMF exposure, compared with no EMF. I got the same result the second time I did the experiment, and then the third time, the fourth time – every time I did the experiment!

I parted company with Roentgen and wrote and submitted a patent application to cover my invention of the method for detecting changes in brain electrical activity in human beings when they are exposed to any stimulus, including weak electromagnetic fields. In addition to wanting to provide for my family, I had another motivation. When I had donated my services to the Public Service Commission in New York, I received only abuse in return. The lesson I had taken away from that experience was that no one appreciates anything they get for nothing. If people wanted to use my method, they would pay; then they would appreciate it. The patent was awarded without the need for me to change even a single word.

My career in science entered a productive stage, like a tree bearing fruit. I began publishing papers in neuroscience journals describing how EMFs consistently affected brain electrical activity in human subjects, a phenomenon that was as reproducible as gravity. Mindful that knowledge was the product of a method, my first publication in the line was a detailed description of my method. Unlike my patent, which necessarily was written in the complex language of that genre, my methods paper was more in the style of a recipe for baking a cake – first you do this, then you do that, and so forth, and ultimately when you do an ordinary statistical test on the data you will see that the subject detected the EMF.

A drumroll of publications followed in which I described the details of my reproducible phenomenon. EMFs of almost any strength could be detected by human beings, down to levels so low I was confident that essentially every person in the United States experienced such a field every day. I did experiments in which I applied only the electric component of the electromagnetic field and showed that the electric rather than the magnetic component was the ultimate cause of the effect. I discovered that EMFs produced not one effect on the brain, but rather three effects. There was an effect that was triggered by the onset of the EMF in which the brain responded to the change from no field to the presence of a field, there was another effect caused by the offset of the EMF, and there was a third effect associated with the presence of the EMF, that was distinct from the onset effect and continued after the onset effect had died away until the EMF was removed.

I discovered that the effects of the EMF were different depending on whether disease was already present in the subject's brain. People who had multiple sclerosis, for example, responded differently to the EMF than did normal subjects. I showed that it made no difference whether the EMF was low-frequency, like that produced by powerlines, or high-frequency like from cell phones. Each of my results was produced using the method that I patented, each was published, and each was as reproducible as gravity. In experiments involving rabbits, rats, and fish, which permitted me to make measurements that could never be made in human subjects, I learned much about how the body made meaning from EMFs. I discovered that the likely location of the special cells responsible for detecting and/or processing the EMFs was probably in the hindbrain. I went ever farther; at the level of physical theory I analyzed the transduction process and showed how the body could detect EMFs.

Despite my success in the laboratory, I never forgot what Richard Phillips had said to me many years ago, that my work would never be noticed, and even if it was, it would never gain acceptance, and even if it did, it would have no practical consequences in the world. He was gone, but his lieutenants at Battelle had survived and they joined the herd owned by Motorola and the other companies that made and sold cell phones.

The herd published experiments that mimicked all the errors I had seen so often since my education in how science really worked had begun. Krause analyzed an enormous amount of meaningless data and found nothing, a technique that Richard Phillips had pioneered. Fritzer compared the brain waves between two different groups of subjects and should have known that the natural variations were far greater than any potential impact of the EMFs, which was another trick used by Phillips. Inomato-Terada slammed the brains of his subjects with magnetic fields so intense that their muscles contracted, and then measured whether the contractions were affected by puny mobile-phone EMFs, like Phillips who had squeezed big rats into little cages which effectively hid the smaller stress effect produced by EMFs. Roschke played the game that was most common among the investigators who favored Motorola's point of view, which was to average the results from subjects whose measurements went up with the results from subjects whose measurements went down, and then report there were no changes, on average. I had thought no one was better at that game than Phillips, who used it to prove to the satisfaction of the power industry that male and female mice in the three generations were not affected by EMFs, even though the opposite was true. In the twenty-first century, however, the method of obfuscation by averaging was improved. Wagner averaged the data along three dimensions so a positive effect was triply unlikely. Kleinlogel did it twice, in two identical experiments. Hamblin also did it, and then professed surprise she found no effects. Nothing these investigators did was worth doing, but they did it well, at least from Motorola's perspective.

Behind them stood the other types of mistaken experts that I also knew well. Michael Repacholi was as committed in his belief in the intrinsic safety of EMFs as had been Herman Schwan. Motorola and the other companies appreciated Repacholi's value, and with their help he became a kind of EMF czar at one of the agencies of the World Health Organization. From that bully pulpit he purported to teach the world about EMFs.

At a meeting where he was the honored guest I listened to Repacholi deliver a paean to the activities at the WHO that he guided and controlled. He proudly claimed to have stemmed the tide of concern regarding the propensity of EMFs to produce cancer and other diseases, and to have established the principle that cell phones were harmless. He touted legislation he had drafted, and expressed hope it would be enacted by governments around the world so that the fact that EMFs were safe would be enshrined in law.

From the audience I asked him, "Can you tell me exactly who at WHO has made the assessment of EMF safety, and on what basis?"

From the podium, he began his answer by describing a formal process for considering scientific evidence, but without naming names or giving reasons. I interrupted him. "Perhaps I wasn't clear. I didn't ask about the procedural process. I would like to know the names of the human beings who make the judgment regarding the safety of EMFs, and why they made them."

Again his answer dealt only with process, and he did not name names or give me reasons. I interrupted him again and said, "I am trying to get at the names of the individuals, the human beings who made the judgment that are contained in the WHO reports. Will you please tell me their names?"

For a third time he refused to answer.

Standing beside Repacholi at that meeting was his sponsor, Mays Swicord. When I first knew him he worked for the Food and Drug Administration where his duty was to protect the public against the risks of environmental EMFs. Over and over he had denied they posed any health risks, but he never explained his reasons. Swicord's speeches made him an ideal choice for Motorola, and he went to work for that company, delivering his message all over the world. No place was too far for this committed man, even Armenia.

Many of the worst evils that occurred during my lifetime, Auschwitz, Treblinka, Darfur, the Soviet Gulag, all had in common the fact that their psychological impression on right-thinking people occurred slowly, over time, and not instantly like the explosion of a bomb. It's as if the horror of some things is so great, a true appreciation can come only slowly. The work of Elisabeth Cardis and her colleagues belongs in this class. She was the head of another agency of WHO, the one that was supposed to warn people about what causes cancer. Together with others who sold their souls, she conducted a huge questionnaire study of people all over the world who used cell phones and people who got brain cancer, with the aim of proving that warnings to the public that there was even a possibility cell phones caused cancer were completely unnecessary. Those monsters didn't see that their work was founded on the despicable assumption that involuntary human experimentation was a proper source of human knowledge, and a proper basis for determining public policy.

Motorola needed a risk assessor to synthesize the evidence of doubt Cardis and her collaborators had manufactured, and who better than Dwight Spieler and Asher Sheppard, who both had the ability to deceive. At a symposium, Dwight Spieler gave his perspective on the problem of public perception of possible health risks, which he summarized using language from someone else who also had a knack for persuading an audience. There are known knowns, he said, which are things that we know that we know. There are known unknowns, which are things that we now know we didn't know. There are also unknown unknowns, which are things that we do not know we don't know, and each year we discover a few more of these unknown unknowns. His message was that cell-phone EMF health risks were simply one more unknown unknown, and that such remote contingencies matter only to the over-emotional and under-rational.

Asher Sheppard, who supported his family by means of his income from the Asher Sheppard Consulting Company, spoke on behalf of Motorola's interest at many meetings, both public and private, where he continued his life's work which was to destroy any scientist who furnished evidence antagonistic to the interests of his clients, as he had tried to destroy me thirty years earlier when he wrote a secret report for Judge Matias. Sheppard traveled widely and attacked every scientist Motorola perceived as an enemy; he was always careful to do so only in a forum that was friendly to him, and only under conditions where his victim had no procedural right to respond.

This is the end of my journey. I've lived through two EMF cycles, and I won't survive long enough to see a third. You probably will. I want to tell you why I am optimistic that the cycling will eventually end, even if not while you are alive.

EMF experts sound erudite, but when they make mistakes they do so in characteristic ways, so they always betray themselves. The environmental science of EMFs consists of both knowledge and justice, so each element should be reflected in the character of anyone who is truly an expert. We can classify the different kinds of mistaken experts based on what they lack.

There are industry experts who make their living by proving negative judgments, irrespective of what environmental science requires. These experts love money, which they seek above everything else. They see themselves as completely free to say anything in pursuit of money and prestige. Their minds are formless in the sense that they have no idea what truth is.

If an expert stakes a claim to knowledge that cannot be inferred from evidence by the process normally used to make meaning, he goes past science and into metaphysics. The mind of such a person is hardened against any empirical evidence or inferences that point away from what he believes. He regards any such evidence as necessarily flawed and feels no need to dig into it. Such experts emphasize belief over knowledge.

If an expert who works for the government becomes too passionately loyal to his bureaucracy, he forgets about justice. From the outside he appears to be following the proper method for interpreting data. On the inside, however, he triages away all inferences that may impede the mission of his agency, irrespective of whether that is the proper result of the normal process for making meaning from data. The testimony of such an expert is not reliable, like a lover who is asked to judge the beauty of his beloved.

Other experts devote themselves to pinpointing the source of epidemics like cholera or lung cancer, which they do by the method of questionnaires. But some questionnaire experts become disoriented and ask questions like "Do cell phones cause cancer?," which cannot be answered using their method, as gold cannot come from base metals.

These are the four kinds of mistaken experts: toadies, casuists, lovers, and alchemists. Where do they come from? Not from under rocks or out of thin air, but from the same stock that gives rise to experts who maintain the proper balance. In each instance, I suppose there was a crisis where the expert made the wrong choice but nevertheless was rewarded with a publication in a prestigious journal, special attention in the media, a promotion, or some financial benefit. Thereafter, it became progressively easier to put environmental science in the background.

With so many mistaken experts, will EMF environmental science survive? If you look, you can see that it continues to grow and develop, despite the mistaken experts. Think about this. EMFs don't cause disease as a result of their frequency, intensity, or wave shape. Rather, EMFs introduce change into the body that is neither necessarily good nor bad, but is rather the kind of change that naturally occurs. The body is destroyed by its own obliquity, which is disease, not by the detection of EMFs. The body cannot be destroyed by EMFs because EMFs are not disease, which is the evil that is naturally connected with the body. In the same way, the bad experts can't destroy environmental science, because they don't make it more unjust or ignorant or greedy.

Injustice, ignorance, and greed are ultimately fatal to the experts themselves, not to environmental science. But don't make the mistake of supposing that the unjust and foolish expert, even when he is caught in the act, is immediately destroyed by what makes him bad. More frequently he can continue to act that way, and even be acclaimed and respected in many quarters for doing so. The mistaken experts kill other people but they themselves often live for a long time. In the end, however, they corrupt themselves. The environmental science of EMFs won't be destroyed by the mistaken experts. I think it will fare well.

Epilogue

I wanted to know things and I thought every question had a correct answer. Learning why Sputnik didn't fall to earth helped turn me toward science. I stole an ammeter, and for months I suffered guilt for having broken the rules of science, even though I wanted nothing more than to follow them. I forgave myself, but only after deciding to never use the meter and promising to never again break the rules. Nobody could tell me why my classmate got cancer, but I assumed I just hadn't asked the right person. By the time the Jesuits were finished with me I was certain I understood what truth was and that, at least in the laboratory, I knew how to find it: use physics. When I began working for Dr. Becker, my fear that I couldn't perform at his level was soon replaced with curiosity regarding how he learned what he knew. His method was clear enough, taking and interpreting data, but how he chose from among its possible meanings mystified me. After an incubation period of ten years during which I earned graduate degrees in physics and law and learned how to do research, I began speaking about hazards of environmental EMFs, and was criticized from different directions. Experts said the results I found in my research violated the laws of nature, and had been refuted by other investigators. The constant refrain of the chorus that evaluated my grant applications was "no mechanisms," which in their parlance meant "bad ideas." Lawyers attacked me, saying that the implications for human health I drew from the EMF studies were not generally accepted.

My opponents have been persuasive, but little of what they said is true, and their biggest falsehoods are amazing; that nature is not free, like a marionette, that God causes disease so the best scientists can do is find cures; that involuntary human experimentation is desirable whenever other forms of research are too expensive. My quarrels with physicists, the institute of health, and the mouthpieces for the masters of energy produced a heap of calumny that is difficult for me to counter. Nevertheless I am confident in my case; the best proof that I am closer to the truth than my opponents is that I survived to tell this story even though they are powerful and I am only a gadfly.

To judge my case fairly you must take into consideration a prevailing cultural prejudice, that good science is always perfect and perfect science is always good. The disseminators of this tale are numerous and they have been at work for a long time. You accepted their opinions when you were young and more impressible than you are now, and because you heard no other point of view. These mostly unidentifiable myth-makers convinced themselves and then persuaded you that science was man's highest aspiration and most transcendent accomplishment. I cannot deal with them since I do not have them here to cross-examine them. So I ask you to recognize that I had two kinds of opponents, one age-old and the other recent, and I hope you will see the necessity of my choice to concentrate on answering only the latter, and that you will not judge my case under the influence of the old lies.

When Herman Schwan told me environmental EMFs could not cause disease I took what he said as a riddle, considering that I had simulated environmental EMFs in my laboratory and found they affected the metabolism of animals, a fact I interpreted to mean EMFs probably also affected the metabolism of people, from which I drew an inference that at least some of the metabolic effects would be harmful and that, if so, environmental EMFs could cause disease. I was confident in my belief that the kind of science I had just mastered, the platonic form of physics, could lead to an explanation for everything that happened in the world, so I tried to resolve the riddle according to the rules, by means of definitive experiments that produced consistent results. But no two experiments agreed exactly. At first I thought the inconsistencies arose from experimental errors, conspicuous or otherwise – reflections in the mirror of my own inadequacy. But everyone else's experiments – neglecting those of the herd that belonged to the masters – were also inconsistent. After talking and listening to many people, some honest and some not, I realized that the so-called inconsistencies were themselves consistent and therefore real because if the same thing happens each time an experiment is repeated, then nature has spoken and it remains only to ferret out the meaning of the answer, like a Delphic oracle. Intoxicated with physics, the herd, and other experts who sympathized with them, argued that all reliable responses were robotically reproducible, a state of affairs that is antithetical to the behavior of every living thing, from the tiniest virus to man himself, all of which desperately fight to stay in being and, in doing so, exhibit behaviors that are never precisely predictable and are often inconsistent with those of others of its species. No solid, liquid, gas, or man-made object struggles for existence, so each behaves today as it did yesterday and will tomorrow. Among EMF studies, inconsistency is to be expected.

The law governing human reaction to EMFs is particular to each person and not, in any important way, general like the laws of physics. Life does not point beyond itself to pithy equations. There is no deep theory behind biophenomena. The biophenomena are the theory, so an answer regarding whether EMFs cause harm can't emerge from the same process that successfully explained the behavior of air, water, fire, and earth. Attempts to do so are based on a kind of lawfulness, embodied by equations, that is alien to life.

Those who study life have within their reach the possibility of understanding something about the relation between EMFs and disease. But success depends on recognizing proponents of misleading concepts. One of the heavyweight misleaders is our nation's institute of health, which is where anyone who lacks some other patron must go, sooner or later, for money to perform research. Approaching this august body one encounters not a secular organization but something akin to a church. There are dioceses called study sections, cardinals called executive secretaries, and a titular head who is not expected to explain himself, like an infallible pope. The institute has a foundational myth, that science is distinct from values, a theology based on belief in the existence of mechanisms, commandments regarding how the scientific enterprise ought to be conducted, a concept of truth holding that data speaks for itself, saints like Dennis Thibodeaux, non-democratic traditions such as a star-chamber process for evaluating proposals, darkness in its history as well as a reputation for doing good, and a constant need for more money to help spread its message. Occasionally I was successful, and got to wet my nose, but as a suspicious character who practiced a strange religion, mostly I was humiliated. Nevertheless, based partly on research the institute funded only because it didn't know what I was doing, I finally came to understand the limitation on what was knowable about EMFs. Each living thing is a source of purpose, and follows its own inner law which governs its own unique becoming. We can learn something about the nature of the laws by observing the behavior of a fair number of individuals and searching for a pattern. But we cannot encompass all possible behaviors because we cannot observe every individual, or even one individual in all possible circumstances, so the possibility of surprise is ever present, an uncertainty that must be accepted with a sense of resignation. This inseparability of law and subject restricts an explanation of the reactions triggered by the stress of EMFs to a description of an archetypal form of the process, which continuously develops. Such knowledge is naturally imperfect.

The institute professes to believe in the primacy of mechanisms, the opposite of dynamical changes, so its suspicion of me was inevitable. Were the institute to prize observation over belief, and require its grantees to debate rather than acquiesce, reasonable people would see that when the effects of EMFs are viewed in the dimension in which they actually occur, time, otherwise invisible patterns can be perceived, even though none is precisely predictable. This way of doing science yields living knowledge rather than the mostly sterile work product of the institute's grantees, which is evaluated only by their brothers and sisters on the study section from whom they received the public's money and to whom in due course they return the favor. By vetoing the study of change in favor of slavish devotion to mechanisms, the institute conveniently avoids political difficulties since, unlike mechanisms, change has a cause. Thus the institute's research always leads away from the idea that disease is a reaction to what is in the environment and toward the idea that disease is a curse from God, who cannot be sued.

I was swept up in the politics of the institute and went to meetings where I expected to debate some of my opponents. But the institute stacked the jury with experts from the herd, and I lost every vote, some by lopsided margins. This was how the institute was able to pronounce EMFs safe, a decision that was not the product of a scientific method but rather was arrived at on religious and political grounds.

At first, Judge van Orsdel's rule of evidence, that for something to be a scientific fact all scientists must agree it is true, was only a tautology with no import because the scientific enterprise generated no controversies at law, only certainty that excluded reason-based disagreement. But asking "Do EMFs cause cancer?" was an implicit demand for a new kind of knowledge, something that is by its nature dependent on the existence of minds and the opinions held in them, that is uncertain, and that is rooted in the idea of justice. The judge's rule, a perfect weapon for aborting this kind of knowledge, was employed with impressive efficiency by the masters who run the Electric Power Research Institute, Motorola, and the other EMF-producing companies to insure they become as profitable as they can. The masters organized worldwide research efforts, established committees and commissions, lobbied governments, and put forth experts who said they could explain EMFs but who actually knew nothing about the subject. On this uneven playing field, the masters created a socio-cognitive reality that prevented general acceptance of any role for EMFs in promoting human disease. Their epistemological achievement did not quiet all opposing voices, but in many circles became entrenched.

The masters did not put forth their real case or speak with their own voices, and so avoided openly avowing their morality, that justice is a concept invented by weaklings who cannot otherwise compete successfully. This explains why the masters themselves remain largely invisible even though their committees, commissions, experts, and lawyers are well known. These surrogates accused me of creating hysteria among ordinary people who worried about the hazards of EMFs, saying that people should guard themselves against being deceived by the shock of my message, but what they really feared was the force of its truth.

Conceived of as something corporeal, knowledge of harm from EMFs is not possible. Conceived of as an idea, we can know about the harm they cause. This realization first occurred to me in a flash of insight and was sharpened in many battles. Ordinary people are as capable as experts of knowing about such harm because everyone wants justice, and everyone values something. Different people prefer pleasure or health in different proportions. For those who prefer a greater proportion of health, environmental science is a good thing.

When the Supreme Court unanimously overruled Judge van Orsdel, saying that scientific knowledge was not a matter of authority or popularity, but rather the product of a method, the environmental science of EMFs was born. It is still a child, not mature enough to take its rightful place in the world, where it has powerful enemies but friends who are inchoate because they do not understand how they would benefit. Environmental science will not grow to adulthood until a keen lawyer appears, someone who loves knowledge, respects the societal purpose of science, understands the truth about EMFs and, equally as important as all these other things, makes the effort needed to grasp the sophistical defects in the arguments invented by Patty Ryan and his brethren in the bar and taught to the herd, because the inferiority of their case must be understood before it can be exposed. Until we see this bold warrior, whose identity is presently unknown although

I can confidently say it is not Peter Richman, the masters will continue to prevail. Although there is no possibility of beating them today, the masters cannot delay forever the day they are called to pay for their evil.

I have always heard words like "magnificent achievement," and "glorious" used to describe science, but most of the achievements I saw in the world of EMFs were not magnificent and had no glory because their chief purpose was to support the masters who control EMF science as if it were a kite, first pulling on one string and then the other to spill or capture air so that the kite moves up or down, left or right, as desired. My heroic idea of science is gone, replaced by a more sober view, shaped by my experience. Words such as "inconsistent," "controversial," "mechanisms," or "general acceptance" sound obscene to me. Only uncertainty seems real, and this sense of reality is what I am. I found Andrew Marino.

Index

60 Minutes 15, 124, 127, 154, 256
Abrams, Dr. Albert 73, 94, 135
Adair, Robert K. 294, 349
Adey, Ross 98, 164, 257, 291, 373
* "Afelis, Paul" 372, 386-88
Agent Orange 331-32, 393, 417
Albright, Jim 194, 207, 221, 238, 425
Albright, Merrilee 194
Alice in Wonderland 278
American Association for the Advancement of Science (AAAS) 324
American Cancer Society 436
American College of Legal Medicine 324
American Embassy in Moscow 97, 163, 165
American Insurance Association 324
American Medical Association 325
American Physical Society (APS) 345, 351
American Tort Reform Association 325
Anders, Frank 2, 193-94
Anderson, Jack 96-98, 163
Angelos, Mike 350
Annals of the New York Academy of Sciences 99
"Antigone, Sophie" 247-49
Appleton, Budd 165-67
Arthritis Foundation 346
artificial ligaments 221-23, 227, 417
Aspirin Foundation 426
auras 181-82, 345

Bachman, Dr. Charles 39-41
Banks, Robert 128
Barnes, James 372, 379, 386
Baruch, Bernard 51-52, 370
Bassett, Dr. Andrew 44, 47, 50-53, 70-71, 75, 92-94, 99, 155, 169-74, 176-80, 193, 370, 374, 429
BC Hydro 247-49, 254-55. *See also* British Columbia Hydro
Becker, Dr. Robert 2, 16, 40, 41, 43, 45-53, 57, 65, 69-71, 74-75, 77-80, 85, 87-88, 90-96, 98, 104-107, 111, 113-14, 117, 119-21, 127, 132-33, 149-57, 159, 169, 173-77, 179, 183-84, 186-87, 189-93, 207, 219, 236, 257, 265, 277-78, 287, 300, 337, 341-42, 350, 358, 425, 431, 443
Bendectin 319-20, 322, 325, 328, 330, 414
"Benecke, Edward" 311-13, 315, 317-18
Berlin, Edward 121, 123, 125, 429
Berndt, Joe 103-104, 120, 429
BHT (butylated hydroxytoluene) 80-82, 89, 100, 113-14, 228, 319
Bioelectromagnetics 136
Bioelectromagnetics Society 136, 139, 148, 265
biological uncertainty 345, 399
Black, Jonathan 91
Blackmun, Justice Harry 2, 319, 321-23, 326-27, 330, 426
Bloembergen, Nicolaas 293
blue-ribbon committees 248, 276, 280, 284, 349, 383
Bockman, Richard 294, 302
Boltzmann, Ludwig 431-32
bone nonunions 169-70, 173-75, 178-80, 191

* Quotation marks around a name indicate an invented name, the character is a composite, or a privacy interest is being protected (please see the Preface).

Boone, Pat 373
"Boutbad, Edward" 317-18
Boutwell, Roswell 294, 303
brain cancer 354, 374, 378-79, 382-85, 390, 402, 404-405, 408, 412-13, 421, 424, 427, 440
brain waves 251, 349, 364-65, 379, 381-83, 428, 430, 433, 435, 438
Brennan, Justice William 245
Brighton, Carl 70-71, 75, 91, 95, 99, 155, 169, 174, 176-78
British Columbia Hydro 247-48. *See also* BC Hydro
Brown, Frank 86
Brown, Fred 2, 43, 69
Brown, Harold 164
Burton, Dr. Richard 30-31
Busch, Harris 257
Bush, George H.W. 222, 235

Carbala, Vince 36
carbon fibers 2, 7, 219-26, 229-34, 236-37, 275, 417
Cardis, Elisabeth 440
Cardozo, Judge Benjamin 2, 65, 196-97
Carey, Hugh 111, 121-22, 125, 290, 429
Carson, Rachel 54-56
Carter, Dixie 273
Carter, Jimmy 153
cataracts 159-64, 166-67
cell phone 2, 393, 401-405, 408, 411-15, 420-24, 426, 430, 435-36, 438-40, 442
Central Intelligence Agency 96-97, 159, 162. *See also* CIA
"Certain, Edward" 311
Chamber of Commerce 324
Charles, Bob 23, 197
Charles, Father 38
Chernoff, Neil 372

"Christopher, John" 8, 363-65, 369-78, 381-86, 389-90
CIA 163. *See also* Central Intelligence Agency
Clark, Richard 124, 198
Cleland, Max 132, 155, 191
Cole, Phillip 292
Columbia University 23, 44, 51-52, 171-74, 179, 193, 294, 303
Commoner, Barry 186
Cone, Clarence 94
Congress 156, 159, 164, 208, 322-24, 358-59, 363, 369-70, 390
Congressional Record 77
"Cook, Norman" 276, 282
"Cordaro, Jack" 318, 320-21, 323-27
Cormack, Allan 294
"Cottom, Edward" 312
Court, Margaret 91
Cousins, Norman 157
Cowen, Murray 184-85
Cranston, Alan 155, 191
Cronkite, Walter 56
cross-examination 82, 114, 115, 117-19, 144, 241-42, 252, 254, 262-63, 268, 294, 312, 317, 329, 363, 422, 425, 444

Dandy, David 229-30
Darwin, Charles 426
Daubert decision 332-33, 407, 417, 425
Defense Department 159-62, 164-66, 182
de Lafayette, Marquis 71
Delgado, Jose 357
DeMarco, Denny 23, 197
Department of Defense. *See* Defense Department; *see also* DOD
Department of Energy 15, 129-30, 132-34, 136, 142, 149, 188, 284, 372
Derek, Bo 430

deterministic chaos 188
Dirac, Paul 13, 53-54, 199-200
Distinguished Gentleman, The 273
DOD. *See* Department of Defense; *see also* Defense Department
Doyle, Arthur Conan 73
Dubos, René 215, 217

EEG 432-434. *See also* Electroencephalogram
Egan, Edmond A. 294
EIC 248-49. *See also* Environmental Information Corporation
Einstein, Albert 37, 56-57, 113, 199-201, 289, 294
Eisenhower, Dwight D. 416, 424
Electric and Magnetic Fields Research and Public Information Dissemination Program. *See* RAPID
Electric Power Research Institute (EPRI) 128-31, 133, 136, 145, 186, 188, 249, 257, 280, 284, 286, 347, 357-58, 360, 372, 386, 390-91, 425, 446
electroencephalogram 348. *See also* EEG
Electromagnetic Fields and Life 86
Electromagnetism & Life 430
Ensor, Sam 16, 25-28, 45
Environment 186
Environmental Information Corporation 247. *See also* EIC
Environmental Protection Agency 103, 152, 185, 280, 316, 318, 357, 369, 372, 405, 428. *See also* EPA
environmental science 186, 393-98, 400-401, 403, 406-409, 412, 414, 419, 424-26, 441-42, 447
EPA 208, 357, 428. *See also* Environmental Protection Agency

epidemiology 258, 261, 405
"Erdgas, Anthony" 17, 249-50, 252-53, 255, 269-270, 430
"Erica" 370-74, 378, 384-85, 389
"Faul, Dr." 393-94, 399, 411, 420
FCC. *See* Federal Communications Commission
FDA. *See* Food and Drug Administration
Federal Communications Commission 14, 316
Feynman, Richard 85
Flugum, Robert 129-36, 142, 146, 279, 290, 429
Food and Drug Administration 81-82, 113, 178-80, 221, 234, 316, 325, 372, 406, 440
Franklin, Benjamin 71
Frasier, George 35
Frey, Allan 93, 257, 411
Friedman, Howard 87, 99
Fritzer, Gunther 438
Frye, James 315-16
Frye rule 132, 316, 319-26, 330, 332
Fudge factors 28

Galileo 57, 216
Galvani, Luigi 71-72
Garrett, Alfred C. 72
"Gelling, Adolph" 286
Gelling, Paul 36, 201
Gelmann, Edward 294, 302
General Crushed Stone Company 60, 62-63
Gilpin, Dickie 21
Glashow, Sheldon Lee 293
glaucoma 2, 88-89, 99, 107, 127
Glenn, John 77
Goadby, Albert 239-45, 273
Goethe, Johann Wolfgang von 56, 350, 352, 431
Graham, Billy 144

Greene, Melvin 292
Greenway Avenue 21-23, 95
Gustavson, Karl 248-249, 254, 430
Gyuk, Imre 372

Hamblin, Denise 439
Handler, Philip 2, 149-52, 154, 156-58, 194-96, 211, 265, 374, 383, 429
Harrington, Dan 93
"Harris, Rice" 278, 283
Hart, Franklin 62, 64, 82-84, 201, 429
"Harvey, Robert" 277, 282
Hastings, Woodland 151, 153, 429
Hays, Marguerite 152-55, 191-92, 429
Hegel, G.W.F. 30-31, 54
Hexcel Medical 229-30
"Holy Trinity" of science 216
Hope, Bob 144
Hourahane, Don 219-21, 225, 231
"Hutson, Anthony" 311

immune system 88, 90, 224, 228, 280, 361, 362, 364, 370, 375-78, 394-95, 432-33
Inomato-Terada, Satomi 438
International Brotherhood of Electrical Workers 33

Jacobson, Cecil 97
Jaffe, Lionel 78, 94, 153
Jefferson, Thomas 71
Jenkins, David 219-20, 229-30
Jesuit 29-30, 38, 148, 197, 212, 443
Johnson & Johnson Corporation 170-71
"Jonson, Linn" 312
Journal of Equine Veterinary Science 226
Journal of Ophthalmology 99
Journal of the New York State Medical Association 80
"Juko, Jacob" 375, 378-79, 382
Justesen, Don 98, 136, 164-66, 265, 267-70, 372, 379, 429

Kahn, Alfred 120-21
"Keine, Phillip" 256-58, 260-63, 269-70, 411, 421-22, 429
"Keogh, Patrick" 412-13, 422-25
Kierkegaard, Søren 30
King, Billie Jean 91, 96
King, Larry 373
Klein Independent School District 278
"Klein, Judy" 274-76, 279, 284-86
Kleinlogel, Horst 439
"Koechel, Greg" 311
Koslov, Samuel 162-65, 168, 290
Krause, Christina 438

Labarthe, Darwin 294
Laird, Melvin 105
Lasalle, Bernard 77
Lavin, Linda 38, 59. *See also* Marino, Linda (Lin)
legal liability 67, 318, 415, 422
"Lindow, Christopher" 312
Lorenz, Edward 188, 433
Louisiana State University Medical School 193, 207
"Lugner, Monica" 372, 386-89
Lynam, Jimmy 21

MacArthur, Douglas 406
magnetism 33, 40, 71
"Mandel, Rose" 375-76, 378
Marino, Linda (Lin) 5, 59-60, 64, 102, 203, 207. *See also* Lavin, Linda
Marton, John 247
Matias, Thomas 115-25, 196-98, 279, 429, 441
Mattill, John 186, 429
Maxwell, James Clerk 289, 343-44, 421
McCabe, Judge Peter 294-98, 303-306
McCarron, Dave 23

"McGurk, Jack" 284, 291-93, 295-96, 298-302, 304
McIntyre, Judge John 248-50, 252-55
Meadows, Tom 121-22
mechanism(s) 92, 118, 140, 162, 189, 215-16, 237, 278, 293, 297, 359, 374-78, 433, 443, 445-46, 448
Merrill Dow Corporation 319, 322, 325
Mesmer, Franz 71, 135
Mevissen, Meike 375
microwave(s) 33-34, 48, 93, 97, 159-67, 187, 193, 248, 256, 265-67, 313-14, 317-19
"Milkin, Dr." 395, 401, 411
Miller, Bennett 133, 429
Miller, Morton 2, 114, 118-20, 122, 128-30, 134, 150-51, 157, 186, 357, 424
Millikan, R.A. 73
Minkin, Cedric 92, 94-95, 429
Mishra, Nirmal 222-26, 228-31, 235, 237, 429
Mobile Phone Manufacturers Forum 396, 425
Mondale, Walter 15, 133
"Montpelier, Tim" 112, 128, 139, 145-46
Moore, Dudley 430
"Morris, Dr." 395, 398, 411, 413, 418
"Morrison, James" 375-76
Motorola Corporation 412-13, 415, 419, 422-26, 428, 436, 438-41, 446
Moulder, John 349
Murphy, Eddie 273

National Academy of Sciences (NAS) 57, 149, 151-54, 156-58, 196, 276, 324, 383, 396
National Aeronautics and Space Administration (NASA) 2, 40, 193
National Institute of Environmental Health Sciences (NIEHS) 358, 361
National Institutes of Health (NIH) 45, 78-79, 127, 133, 185-86, 207-14, 216-18, 238, 249, 310, 320, 337-38, 347, 359-60, 363, 369, 370-73, 375, 377, 389-91, 424
National Science Foundation (NSF) 139
natural killer cells 359, 365
Nature 44, 49, 77
Navy 35, 96, 104-106, 114, 119, 149-54, 156-57, 159-60, 162, 164-65, 173, 267
Nelson, Gaylord 150, 154, 157
Newcomer, Judge Clarence 240, 242-45
New England Journal of Medicine 325
Newton, Isaac 28, 53, 289, 344, 431
New York Academy of Sciences 90-91, 95
New York Power Authority 274, 277, 291, 293, 304
New York Public Service Commission 128, 184, 196, 273, 437
Nietzsche, Friedrich 16, 30, 148, 197
Nixon, Richard M. 79
Nobel Prize 54-55, 85, 296, 298, 337
nonlinear 433-34
"Novak, Hank" 275, 282

Occupational Safety and Health Administration 406
Ohms 273
Olden, Kenneth 389-91, 429
oscilloclast 2, 73, 94, 315-16

Palsgraf decision 65, 100, 415
Pasteur, Louis 52
PAVE PAWS radar 354

Pearl Harbor 160, 162, 165
Pennsylvania Utilities Commission
 239
Perkins, Elisha 72, 135
Perry, Stephen 183
Peterson, Peter 291-92
phantom leaf effect 181-82
"Phant, Phu" 372
Pharmaceuticals Manufacturers
 Association 325
Philadelphia Electric Power Company
 239-45, 274
Phillips, Richard 128-36, 134-36, 143,
 188, 251, 255, 265, 269, 280,
 345, 356, 429, 438-39
"Pick, Charles" 372, 375-77, 381-82,
 385-86
piezoelectric effect 50, 94
Pilla, Arthur 171-80
Plato, *Dialogues* 11, 23
Pollack, Herbert 97, 163
Pound, Robert V. 293
"Power, Dwain" 311
Presman, Aleksander 86, 95
Prigogene, Ilya 215, 217
product liability 322

"Quarles, Robert" 310
questionnaire studies 183-84, 241,
 256-58, 261, 283, 287, 293,
 315-17, 331-32, 346, 388, 404-
 406, 427, 440

radar guns 309-14, 317-18
Radio Corporation of America (RCA)
 33, 35, 74
RAPID 361-63, 369-70
Rather, Dan 154
Ray, Karen 186
Readers Digest 239
Recurrence quantification analysis
 (RQA) 434-35, 437
regeneration 40, 44-46, 53, 57, 69,
 77-78, 94, 155-56, 169, 176,
 191-92, 342
Rehnquist, Justice William H. 322
Reiter, Russel 372
Repacholi, Michael 2, 294, 304, 349,
 439-40
"Richman, Peter" 393, 411, 413, 423-
 25, 448
"Richman, young" 393, 396-401, 403,
 407, 411-14, 416-19, 422-25
Riggs, Bobby 91
risk assessment 17, 139-45, 147, 290
"Robinson, Sherry" 255
Rockwell, Norman 72
"Rodriguez, Carl" 275, 281, 284-86
Roentgen, Wilhelm 57, 73, 434, 437
Romano, Joseph 239-45
Romero Sierra, Cesar 93
Roschke, Joachim 438
"Rosenbalm, Terry" 310
Rose, S. Meryl 53
"Ruhig, Martin" 372, 384, 390-91
Russia (Russian) 24, 46, 86, 97, 104,
 112, 115, 133, 241, 265-66,
 416. *See also* Soviet Union
"Ryan, Patty" 2, 17, 257-59, 264, 283,
 292-307, 333, 349, 365, 423,
 425, 447

salamanders 40, 43-45, 77-78, 182,
 184, 345
Sandpiper Elementary School 255-
 56, 264
Sarius, Wayne 132-33
Saturday Evening Post 73
Saturday Review 156-58
Savitz, David 247, 257, 259-61, 372
"Scarafiotti, David" 310
Scheifelbein, Susan 156-58
Schlesinger, James 133
Schmidt, Mr. 34, 81, 201, 429
Schrodinger, Erwin 54-56, 199, 338,
 342

Schwan, Herman 34, 74, 80, 113-14, 117-20, 122, 129-30, 133, 142-43, 149-51, 157, 237, 349, 429, 439, 444
Science 44, 129, 133, 153, 172, 182, 194
Scientific American 53
scientific evidence 81, 128, 142, 228, 230, 247, 250, 262, 267, 278, 282, 314, 316, 319, 322, 325, 330, 332, 373, 408, 439
Seaborg, Glenn T. 294
Semmelweiss, Ignaz 52
Shah, Kanu 106-107, 113
Sheppard, Asher 123-24, 139, 279, 372, 429, 440-41
Simpson, Bob 106-107, 111-12, 115-21, 123, 262, 292, 393
Sinclair, Upton 73
Singer, Marcus 53
Sinyukhin, A.M. 53
"Smith, John" 262-64
Smith, Stephen 257
South Africa 219-22, 225-26, 266, 274
Soviet Union 162-63, 187, 440. *See also* Russia
Spadaro, Joe 174
Spaeder, Joe 24, 197
spectral analysis 349, 432
"Spieler, Dwight" 129, 139, 142, 146-48, 429, 440
Sputnik 2, 24, 46, 197, 275, 443
standards of legal proof 101
Starr, Chauncey 2, 128-29, 133, 145-48, 186, 279, 284, 286, 357-58, 429
State Department 97, 163
State University of New York Press 186
"Stein, Dr." 394, 399-400, 407, 421
St. Joseph's College 24
Stolwijk, Jan 294, 303
stone quarry 2, 61, 82, 419

stress (stressor) 51, 88, 96, 99-100, 113, 127-28, 135-36, 154-57, 177, 185, 189, 253, 277-78, 280, 283-84, 287, 328, 341, 345, 347-48, 351, 359, 383, 438, 446
Strover, Angus 219-21
"Stuart, Aaron A." 294-95, 300
"Stum, Eva" 372, 386, 388-89
"Sudduth, Lawrence" 310
suicide and EMFs 183-84, 280, 357
Supreme Court 156, 245, 318-19, 327, 332, 411, 417, 447. *See also* U.S. Supreme Court
Swicord, Mays 419, 440
Syracuse University 36, 39, 123, 343, 350
Szent-Gyorgyi, Albert 2, 13, 53, 337-38, 341-42, 365, 431

"Tauschen, Hans" 139, 146, 148, 429
Technology Review 80, 186
Terrace, Herbert 294, 303
"Thibodeaux, Dennis" 207-208, 210-11, 215, 217-18, 237-38, 370, 445
toxic torts 293, 305, 321, 393, 414-15
transduction 364, 438
"Traveler, Terrance" 310
Trischka, John W. "Trashcan" 36, 48, 199, 201, 344, 429
Truman, Harry 406
Tucker, Margaret 294, 302
Tyler, Paul 2, 96, 104-106, 149-50, 159, 161, 164-65, 167-68, 290

U.S. Central Intelligence Agency. *See* CIA; *see also* Central Intelligence Agency
U.S. Chamber of Commerce. *See* Chamber of Commerce
U.S. Congress. *See* Congress
U.S. Department of Defense (DOD).

See Defense Department
U.S. Department of Energy.
    See Department of Energy
U.S. Embassy, Moscow. *See* American
    Embassy in Moscow
U.S. Environmental Protection
    Agency. *See* Environmental
    Protection Agency
U.S. Federal Communications
    Commission. *See* Federal
    Communications Commission
U.S. Food and Drug Administration.
    See Food and Drug
    Administration
U.S. Navy. *See* Navy
U.S. Occupational Safety and
    Health Administration.
    See Occupational Safety and
    Health Administration.
U.S. State Department. *See* State
    Department
U.S. Supreme Court 245, 318-19. *See
    also* Supreme Court
U.S. Veterans Administration. *See* VA;
    see also Veterans Administration

VA 149, 155, 291, 358. *See
    also* Veterans Administration
"Vanderhide, Henry" 208, 210-11
van Orsdel, Judge Josiah 316, 330,
    446-47
"Vesta, Michael" 310
Veterans Administration 39, 41, 45,
    78, 82, 98, 127, 132-33, 149,
    151, 152, 164, 184, 191-93,
    350. *See also* VA
Volta, Alessandro 71-72

Wagner, Peter 439
Waite, Ralph 273
Wallace, Frank 115, 121-24, 429
Wallace, Mike 2, 15, 124-25, 198, 256
Wallner, Father 16, 29, 31, 148, 197,

215
Washington Post 77, 96, 155
Watson, James 12, 298
Webber, Chuck 433-34, 437
Weinstein, Judge Jack 404
Wertheimer, Nancy 257, 259, 261,
    347, 372, 386, 387
Westinghouse Corporation 34-36,
    199
"William" 37, 341, 350
Wilson, Richard 293
Winship, John 100-102
World Health Organization (WHO)
    291, 294, 304, 396, 439-40

Yalow, Rosalyn 294

Zaner, Ken 294, 303
"Zapper" Corporation 310, 312, 313,
    317-18
Zaret, Milton 2, 7, 159-67, 265
Zbilut, Joe 433-34
Zhirmunskii, A.V. 53
"Zum, James" 311